MINERAL NUTRITION
OF PLANTS:
PRINCIPLES AND
PERSPECTIVES

Dennis Robert Hoagland
1884–1949

Pioneer of the modern era of research on the mineral nutrition of plants. (Reproduced by permission of the Director, The Bancroft Library.)

MINERAL NUTRITION OF PLANTS: PRINCIPLES AND PERSPECTIVES

Emanuel Epstein

*Department of Soils
and Plant Nutrition
University of California
Davis*

John Wiley and Sons, Inc.
New York · London · Sydney · Toronto

PREFACE

With this book I mean to fill a need that has long been felt by students and researchers in many areas of the biological, agricultural, and environmental sciences—the need for a textbook on the principles of the mineral nutrition of plants. This subject is of fundamental importance in biology, for it deals with the processes by which the living world acquires from the environment the mineral elements essential to life. In addition to this intrinsic interest is the importance of plant nutrition as the science basic to agriculture, forestry, range management, stewardship of the evironment, and other activities devoted to the raising of plants and animals and the welfare of human beings.

In writing this book I have kept in mind several kinds of readers: undergraduate and graduate students of plant physiology and plant biochemistry; research workers in plant nutrition, plant physiology and ecology; students and researchers in the fields of soil science, agronomy, horticulture, vegetable crops, forestry, range management, environmental sciences and other disciplines, who may have occasion to use such an exposition of the principal facts and current thinking in the field of the mineral nutrition of plants.

With these readers and their needs in mind, I have adopted the following guidelines for the writing of this book.

(1) The student should be able to *read* the book in its entirety: he should not be reduced to merely consulting it for specific points of information. This means that the book shall not attempt encyclopedic completeness.

(2) As a corollary to the point just made, the book must be highly selective, and of necessity much interesting and valuable material has been omitted from it. Inevitably, the choices made among the existing materials depend to some extent on chance, on what is and what is not readily available, and on the vagaries of memory and associations. Deliberate selections were made with a view to including a large variety of experimental plant materials and experimental approaches. In regard to old material versus

v

new, preference is given to the new when, as is often the case, it is experimentally more precise and more sophisticated in interpretation than earlier work; the history of the subject is covered in the first chapter, however, and classical contributions are mentioned throughout the book.

(3) The book deals in the main with the mineral nutrition of higher green plants. However, in discussing cellular processes I have drawn on findings from work with other organisms as well, for the emphasis in the book is on basic processes, not on any particular group of plants.

(4) Mineral plant nutrition, in the widest sense, may be taken to include virtually all of plant physiology and plant biochemistry, including the intermediary metabolism of compounds containing nitrogen, phosphorus, and sulfur, as well as a good deal of soil science as it relates to soil–plant relationships and much of soil microbiology to boot. This book focuses more sharply on those mineral nutritional activities of plants which represent the "initial" or "acquisitive" aspects of physiology and metabolism. The reason for this emphasis is that these aspects of mineral plant metabolism have come to be identified as the specific subject matter of mineral plant nutrition. They are often given short shrift in books on general plant physiology and biochemistry, which stress instead the intermediary aspects of metabolism.

(5) The book is intended for students and researchers and is therefore documented by references to the literature, so that its statements can be scrutinized and verified, and so that the reader is enabled to enlarge his view of any topic on which he wants more information than is given here.

(6) I have not hesitated to indicate preferences for certain views among conflicting ones because I believe that the writer of a book of this nature must give his readers considered judgments instead of flinging at them a random grab-bag of conclusions and conjectures culled from the literature.

A book like this, though written by one person, is nevertheless the product of a grand cooperative enterprise—the enterprise of biological and agricultural science. I am unable to express individually my true appreciation to my colleagues all over the world who helped by sending reprints, answering inquiries, and supplying material for inclusion in the book. The importance of the superb collection of material in the library of the University of California, Davis, is beyond reckoning. My friends and co-workers in the Department of Soils and Plant Nutrition were ever ready to discuss moot points with me, and so were colleagues all over the campus on whom I called for counsel and judgment. My students, both undergraduate and graduate, and post-doctoral fellows were an inspiration; this book would not have come into being without them. My own research during the years when the book was taking shape was supported by the

National Science Foundation, the Office of Saline Water, United States Department of the Interior, and a Fulbright grant for a period of research and study with colleagues in Australia. To all these, and to Peggy, my thanks.

University of California, Davis   EMANUEL EPSTEIN
August, 1971

CONTENTS

PART I

THE ELEMENTS OF PLANT NUTRITION

INTRODUCTION
AND HISTORY

1. THE SCOPE OF MINERAL PLANT NUTRITION

The processes of plant nutrition are those having to do with the acquisition of nutrient elements by plants and with the functions of the elements in the life of plants. As a science, plant nutrition is a specialty within the general subject of plant physiology. The parts of it that deal with metabolic and biochemical functions of the chemical elements link it with other aspects of plant physiology and plant biochemistry. On the other hand, those processes—physical, chemical, physiological, and biochemical—having to do with the interactions of the plant with its chemical media, with the initial acquisition of chemical elements, and with their distribution within the plant are the particular province of plant nutrition.

2. "MINING" THE ENVIRONMENT

Plant nutrition is of unique importance in the realm of life on earth and in the affairs of man. All living things consist of atoms of the chemical elements. The ultimate reservoirs of these elements on earth are the rocks, the oceans, and the atmosphere. Rocks weathered to soil, the oceans and the water they furnish through the atmosphere to lakes, streams, and soils, and the atmosphere itself—these simple compounds, aggregates, solutions, and gases are the ores which the living world mines for the elements that go into the make-up of palm and pine, of rice and redwood, of mice and men.

Not all living things, however, participate in this primary mining of the raw materials of life. Only green plants and certain microorganisms are able to extract simple inorganic compounds and ions from the environment without having to rely on complex energy-rich compounds previously synthesized by other living organisms. These self-sufficient organisms are called "autotrophs," in contrast to the "heterotrophs" which must draw on organic metabolites "prefabricated" by autotrophs. In the economy of the living world, by far the most important agents in the primary acquisition of energy and chemical elements from the external environment are the photosynthetic plants—the algae in oceans, streams, and lakes, and the green plants on land.

It is through the activities of these plants that carbon and nitrogen, potassium, phosphorus, sulfur, magnesium and other essential nutrient elements are initially abstracted from the inorganic environment and incorporated into living cells and tissues. After this initial acquisition by living plants, nutrient and other elements may then find their way into the cells

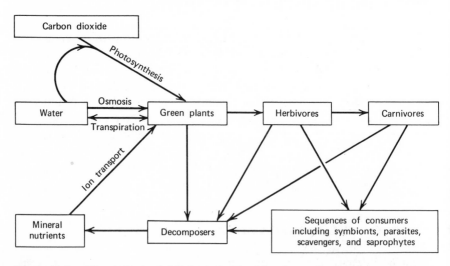

Figure 1-1. Flow chart of nutrients in the biosphere. A complete ecosystem could exist without the organisms and processes indicated by the three boxes— herbivores, carnivores, and sequences of consumers. Not included in the diagram are the return of carbon dioxide to the free carbon dioxide pool, effected by all organisms through respiration as well as abiotically through the burning of organic materials, and the release of oxygen in photosynthesis and its uptake in aerobic respiration.

of consumers—decomposers and herbivores, from herbivores to carnivores, and via decomposers back to the non-living reservoirs of inorganic matter (Figure 1-1). The entire living world depends upon plants and their ability to direct the movement of inorganic substances at the boundaries between their own cells and the environment on which they abut.

The major part of the total biomass of the world is terrestrial (Bowen, 1966). For terrestrial life, higher green plants are by far the most important autotrophs; all the rest of us depend on them. "All flesh is grass": the web of life is woven on a loom of land.

TRANSLOCATION

The cells of most algae and other aquatic plants are all exposed to the aqueous medium in which they float, and they take up water and solutes directly from it. Most terrestrial green plants, on the other hand, have tissues that are far removed from the soil which is the source of water and inorganic nutrient elements. The top branches of a redwood tree may sway 300 feet above the soil from which its roots absorb water and dissolved nutrients. In land plants there are elaborate structures and mechanisms which effect the long-distance transport or "translocation" of water and solutes within the plant body. Hence, translocation of substances is another major subject of plant nutrition.

FUNCTIONS OF ELEMENTS

If one of the chemical elements essential to the life of a plant is present in the environment of that plant in insufficient amounts or in chemical combinations which render it poorly available for absorption, the deficiency of this element in the cells of the plant will bring about derangements in its metabolism. Eventually, these metabolic disturbances manifest themselves in the development of visible symptoms such as stunted growth, yellowing or purpling of leaves, or other abnormalities. These symptoms of nutrient deficiency are more or less characteristic for a given element, but depend also on the severity of the deficiency, the particular species or strain of plant, and many environmental factors.

Occurrence of these diseases and the identification of nutrient deficiencies as their cause, as well as the development of toxic conditions resulting from excessive absorption of certain elements, focus attention on the functional roles of elements in the metabolism of plants. Certain elements—

carbon, oxygen, hydrogen, nitrogen, for example—are major constituents of the principal classes of compounds that make up living plants; these four elements, in fact, constitute about 95 per cent of the fresh weight of most living plants. These are the elements of water and those which make up the bulk of the carbohydrates, proteins, and lipids of plant cells. Among the elements accounting for the remaining few per cent of the weight of the plant are many which, though they may range in concentration from just one or two per cent down to a few parts per billion, are just as essential to the functioning of the plant as the elements that make up the bulk of plant matter. The study of the functions of the elements as components of structural compounds and dynamic metabolites is the third main topic of plant nutrition, in addition to the processes of absorption and translocation.

5. HEREDITY AND ENVIRONMENT

Finally, all physiological performances of plants, including nutritional ones, are a function of their innate genetic constitution and of the environment in which they live. The interplay between these two factors in the nutritional physiology of plants is especially fascinating because of the great diversity of soil, the main mineral medium of higher plants. Plants are adapted genetically and physiologically to this diversity. Genetic and ecological aspects of the mineral nutrition of plants are discussed in the last part of this book.

6. HISTORY OF PLANT NUTRITION RESEARCH

In antiquity there had been built up a substantial amount of agricultural "know-how" but there was no experimental plant science of essentially the kind we have today. The advance of science was hindered by the general acceptance of a speculative scheme advanced by the Greek natural philosopher, Aristotle (384–322 B.C.) who held that all matter consisted of the four "elements," earth, water, air, and fire. Although another Greek philosopher, Democritus (460–360 B.C.), had advanced an atomic theory of matter much closer to modern conceptions, the speculations of Aristotle prevailed for about 2000 years until the beginnings of modern experimental science in the 16th century.

J. B. van Helmont (1577–1644) was a Belgian physician to whom belongs the honor of having conducted the first quantitative experiments in plant nutrition. In a famous experiment he investigated the source of the materials that plants are composed of (Gabriel and Fogel, 1955). "That all vegetable [matter] immediately and materially arises from the element of water alone I learned from this experiment. I took an earthenware pot, placed in it 200 lb of earth dried in an oven, soaked this with water, and planted in it a willow shoot weighing 5 lb. After five years had passed, the tree grown therefrom weighed 169 lb and about 3 oz. But the earthenware pot was constantly wet only with rain or (when necessary) distilled water; and it was ample [in size] and imbedded in the ground; and, to prevent dust flying around from mixing with the earth, the rim of the pot was kept covered with an iron plate coated with tin and pierced with many holes. I did not compute the weight of the deciduous leaves of the four autumns. Finally, I again dried the earth of the pot, and it was found to be the same 200 lb minus about 2 oz. Therefore, 164 lb of wood, bark, and root had arisen from the water alone."

This was an excellent experiment—well planned, carefully done, and accurately described. If the conclusion was totally wrong we can hardly blame the experimenter, who was a pioneer in a new science. Leonardo da Vinci (1452–1519) had performed a similar experiment but his account of it lay unpublished in his notebooks (Bodenheimer, 1958), and Robert Boyle (1627–1691) in "The Sceptical Chymist" described experiments of the same kind, from which he drew the same conclusion (Russell, 1961).

The importance of mineral matter for the growth of plants was recognized by John Woodward (1665–1728), a Professor of Medicine in London. He concluded that most of the water that enters into plants "passes through the pores of them and exhales up into the atmosphere; that a great part of the terrestrial matter, mixt with the water passes up into the plant along with it; and that the plant is more or less augmented in proportion as the water contains a greater or smaller quantity of that matter." These conclusions were based on experiments in which he found that plants grew better in water containing dissolved solids than in distilled water.

Stephen Hales (1677–1761), an English clergyman, published in 1727 a book entitled in the elaborate manner of the time, "Vegetable Staticks: Or, An Account of some Statical Experiments On The Sap in Vegetables: Being an Essay towards a Natural History of Vegetation. Also, a Specimen of An Attempt to Analyse the Air, by a great Variety of Chymio-Statical Experiments; Which were read at several Meetings before the Royal Society" (Hales, 1727). This is the first book describing and discussing a substantial number of experiments in plant physiology. Hales had a passion for exact measurements, and the "staticks" in the title of this book refers to quantitative experiments and determinations. He was particularly interested in the "sap" of plants and made measurements of amounts of water absorbed and transpired. He related these amounts to the area of the root surface through which the water was absorbed, and the areas of the leaf surfaces through which it was transpired, and calculated relative velocities of water movement through unit area of root surface and unit area of leaf surface. He performed many experiments on root pressure which he described so well in writing and by means of illustrations that a modern experimenter would have no trouble repeating them. Stephen Hales may properly be called the first plant physiologist. The chief award given by the American Society of Plant Physiologists is called the Stephen Hales Award in his honor.

Stephen Hales and others before him had had an inkling that the air contributed something to the substance of plants, but the true nature of this contribution could not be understood as long as there was no realistic conception of the composition of the air and the nature of combustion. Like others of his time, Stephen Hales believed in the "phlogiston" theory according to which all combustible materials were compounds of "phlogiston." On burning, phlogiston is expelled, and the "calces" (ashes) are left behind. These ideas gave way to essentially modern views toward the end of the 18th century.

J. Priestley (1733–1804) prepared oxygen by heating mercuric oxide, but remained an adherent of the phlogiston theory all his life. In spite of the handicap of working with a faulty chemical theory that was being overthrown in his own time Priestley made important contributions to chemistry and physiology. He observed that green plants emitted the same gas that was released by heated mercuric oxide, and thus took the first step in clarifying the process of photosynthesis. Jan Ingen–Housz (1730–1799), a Dutch physician who spent much of his life in England, repeated Priestley's experiments and made the important discovery that light was essential for the evolution of oxygen by green plants. Like Priestley, he had totally wrong ideas of the meaning of his discovery.

The first investigator to gain a reasonably valid view of photosynthesis was Jean Senebier (1742–1809), Swiss clergyman, librarian, and scientist. He found that the amounts of oxygen evolved by green leaves kept in water were proportional to the concentration of carbon dioxide dissolved in the water. Like his predecessors, he did not use the terms oxygen and carbon dioxide but supposed that light, combined with some substance of the green leaf, decomposed "fixed air" (carbon dioxide) into "dephlogisticated air" (oxygen). Nevertheless, he carried the investigation of photosynthesis about as far as it could be taken while the phlogiston theory was still dominant in the thinking of chemists and physiologists. This theory, however, did not survive the 18th century.

The end of the century was revolutionary not only in affairs of state and society, but in science as well. The French chemist, Antoine L. Lavoisier (1743–1794) administered the coup de grace to the phlogiston theory and laid down the tenets of modern chemistry. In chemical reactions, he proposed, no elements are created, none are transmuted into others, and none are destroyed. Chemical reactions consist of changes in the combinations of elements. Lavoisier devised the chemical nomenclature of elements which is, essentially, our modern form. He laid the firm foundation on which modern chemistry, physiology, and biochemistry are built.

The first scientist to make full use of the new chemistry of Lavoisier in research on plant nutrition was the Frenchman, Theodore de Saussure (1767–1845). His work, followed by that of others, greatly extended knowledge concerning the significance and the absorption by plants of elements derived from the soil. He first compiled his work in a book under the title "Récherches chimiques sur la Végétation," published in 1804, and continued to make important contributions till shortly before his death in 1845.

De Saussure combined knowledge of the new chemistry with careful experimentation and with equally careful interpretation of the results obtained. He was fully aware that explanations "in a subject as new and as complicated as this one . . . are doubtless very often venturesome," and resisted the temptation to draw inferences well beyond the reach of the experimental results. Yet when he felt that he was on safe ground he drew firm conclusions. De Saussure was often on safe ground and recognized many features of plant nutrition that we now take for granted. He grew plants of lady's thumb, *Polygonum persicaria,* in solutions of single salts and of some organic solutes as well, and observed that these various dissolved substances were not all absorbed by the plants in equal amounts. Thus he may be called the discoverer of selectivity in the absorption of solutes by plants. He also laid down the principle of essentiality. While admitting that certain elements absorbed by plants might not be essential he insisted that some were indispensible. His published tables giving the chemical composition of the ash of various species of plants were the first compilations of such data.

De Saussure's conclusions concerning the importance of elements derived from the soil were much debated during the first decades of the nineteenth century. The German investigators, Carl S. Sprengel (1787–1859) and A. F. Wiegmann (1771–1853), obtained evidence which led them to support de Saussure's views. Sprengel's work on the importance of nutrient elements absorbed from the soil was particularly significant. He wrote that a soil may be favorable in almost all respects, "yet may often be unproductive because it is deficient in *one single element* that is necessary as a food for plants." Here was a clear statement of the "law of the minimum" wrongly credited to Liebig by many authors—including Liebig himself.

Research on plant nutrition and soil-plant interrelations of the early and middle nineteenth century reached a high point in the work of the Frenchman, Jean–Baptiste Boussingault (1802–1887). He is generally credited, and rightly, with having laid the foundation of the new agricultural science. Much of the material in his writings strikes the modern reader as essentially contemporary in outlook and approach. Unlike his predecessors, he was not content with studying the elemental composition of crop plants but stressed the balance between the amounts of each element absorbed by the crop and the amounts subtracted from the soil and fertilizer medium in which the plants grew. He published numerous tables giving the chemical composition of crops, and calculated the amounts of the various elements removed per hectare. He studied the effects of fertilizers and soil amendments upon the balance of elements between crop and soil. These researches were the forerunners of the innumerable studies published ever since with titles like "The effect of . . . on yield and composition of . . ."

Boussingault must be given credit for providing the first "hard" evidence for the fixation of atmospheric nitrogen by leguminous plants, despite the fact that earlier students had suspected that legumes possess some special power of acquiring nitrogen, and despite the fact, too, that Liebig so savagely questioned his results that Boussingault himself was moved to doubt. What led Boussingault to his conclusion was the finding that red clover and pea when grown in a soil containing no available nitrogen, gained in addition to carbon, hydrogen, and oxygen measurable amounts of nitrogen, whereas wheat and oats, under identical conditions, failed to gain any nitrogen. The nitrogen gained by the legumes "is derived from the atmosphere; but I do not pretend to say in what precise manner the assimilation takes place." For a nice account of Boussingault and an appreciation of his contribution to early research on the nitrogen problem see Aulie (1970).

Justus von Liebig (1803–1873) of Germany was the foremost organic chemist of his time. The British Association for the Advancement of Science invited him to prepare a review of the subject for presentation at its meeting in 1840. This invitation resulted in his writing a book entitled "Organic Chemistry in its Applications to Agriculture and Physiology." The book went through many editions, was translated into several languages, and became extremely influential. At the time of its initial publication in 1840, Liebig himself had done virtually no work of his own in the fields of agricultural chemistry and plant nutrition. He did not let this deter him from claiming to be the first investigator to study "the application of chemical principles to the growth of vegetables" since Sir Humphry Davy (1778–1829), British chemist. Liebig's own book was the best refutation of this claim, because it represented a summary and compilation of the work of de Saussure, Sprengel, Boussingault, and many others.

His own judgments and interpretations were often faulty, and in many cases represented a denial of sound conclusions already established by other investigators. For example, he attacked Boussingault's conclusion that leguminous plants do, and non-leguminous plants do not, derive nitrogen from the atmosphere. He claimed, instead, that all plants absorb nitrogen in the form of ammonia from the air. The ammonia he considered to originate from the decay of organic matter. Almost half a century after de Saussure's clear-cut evidence for selective solute absorption, Liebig wrote: "All substances in solution in a soil are absorbed by the roots of plants, exactly as a sponge imbibes a liquid and all that it contains, without selection." Also, his claim "that any one of the alkaline bases may be substituted for another, the action of all being the same," was mere conjecture, not credible even at the time it was written.

Liebig's main contribution to plant nutrition was that he finally did away with the humus theory according to which organic matter of the soil is the source of the carbon that plants absorb. He considered that soil contributes soluble inorganic constituents. This was not a novel conclusion, for de Saussure, Sprengel, and Boussingault had concluded likewise, but Liebig's assertive and authoritative manner of writing and the vigor with which he propagated his views finally won acceptance for the "mineral theory of fertilizers." After publication of the first edition of his book in 1840 he and his many students and collaborators turned to laboratory work on mineral constituents of plants, and as a result many of the ill-considered statements in the book disappeared from later editions. Improved analytical methods were devised, and Liebig gained a much more accurate knowledge of the mineral composition of plants than his predecessors had obtained.

For several decades after the first publication of Liebig's book in 1840, work on the role of atmospheric nitrogen in plant nutrition continued and many investigators gained the conviction that leguminous plants were able to fix free nitrogen from the air. The subject remained a baffling one, however, and successful demonstrations of nitrogen fixation by leguminous plants were doubted because other experiments failed to show any such effect. Finally the German investigators Hellriegel and Wilfarth announced in 1886 their discovery of the role of bacteria in the nodules of the roots of leguminous plants. The Russian botanist, Woronin, had discovered earlier that the root nodules of legumes contain bacteria, but their role remained obscure. Hellriegel and Wilfarth grew peas in sterile soil, as controls, and in soils inoculated with leaching from a soil in which peas did well. The controls in uninoculated soil developed no nodules and failed to develop. The plants in the inoculated soil developed root nodules and made good growth. The experimenters concluded that legumes fix atmospheric nitrogen, but only when infected by symbiotic bacteria. Non-leguminous plants do not fix free nitrogen but depend entirely on combined nitrogen in the soil. This work provided the final confirmation of conclusions first reached by Boussingault in 1837.

Science is a cooperative and cumulative endeavor, and many discoveries, like that of biological nitrogen fixation, cannot be credited fairly to any one investigator or team. P. W. Wilson (1957) of the University of Wisconsin College of Agriculture has written a delightful account of the ups and downs of the discovery of biological nitrogen fixation, for a while a veritable comedy of errors at times; yet eventually, through the work of many men, the facts emerged and the reality of the process was established.

By the end of the 18th century, Senebier had established the main facts of photosynthesis, and by the end of the 19th century, an inventory of the principal facts of biological nitrogen fixation was at hand. One other aspect of plant nutrition, the absorption of nutrient salts by roots, was considerably clarified in the course of the 19th century. Liebig had concluded that the water of the soil, the soil solution, was devoid of potassium, phosphate, and ammonium. According to his view, plants would languish and die if these nutrients were supplied to them in solution. He visualized instead a direct transfer of nutrients from the surface of soil particles to the surface of roots in intimate contact with them. This view was elaborated later in a more sophisticated form by H. Jenny and his associates in Berkeley, beginning in the 1930's.

In 1860 the German botanist Julius von Sachs demonstrated that the solid phase of the soil can be dispensed with entirely in the nutrition of plants. He prepared solutions of salts supplying the major essential mineral nutrients and containing (without his knowing it) adequate trace amounts of essential micronutrient elements present as contaminants in the major salts. In these nutrient solutions he grew plants to maturity. W. Knop in the early 1860's devised another nutrient solution, and growing plants in solution culture has been a favorite research technique in plant nutrition ever since. These researches, along with studies of the chemical composition of the soil solution, have emphasized the importance of the soil solution as a source of major nutrients available for absorption by roots.

Readers who wish to delve into the history of this subject will find the following references useful. Bodenheimer's (1958) history of biology, as its title indicates, covers a broad subject but serves to set the scene. It contains source readings. The short history of plant science by Reed (1942) is interesting and deals with aspects of mineral nutrition, among others. The source book compiled by Browne (1944) has excerpts from original writings which make good reading. The first chapter of the book by Russell (1961) gives a brief survey of the history of research on plant nutrition and soil–plant relationships.

Developments that have taken place in the present century are the immediate precursors of current contemporary work. They will not be looked on as "historical" but will be mentioned in connection with the various specialized topics of which they are a part.

REFERENCES

Aulie, R. P. 1970. Boussingault and the nitrogen cycle. Proc. Am. Philosoph. Soc. 114:435–479.

Bodenheimer, F. S. 1958. The History of Biology: An Introduction. Wm. Dawson and Sons, Ltd., London.

Bowen, H. J. M. 1966. Trace Elements in Biochemistry. Academic Press, London and New York.

Browne, C. A. 1944. A Source Book of Agricultural Chemistry. Chronica Botanica 8:1–290.

Gabriel, M. L. and S. Fogel. 1955. Great Experiments in Biology. Prentice-Hall, Inc., Englewood Cliffs.

Hales, S. 1727. Vegetable Staticks. Available in a 1969 reprint, M. A. Hoskin, ed., American Elsevier, Inc., New York.

Reed, H. S. 1942. A Short History of the Plant Sciences. The Ronald Press Company, New York.

Russell, E. J. 1961. Soil Conditions and Plant Growth. 9th ed. John Wiley and Sons, Inc., New York.

Wilson, P. W. 1957. On the sources of nitrogen of vegetation; etc. Bact. Rev. 21:215–226.

2

BIOGEOCHEMICAL
CYCLING OF ELEMENTS
IN NATURE

1. INTRODUCTION

Plant nutrition, as we have seen in the previous chapter, is intimately concerned with the traffic of elements and compounds between the totality of living things (the biosphere) and the non-living surroundings which form its setting and substrate. Collectively these two entities are called the ecosphere (Cole, 1958). In particular, plant nutrition deals with the initial acquisition by the biosphere of the elements and simple inorganic compounds that are the raw materials from which living matter molds more living matter. The non-living substrate of the biosphere consists of the rocks and minerals of the earth (the lithosphere), of the surrounding envelope of nitrogen, oxygen, and other gases (the atmosphere), and of the hydrosphere (the bodies of water that cover 71 per cent of the earth in the form of oceans, lakes, and rivers, and the water in rocks and soils and ice and snow). The hydrosphere invades the atmosphere as water vapor which recondenses and returns to earth and sea as rain, snow, and hail. This traffic of water is called the hydrologic cycle.

2. CHEMICAL COMPOSITION OF THE ECOSPHERE

The chemical composition of each of these "spheres" varies greatly from place to place, but average values have been determined and are given in the tables below. The composition of the atmosphere changes with increasing height, both in density and in the relative concentrations of various molecules. However, convection currents keep its composition fairly constant in the layer, 20,000 m thick, nearest the earth (the troposphere), and this is the part of the atmosphere that directly affects living things. The composition of the troposphere is indicated in Table 2-1. The oceans

TABLE 2-1.   Average Composition of the Troposphere[1]

Gas	Composition by volume, ppm	Composition by weight, ppm	Total mass in geograms, 10^{20} g
N_2	780,900	755,100	38.648
O_2	209,500	231,500	11.841
A	9,300	12,800	0.655
CO_2	300	460	0.0233
Ne	18	12.5	0.000636
He	5.2	0.72	0.000037
CH_4	1.5	0.9	0.000043
Kr	1	2.9	0.000146
N_2O	0.5	0.8	0.000040
H_2	0.5	0.03	0.000002
O_3[2]	0.4	0.6	0.000031
Xe	0.08	0.36	0.000018

[1] After Mason (1958).
[2] Variable, increasing with height.

cover 70 per cent of the surface of the earth and make up 98 per cent of the hydrosphere. The average concentration of the major ions in the oceans, and for comparison, in lake and river waters, is given in Table 2-2. The average composition of the lithosphere—the rocky outer crust of the earth—is given in Table 2-3.

These "spheres" form the ultimate reservoirs of chemical elements which make up the living world, the biosphere. The chemical composition of the immediate substrates of growing plants may be very different from the average values given in Tables 2-1 to 2-3. "Fresh water" pumped from wells drilled into the salt-rich floor of the Central Valley of California,

TABLE 2-2. Composition of Sea Water and Lake and River Waters[1]

Ion	Concentration, mM	
	Sea Water	Lake and River Waters
Na^+	457	0.39
K^+	9.7	0.36
Ca^{2+}	10	0.52
Mg^{2+}	56	0.21
Cl^-	536	0.23
SO_4^{2-}	28	0.21
HCO_3^-	2.3	1.1

[1] Data on sea water after Goldberg (1963), calculated on the basis of the total mass of the mineral element (and of C, for HCO_3^-) per liter of sea water, disregarding other ionic or molecular species in which the element may occur. Data on lake and river waters are mean values for North America, after Goldman and Wetzel (1966).

TABLE 2-3. The Commoner Chemical Elements in the Lithosphere[1]

Element	Per cent of weight
O	46.60
Si	27.72
Al	8.13
Fe	5.00
Ca	3.63
Na	2.83
K	2.59
Mg	2.09
Sum	98.59

[1] Earth's crust to a 10 mile depth. After Mason, 1958.

for example, is a far cry from the "fresh water," of almost distilled-water purity, fed by snow-melts into the rushing Merced River 100 miles to the east. Rock undergoes a series of chemical changes as it weathers into soil. Even the atmosphere, the most constantly and most thoroughly mixed "sphere" of the three, varies considerably in chemical composition from place to place and from one time to another.

3. ENERGY

The acquisition of elements from the external environment and their synthesis into the manifold array of compounds found in living plants requires energy. The chemical elements are derived from the terrestrial environment; on the other hand, virtually all of the energy that drives the biosphere is of solar origin and reaches the earth as sunlight. About one-third of the incident energy is reflected back into space, the rest heats the earth and is eventually reradiated into space in the form of heat. Heat from the sun cycles water from oceans and icecaps to land, warms the earth and its waters, and not doing so evenly at all times and all places, creates seasons, climates, and weather.

Finally, a tiny percentage of all the solar energy incident upon the earth, about 0.05 per cent of it, powers the activities of the biosphere (Delwiche, 1967). The mechanism by which this electromagnetic energy is captured in the form of chemical compounds is photosynthesis—the synthesis of carbon compounds brought about in the chlorophyll-bearing cellular organelles, the chloroplasts, of green plants. The bodies of plants, synthesized from the raw materials of the ecosphere by the use of energy from the sun, then become the source of both building materials and energy for all the rest of the biosphere (Cole, 1958).

4. ELEMENTAL CYCLING

The traffic of materials between the ecosphere and the biosphere is not all one way. In fact, if it were, life would have ceased long ago, for the supply of essential elements on earth is finite. If these materials were not returned from the biosphere to the ecosphere their supply would become exhausted by being locked up in the bodies of organisms. For example, all the carbon dioxide in the atmosphere at any one time would be tied up through photosynthesis by green plants in a few years were it not for the fact that just about as much carbon dioxide is returned to the atmosphere each year as is fixed by photosynthesis (Bowen, 1966). The return of carbon dioxide to the atmosphere is through the respiration of plants, animals, and microorganisms, and the burning of organic materials—forest fires and prairie fires, and the burning of wood, dung, and fossil fuels by man. Other elements are also subject to such "biogeochemical cycling," and we shall now consider a few of the most important of these cycles.

THE CARBON CYCLE

Carbon is present in the atmosphere in the form of carbon dioxide, CO_2, at an average concentration of 0.3 milliliter per liter. It is a constituent of the hydrosphere as dissolved CO_2 in equilibrium with water:

$$CO_2 + H_2O \leftrightarrows H_2CO_3 \leftrightarrows H^+ + HCO_3^- \leftrightarrows H^+ + CO_3^{2-}$$

The carbon dioxide in the atmosphere is in equilibrium with carbon dioxide dissolved in the oceans, and the concentration of carbonate in solution in the oceans is a function of the solubility of calcium carbonate. This in turn is influenced by the concentration of carbon dioxide (Rankama and Sahama, 1950). The total amount of CO_2 in the oceans is about 50 times that contained in the atmosphere.

Carbon is assimilated by photosynthetic organisms—principally by green plants, including green algae. The first-formed stable products of photosynthesis are phosphorylated compounds of low molecular weight. Beginning with these, the processes of metabolism synthesize the carbohydrates, proteins, lipids and other types of compounds that are constituents of living plants. Part of the assimilated carbon resides briefly in plants in the form of organic compounds, and cycles back to the atmosphere (or, for marine algae, to the ocean) as respiratory CO_2. Other portions of the organic carbon of green plants return to the atmosphere via the respiration of organisms which feed on these plants: animals, bacteria and fungi of decay, and after the death of these their bodies are decomposed by other saprophytic organisms and their carbon is recycled as CO_2 into the atmosphere. Huge amounts of photosynthetic products of earlier geological ages have become fossilized and their carbon has thereby been temporarily removed from the cycle. At present, this heritage of fossil fuels—coal, petroleum, and natural gas—is being burned up at a prodigious rate and returned to the atmosphere in the form of CO_2. Figure 2-1 is a diagrammatic representation of the carbon cycle.

The Carbon Cycle

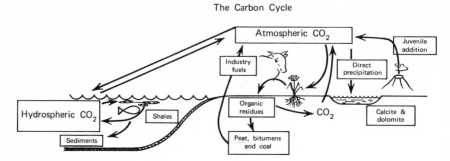

Figure 2-1. The carbon cycle. After Delwiche (1965).

6. THE NITROGEN CYCLE

The large reservoir of nitrogen is free nitrogen gas, N_2, which constitutes 78 per cent of the atmosphere by volume. In this form nitrogen is not directly available to green plants. Entry of nitrogen into the biosphere is brought about chiefly through the activities of nitrogen-fixing micro-organisms, either free-living or symbiotically associated with plants. In the latter case, the nitrogen fixed becomes available for the synthesis of amino groups of amino acids and proteins of the host plants. When plants or free-living, nitrogen-fixing bacteria die, bacteria of decay liberate the amino acids of their proteins, and ammonifying bacteria then release the amino groups in the form of ammonium ions, NH_4^+, which are dissolved in the soil solution. Ammonium may be absorbed by plants in this form, or after conversion to nitrite and then nitrate, through the activities of nitrifying bacteria. These are autotrophic. One group utilizes the energy derived from oxidation of ammonium to nitrite and another that from oxidation of nitrite to nitrate (Wallace and Nicholas, 1969).

Plants of a great majority of all species are not hosts to bacterial, nitro-gen-fixing symbionts, and therefore cannot use free atmospheric nitrogen. Instead, most plants depend entirely on fixed (reduced or oxidized) nitro-gen present in the soil solution as a result of the processes just described—nitrogen fixation, decay, and nitrification.

The cycling of nitrogen that we have described does not involve any return of nitrogen into the great pool of free atmospheric nitrogen gas. Nitrogen may, in fact, cycle from soil nitrate through a chain of transfor-mations back to soil nitrate without ever going through the free nitrogen gas form (see Figure 2-2). Nevertheless, when the protein nitrogen of bac-teria, plants, and animal consumers is released to the soil solution as am-

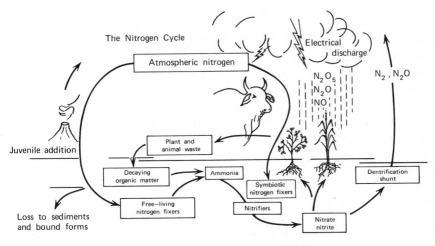

The Nitrogen Cycle

Atmospheric nitrogen

Electrical discharge

N_2O_5
N_2O
NO

N_2 , N_2O

Juvenile addition

Plant and animal waste

Decaying organic matter

Ammonia

Symbiotic nitrogen fixers

Dentrification shunt

Free—living nitrogen fixers

Nitrifiers

Loss to sediments and bound forms

Nitrate nitrite

Figure 2-2. The nitrogen cycle. After Delwiche (1965).

monium, the process constitutes an exit of nitrogen from the biosphere. It may then be reabsorbed by plants in the form of ammonium, or more likely, as nitrate, after nitrification. On the other hand, the nitrogen of nitrate and some other forms of fixed nitrogen may be discharged into the atmosphere by denitrifying bacteria. This loss of gaseous nitrogen to the atmosphere is balanced by biological nitrogen fixation, as discussed above, and to a minor extent by lightning and other atmospheric electrical discharges.

Examination of Figure 2-2 will show that animals do not form an essential link in the nitrogen cycle. The basic processes that nitrogen undergoes can all be carried out by microorganisms and green plants. Delwiche (1965) has given an excellent account of the cycling of carbon and nitrogen in the biosphere.

THE PHOSPHORUS CYCLE

The biogeochemical cycling of carbon and nitrogen is so effective because both of these elements are distributed in the form of gases, CO_2 and N_2, and are available everywhere to those organisms able to bring about their entry or assimilation into the biosphere. The same is not equally true of the elements of water, hydrogen and oxygen, since water is not as ubiquitous as are CO_2 and N_2. No other elements essential to the biosphere are distributed in an equally effective fashion throughout the ecosphere.

The lithosphere is the source and reservoir of phosphorus. The phosphorus content of the lithosphere is 0.12 per cent. Most naturally occurring phosphates are sparingly soluble, like those of calcium, aluminum, iron, and manganese. Phosphate also is held by soil clay minerals as an exchangeable anion, and may be fixed in forms unavailable for absorption by plants. As a result of these features of phosphate chemistry the concentrations of phosphate ions in soil solutions are low, on the order of less than one part per million (Chapter 3). Phosphate enters into the biosphere by being absorbed by plants and microorganisms. Upon decomposition of plants and their animal consumers, soluble phosphate returns to the soil.

Although release of phosphorus from insoluble forms in rock and soils is slow, the sum total of phosphate carried by the rivers into the oceans each year is enormous. Estimates are that 3.5 million tons of phosphorus are lost to the sea yearly, where it precipitates in the form of sparingly soluble calcium phosphate (Cole, 1958). Only a small part of this phosphorus returns to the land, through guano deposited by sea birds, and by man taking fish from the oceans. Thus the supply of phosphorus available for the growth of terrestrial plants is steadily diminishing. Man hastens this process by the manufacture and use of phosphatic fertilizers. Sparingly soluble rock phosphate is converted, through treatment with mineral acids, into much more soluble phosphate fertilizer materials. These are spread

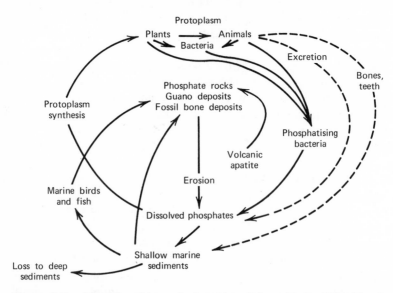

Figure 2-3.   The Phosphorus Cycle. After Odum, E. P. 1971. Fundamentals of Ecology. 3rd ed. W. B. Saunders Company, Philadelphia.

over the land, and are now far more vulnerable to the agencies of erosion than was the original phosphate rock. Much of this material is thus sped on its way to the oceanic burial grounds of phosphorus. Figure 2-3 shows the main features of the phosphorus cycle.

. OTHER ELEMENTS

The cycling of other major mineral elements like potassium, calcium, and magnesium resembles that of phosphorus, in that large quantities of these elements are lost each year to the oceans, without any effective mechanism for their return to land. Plants accelerate the process. They absorb the elements from the soil where they are relatively protected from agents of erosion. Through the fall of leaves and the dropping of branches and fruits, and the eventual death of the plants themselves, the mineral elements that they had mined from the soil are spread on the surface of it; subject to the eroding action of storms and winds, and their dispatch to the oceans is accelerated.

Some elements which, like those mentioned above, do not cycle in gaseous form nevertheless are distributed effectively over the land areas of the world. Chief among these are sodium and chloride, the elements of common salt. Wind whips ocean spray into the air, the water evaporates, and crystals of "cyclic salt" are carried by the air currents into the atmosphere. Eventually, salt reaches land by descending onto it in rainwater. In addition, particles of salt may be deposited directly upon soil and plants, especially in maritime environments where there is much air-borne salt (Boyce, 1954; Cassidy, 1968a,b). It should not be thought, however, that salt from the oceans travels only a short distance from the shore. Cyclic salt circles the globe, and is deposited by rainfall even far inland (Gibbs, 1970; Gorham, 1961).

It is often found that the ratios of mineral elements in rainwater correspond poorly to their ratios in seawater. There are two principal reasons for this. First, the sea water which is carried into the atmosphere does not represent a bulk sampling but rather the surface foam of the windswept sea. This surface film is enriched in organic matter and products of its decay. As a result, its potassium/sodium ratio is higher than that of bulk sea water, and it contains organic nitrogen at a concentration higher than that in the body of the ocean (Wilson, 1959).

The second reason for the lack of correspondence between the ratios of mineral elements in rainwater and the sea is that not only the ocean but soil and vegetation also contribute to the solid material carried by the atmosphere. Dust and plant debris, pollen of plants, and sundry particles of diverse terrestrial origins are swept up and incorporated into the moving air masses above the earth, eventually to return, mainly by being washed down in rain. Such matter, unlike that of marine origin, does not represent a true accession of matter to the land (Gorham, 1961; Tamm, 1958; Wetselaar and Hutton, 1963).

Man's activities contribute much to the sulfur content of the atmosphere in the vicinity of industries which generate sulfur-bearing emanations. The amounts involved are important enough to play a role in the sulfur supply of plants in and near heavily industrialized regions (Gorham and Gordon, 1960; Riehm, 1964). Increasingly sensitive methods for the detection and measurement of minute quantities of various substances are constantly being devised. As they are put to use in atmospheric science many substances not so far detected in rainwater will be found there. Among the

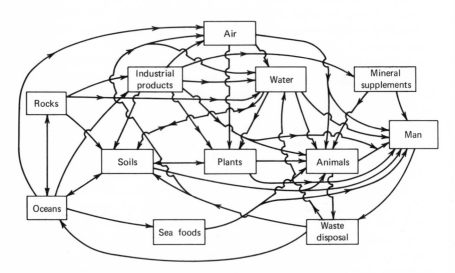

Figure 2-4. The cycling of microelements. After Allaway (1968). Copyright 1968 by Academic Press, New York and London.

substances whose cycling is as yet little known are microelements, including heavy metal elements, some essential, others not essential or even toxic. Figure 2-4 traces the pathways of such elements in the ecosphere. Even organic micronutrients are subject to significant cycling, as shown by the report of Parker (1968) on vitamin B_{12} in rainwater.

FAILURES IN CYCLING

We have described the path of phosphorus in the ecosphere as the "phosphorus cycle" (Figure 2-3). However, to the extent that phosphorus is washed into the sea without return to the land, the cycle is broken and much phosphorus is withdrawn from it at least until geological upheavals raise up the ocean floors again and make the phosphorus deposited there once more available to terrestrial life. This failure of cycling removes from circulation one of the important nutrients of the biosphere.

There is another failure of cycling of quite a different kind. Instead of an important nutrient being lost, harmful substances may accumulate. This happens on an immense scale in the arid and semiarid regions of the world. These regions occupy about one third of the total land area of the earth. About 35 per cent of the area of the United States, not counting Alaska and Hawaii, fall in this category. The soils in these regions, developed under dry conditions, contain large quantities of minerals in which the soils of humid areas are poor, because of the leaching action of rain. Among the minerals present in high concentration in the unleached soils of arid and semiarid regions are sodium chloride, sodium sulfate, and sodium carbonate, and the pH of such soils is often high. The salt content of many of these soils is too high for the growth of crop plants, even where water can be supplied for irrigation. In fact, irrigation may aggravate the condition, for the irrigation water available in arid regions often carries a high burden of dissolved salts—so high that its continued use may build up the salt content of the soil even beyond its naturally high level.

The failure of fresh water to pass through the soils of the arid and semiarid regions of the world causes a failure of cycling of the soluble salts. Instead of being leached away these salts accumulate, often to the point where they render the soils completely infertile.

For excellent general accounts of the mutual relations between the biosphere and its chemical environment the reader is referred to the books by Blum (1968) and Florkin (1960). Perel'man (1967) discussed the subject from the point of view of the geologist, and Bowen (1966) has summarized much quantitative information.

REFERENCES

Allaway, W. H. 1968. Agronomic controls over the environmental cycling of trace elements. Adv. Agron. 20:235–274.

Blum, H. F. 1968. Time's Arrow and Evolution. Princeton University Press, Princeton.

Bowen, H. J. M. 1966. Trace Elements in Biochemistry. Academic Press, London and New York.

Boyce, S. G. 1954. The salt spray community. Ecol. Monographs 24:29–67.

Cassidy, N. G. 1968a. The effect of cyclic salt in a maritime environment. I. The salinity of rainfall and of the atmosphere. Plant and Soil 28:106–128.

Cassidy, N. G. 1968b. The effect of cyclic salt in a maritime environment. II. The absorption by plants of colloidal atmospheric salt. Plant and Soil 28:390–405.

Cole, L. C. 1958. The ecosphere. Scientific American 198(4):83–92.

Delwiche, C. C. 1965. The cycling of carbon and nitrogen in the biosphere. In: Microbiology and Soil Fertility. C. M. Gilmour and O. N. Allen, eds. Oregon State University Press, Corvallis. Pp. 29–58.

Delwiche, C. C. 1967. Energy relationships in soil biochemistry. In: Soil Biochemistry. A. D. McLaren and G. H. Peterson, eds. Marcel Dekker, Inc., New York. Pp. 173–193.

Florkin, M. 1960. Unity and Diversity in Biochemistry. Pergamon Press, Oxford.

Gibbs, R. J. 1970. Mechanisms controlling world water chemistry. Science 170:1088–1090.

Goldberg, E. D. 1963. The ocean as a chemical system. In: The Sea. M. N. Hill, ed. Interscience Publishers, John Wiley and Sons, New York and London. Vol. 2, pp. 3–25.

Goldman, C. R. and R. G. Wetzel. 1966. Mineral nutrients in lake and river waters. In: Environmental Biology. P. L. Altman and D. S. Dittmer, eds. Federation of American Societies for Experimental Biology, Bethesda. Pp. 508–511.

Gorham, E. 1961. Factors influencing supply of major ions to inland waters, with special reference to the atmosphere. Geol. Soc. Am. Bull. 72:795–840.

Gorham, E. and A. G. Gordon. 1960. Some effects of smelter pollution northeast of Falconbridge, Ontario. Can. J. Bot. 38:307–312.

Mason, B. 1958. Principles of Geochemistry. 2nd ed. John Wiley and Sons, Inc., New York.

Odum, E. P. 1971. Fundamentals of Ecology. 3rd ed. W. B. Saunders Company, Philadelphia and London.

Parker, B. C. 1968. Rain as a source of vitamin B_{12}. Nature 219: 617–618.

Perel'man, A. I. 1967. Geochemistry of Epigenesis. Plenum Press, New York.

Rankama, K., and T. G. Sahama. 1950. Geochemistry. The University of Chicago Press, Chicago.

Riehm, H. 1964. Pflanzenernährstoffe in den atmosphärischen Niederschlägen unter besonderer Berücksichtigung des Schwefels. Symp. Internat. Agrochimica 5:453–472.

Tamm, C. O. 1958. The atmosphere. In: Encyclopedia of Plant Physiology. W. Ruhland, ed. Springer–Verlag, Berlin. Vol. 4, pp. 233–242.

Wallace, W. and D. J. D. Nicholas. 1969. The biochemistry of nitrifying microorganisms. Biol. Rev. 44:359–391.

Wetselaar, R. and J. T. Hutton. 1963. The ionic composition of rainwater at Katherine, N . T., and its part in the cycling of plant nutrients. Austral. J. Agric. Res. 14:319–329.

Wilson, A. T. 1959. Surface of the ocean as a source of air-borne nitrogenous material and other plant nutrients. Nature 184:99–101.

3

THE MEDIA OF PLANT NUTRITION

1. THE VARIETY OF NUTRIENT MEDIA

The media or substrates of plants are those portions of the environment that are in close proximity of plants or plant organs and exchange materials with them. There is an astonishing variety of habitats on the earth that furnish media suitable for the growth of plants. The aquatic plants of the oceans and fresh-water bodies are totally immersed, with all their cells in immediate or close vicinity of their aqueous substrate. In estuaries, along beaches, and in marshlands, plants are completely immersed at some times but at other times the blanket of water is withdrawn, exposing the shoots to the gaseous atmosphere. Though they all live in water, aquatic plants face a variety of chemical environments, ranging from ocean water which is 3.5 per cent salt and other still saltier bodies of water to glacier-fed mountain streams and cirque-basin lakes.

The main mineral medium of terrestrial plants is the soil. Soil is an extremely heterogeneous medium in any one place, varying furthermore in important chemical, physical, and biological properties from place to place, depending on parent rock material, climate, topography, age, and biological factors (Jenny, 1941). Plant life, which first evolved in the oceans or in tidal pools, invaded the land in the Silurian Period about 425 million years ago in the form of primitive *Thallophyta*. By the Mississippian Period, 100 million years later, tree ferns, huge horsetails and giant club mosses had colonized large areas of land, and since then evolutionary processes have resulted in the spread of terrestrial plants over the whole face of the earth. Plants take root on exposed rock and in desert soils containing little water, as well as in deep, well-watered alluvial soils and the organic humus of forests. Touched by water, no area of land, no nook and niche of the sun-lit surface of the earth is beyond the colonizing capabilities of green plants.

Like water and soil, the atmosphere is a nutritional medium of plants, by the definition given on p. 29. Photosynthesis by terrestrial plants draws directly on atmospheric carbon dioxide. The oxygen consumed in the respiration of plant tissue comes in part from the atmosphere. Dew condensed from the surrounding air contributes to the water supply of plants in dry regions (Monteith, 1963; Slatyer, 1967; Stone, 1957). Sulfur contained in the atmosphere in the form of sulfur dioxide may contribute to the nutrition of plants (Fried, 1948; Olsen, 1957; Thomas *et al.*, 1950). Air-borne cyclic salt circles the globe, supplying various elements directly by deposition on plants, and indirectly by being carried into the soil in rainwater (Gorham, 1961). These and other aspects of mineral cycling were discussed in Chapter 2.

It is interesting to reflect upon the fact that among living things, only the terrestrial higher plants have parts of their bodies residing permanently in entirely different media—the roots in the soil and the shoots in the atmosphere. This polarity (which we discuss in more detail in Chapters 7 and 8) has necessitated the evolutionary elaboration of structures and mechanisms for a two-way traffic of materials between roots and shoots, instead of the circulatory pattern in animals.

2. SOIL

Soil is the direct mineral substrate of terrestrial plants. All nutrients with the exception of carbon are drawn by land plants mainly from the soil. Among the nutrients which the soil supplies, only hydrogen and oxygen (water), nitrogen, and chlorine cycle via the atmosphere in such a manner as to become widely available to plants and microorganisms capable of assimilating them. Thus a molecule of water absorbed by the roots of a redwood tree on the California coast today may have been yesterday hundreds of miles out at sea. The same is not true of potassium, magnesium, phosphate and other mineral elements. They cannot be adequately imported from afar via the atmosphere but must be in solution in the soil or be released from the solid soil material in the immediate vicinity of the roots (the rhizosphere) if they are to be absorbed. The distances they travel from their points of residence in the soil to the roots which absorb them are often measured in microns, or occasionally more as when water with ions in solution percolates through pores and interstices among the particles of the soil. In any event, compared with the elements that circle the globe in the atmosphere, most mineral nutrient elements in the soil are immobile.

This fact has had a profound bearing on the evolution of terrestrial plants. Since the mineral elements essential for the growth of plants reside in the soil and are for the most part immobile or nearly so, absorbing organs (roots) had to evolve that could penetrate through the whole mass of the soil, bringing the largest possible absorbing surface into intimate contact with the soil matrix. An examination of the soil under turf, or in a forest, or indeed anywhere that plants grow, shows the amazing extent to which the roots ramify through the soil mass.

Dittmer (1937) carefully liberated the soil from the root system of a single rye plant, *Secale cereale,* grown for four months in a box 12 inches square and 22 inches deep (30.5 × 30.5× 55.9 centimeters), using a gentle spray of water, and then measured the extent of the root system. The total area of the root surface abutting on the soil was 6,875 square feet (639 square meters), and the combined length of all the roots was 387 miles (623 kilometers). Chapter 11 of the book on plant ecology by Weaver and Clements (1938) presents a discussion of the growth of roots in relation to the soil environment. More recently, Weaver (1968) and Weaver and Darland (1949) have described the living network of roots in prairie soils. The lacy, tangled fabric of roots of grasses and forbs is shown vividly in the pictures they present (see Figure 3-1). The extent of the growth of roots through the soil medium is also apparent from their finding that after three years' growth the dry weight of the underground parts of grasses in the surface foot (0.3 meter) of soil varied from 5.5 tons per acre (0.4 hectare) for *Andropogon gerardi* (big bluestem) to 1.6 tons for *Bouteloua gracilis* (blue grama).

Hall *et al.* (1953) have devised a radioactive tracer technique for measuring the extension of root systems through the soil with the progress of time. They found that 14 weeks after planting, the roots of a corn plant, *Zea mays,* had penetrated to a depth of more than 20 feet (6 meters), with a radial extent of more than 33 feet (10 meters). Studies of roots in soil and their methodology have been described in a publication of the National Academy of Sciences—National Research Council (1962). The most elaborate installations for the study of root development in soil are "rhizotrons"—underground walkways equipped with windows through which roots can be observed, photographed, and their growth recorded by means of time-lapse cinematography (Taylor, 1969).

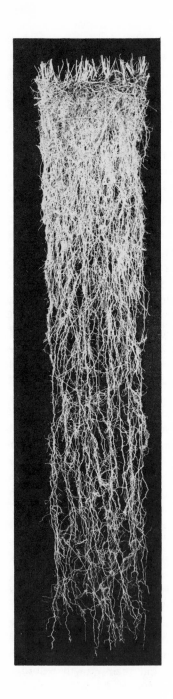

Figure 3-1. Root system of switchgrass, *Panicum virgatum*. Actual size, 12 × 60 inches (30 × 152 centimeters). From Weaver and Darland (1949). The roots actually extend to a much greater depth (J. E. Weaver, personal communication).

Roots, then, represent adaptations that enable plants to mine the soil for essential nutrients. The intimate contact with the soil mass that their function requires is the reason for plants being sessile. Free movement is precluded by the close entwinement that must exist between roots and soil. Photosynthesis (Trench *et al.,* 1969) and nitrogen fixation (Bergersen and Hipsley, 1970) may not be necessarily incompatible with mobile organisms, but the primary acquisition of mineral nutrients from soil is. It follows that terrestrial animals including ourselves have gained the freedom of motion through the evolutionary developments which emancipated us from the function of primary mineral nutrition. This liberation from a stationary existence in turn led to the evolution of sense organs and coordinated body movements; hence the need for a central nervous system and the evolution of the brain. Traced to their remotest origins, the triumphs and the tragedies of our intellect and our emotions are ours because evolution resulted in our reliance for mineral nutrients on those fixed, silent, insensible chemical machines, the plants rooted in the soil.

Compared with a relatively uniform aqueous medium such as lake water or ocean water, soil is an exceedingly heterogeneous substrate. It consists of a solid phase, mineral matter derived from rock and also organic matter in various stages of decomposition; a liquid phase, the soil water, or better, the soil solution, since it contains various solutes; and a gaseous phase, the soil atmosphere which occupies those interstices among the solid particles that are not filled with water.

The soil solution is the most important immediate source of nutrients for absorption by roots. However, it is a very dilute solution which would be speedily depleted by roots were it not for the fact that it is constantly replenished by the release into it of elements from the solid phase. For example, plants were grown for six weeks in soil in which the concentration of phosphate in the soil solution was approximately constant at 1 ppm. From the phosphorus content of the plants at the end of the period it was calculated that the phosphate in the soil solution had been completely renewed ten times each day, on the average (Stout and Overstreet, 1950).

The solid phase of the soil releases mineral elements into solution partly by solubilization of soil minerals and organic matter, partly by solution of sparingly soluble salts, and partly by ion exchange, mainly cation exchange. Clay particles and solid organic matter bear negative charges, so that they act as cation exchange materials. The cations held by these negative charges are mainly calcium, magnesium, potassium, sodium, aluminum, and hydrogen ions, their proportions varying greatly in different soils. The anion of greatest quantitative significance in soils is nitrate, and since it is present almost exclusively in free solution it follows that an equivalent concentration of cations must also be in solution. In a soil capable of supporting the growth of plants additional anions such as phosphate, sulfate, chloride, and others are also inevitably found in the soil solution, and hence equivalent concentrations of cations are of necessity present in this solution. Roots of actively growing plants are therefore bathed by a dilute solution of various salts. The principal cations in solution are potassium, calcium, and magnesium. In soils of arid and semiarid regions, sodium ions often predominate. In acid soils, on the other hand, hydrogen ions and aluminum ions are prominent in the soil solution. Cations in solution exchange freely with cations adsorbed on the solid cation exchange complex (Figure 3-2).

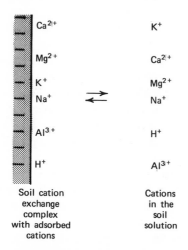

Figure 3-2. Diagrammatic representation of cation exchange.

Soil, then, appears as a system of solid, liquid and gaseous phases of great physical and chemical complexity. As a medium for the growth of plants it is furthermore influenced by the microbial population which resides in it (Barber, 1968, 1969; Burges and Raw, 1967; Gilmour and Allen, 1965; Rovira, 1965). The soil microorganisms live in the film of water coating soil particles and plant roots, and microbial activity is therefore greatly affected by the moisture status of the soil. Plant roots themselves also exert an important effect because they yield sugars and amino acids which serve as substrates for the microflora (Rovira, 1969), so that roots are surrounded by a veritable mantle of microorganisms (Gray, 1967). Thus, microorganisms are part of the immediate microenvironment of roots in soil (Norman, 1961; Parkinson, 1967; Starkey, 1958).

The relative importance of bacteria and fungi in molding the chemistry, physics, and biology of the soil is difficult to assess although it is certain that it differs from soil to soil (Went and Stark, 1968a). While bacteria seem to be concentrated in the immediate proximity of roots, fungi in any soil containing humus form a close network of filaments or hyphae throughout the soil. In sandy soils this may be an important factor in binding soil particles together. Like bacteria, fungi participate in the decomposition of plant litter on and in the soil.

Many fungi form associations with plant roots. The fungal threads or hyphae either form a sheath around the roots or actually penetrate the root tissue. These mycorrhizas ("root-fungi") make for a more intimate association between roots and soil than would otherwise obtain. Inorganic nitrogen, the result of mineralization of organic matter by the mycorrhizas, is passed on to the root tissue of the host plant. The supply of other nutrients to mycorrhizal roots also seems to be facilitated by these fungi. Harley (1969) has published a monograph and shorter accounts (Harley, 1956, 1963) on mycorrhizas and their significance in the life of the plants they infect. Went and Stark (1968b) have stressed their importance in the tropical rain forest, and Björkman (1970) has discussed their stimulating effect on the growth of conifers.

Algae play an important role in many soils. They frequently act as early colonizers on newly barren surfaces, as after volcanic eruptions. Shields and Durrell (1964) have reviewed the relation of algae to soil fertility. An account of earlier work on terrestrial algae is that by Fritsch (1936).

Various aspects of the chemistry of the soil in relation to plant nutrition are discussed by Bould and Hewitt (1963). *Soil Conditions and Plant Growth* (Russell, 1961) is a classic that has gone through nine editions since it was first published in 1912. Black (1968) has written a general textbook on soil–plant relationships, and a book by Fried and Broeshart (1967) emphasizes those chemical processes which occur outside the plant. The book on soil biology edited by Burges and Raw (1967) deals primarily with the organisms in the soil. Many aspects of the interplay between roots and soil, with emphasis on microorganisms and their activities, are discussed by various experts in the symposium volume edited by Baker and Snyder (1965). McLaren and Peterson (1967) have edited a volume on soil biochemistry. The concept of soil in its historical development has been discussed by Simonson (1968), and Eyre (1968) has given a world picture of vegetation and soils. Kellog and Orvedal (1969) of the U. S. Department of Agriculture have written on the arable soils of the world and their potential for agricultural production.

3. SOLUTION CULTURE

Introduction. Soils, being such complex and heterogeneous media, are not amenable to close experimental control. Specifically, the composition of the soil solution bathing the roots is an unknown quantity, varying from one root to the next, and from moment to moment. Active roots withdraw water and nutrients, microorganisms influence the root microenvironment, respiratory carbon dioxide from both roots and microflora changes the composition of the soil solution and the soil atmosphere, and there is constant traffic of materials between the soil solution and the solid phase of the soil.

Since accurate control of relevant variables is a *sine qua non* for meaningful experimentation, scientists interested in the physiology of mineral plant nutrition have turned to simpler, artificial substrates which enable them to control precisely the composition of the solution bathing the roots. The solution culture is by far the most important experimental device for investigations in plant nutrition.

In a typical solution culture set-up (Figure 3-3) the plant is supported in such a way that its roots are immersed in a solution—the culture solution—containing nutrient elements. The solution is kept oxygenated by bubbling air through it from a compressed air line. If all essential mineral nutrients and the pH are maintained at suitable levels in the culture solution plants will grow in such a solution just as well as in the most fertile soil, other conditions such as light and temperature being equally favorable.

Basically, the technique of growing plants in solution culture has changed little since Knop and Sachs first used it in the 1860's (Hoagland and Arnon, 1950). However, refinements in chemical and mechanical aspects of the technique have made it into a powerful tool for the investigation of many aspects of plant nutrition and metabolism. By means of special methods of purification, the concentration of certain contaminant elements can be brought down to the level of 1 part per billion or less. The concentration of dissolved oxygen in the solution can be regulated by bubbling appropriate mixtures of oxygen and nitrogen through the solutions, instead of air. Esoteric compounds such as herbicides can be added at the desired concentrations. It is possible to transfer the plants, with precise timing, from one solution to another of different chemical or isotopic com-

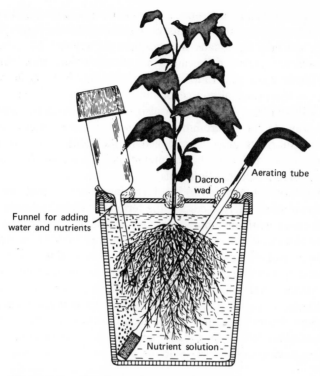

Figure 3-3.   Solution culture set-up.

position. Although the method of solution culture has already passed its centenary its possibilities as a research technique are by no means exhausted. Hewitt (1966) has written a comprehensive account of experimental culture methods used in the study of plant nutrition; it is the standard handbook on the subject.

The Chemical Composition of Nutrient Solutions.   Much time and effort have been expended on the formulation of nutrient solutions. Many compositions have been successful, and there is little likelihood that any particular combination of salt concentrations and ratios will prove decidedly superior to all others, although the search for such "best" or "balanced" elixirs of plant life is a time-honored pursuit (Homès, 1961, 1963; Shive, 1915; Shive and Martin, 1918).

There must be three macronutrient elements present in nutrient media in the form of cations, viz., potassium, calcium, and magnesium. The three macronutrient anions are nitrate, phosphate, and sulfate. All macronutrient elements can therefore be furnished by three salts, for example, potassium nitrate, calcium phosphate, and magnesium sulfate. As a rule, however, four salts are preferred, because this affords greater flexibility in varying the concentrations and ratios of ions if this should be desirable. Although nitrate is the principal source of nitrogen in most culture solutions, as it is in soil solutions, nitrogen can also be furnished as the ammonium ion. In addition to the macronutrient elements at appropriate concentrations, the micronutrients must be supplied in the solution at low but adequate levels, and the pH must be maintained within a suitable range (see next section, pH of Nutrient Solutions).

Hoagland and Arnon (1950) formulated two nutrient solutions which have been very widely used and the term "Hoagland solution" has become a household word in laboratories devoted to plant nutrition all over the world. Hoagland solution 2 contains ammonium ions as well as nitrate, and as a result is better buffered than solution 1. The second solution was modified by Johnson et al. (1957). The composition of their solution, with some further slight changes in micronutrient composition, is given in Table 3-1. Plants of many species have been successfully grown in this modified Hoagland solution. Steiner (1961) and Homès (1961, 1963) have devised schemes for systematically varying the concentrations of the macronutrient elements in culture solutions.

✕ **pH of Nutrient Solutions.** The pH of soil solutions ranges from 4 or below in the most acid soils to 8 and above in the most alkaline soils (Walker *et al.,* 1964). Some plant species are particularly well adapted to growth at either extreme of the pH range, as discussed in Chapter 13, which deals with ecological aspects of plant nutrition. Many species, however, grow well in the pH range from about 5 to 7, and most nutrient solutions have pH values in this range.

The importance of the pH of nutrient media is two-fold. First, the pH influences the oxidation–reduction equilibrium and the solubility of several constituents and the ionic form of several elements. In an aerated solution with a pH of 8, ferric iron, Fe^{3+}, precipitates as the extremely insoluble ferric hydroxide, $Fe(OH)_3$, with the result that iron may not be available for absorption by plants. The state of oxidation and the solubility of other heavy metal ions are also greatly influenced by pH (Hodgson, 1963; Ponnamperuma, 1955). As for anions, the ionic form of phosphate is a function of the pH. At pH 4, phosphate is predominantly $H_2PO_4^-$, but at pH 9, all but 1.5 per cent of the phosphate is in the HPO_4^{2-} form (Larsen, 1967). Neither the chemical nor the physiological behavior of these ionic species is identical.

TABLE 3-1. Nutrient Solution[1]

Macronutrients

Compound	Molecular weight	Concentration of stock solution, M	Concentration of stock solution, g/liter	Volume of stock solution per liter of final solution, ml	Element	Final concentration of element, μM	Final concentration of element, ppm
KNO_3	101.10	1.00	101.10	6.0	N	16000	224
$Ca(NO_3)_2 \cdot 4H_2O$	236.16	1.00	236.16	4.0	K	6000	235
$NH_4H_2PO_4$	115.08	1.00	115.08	2.0	Ca	4000	160
$MgSO_4 \cdot 7H_2O$	246.49	1.00	246.49	1.0	P	2000	62
					S	1000	32
					Mg	1000	24

Micronutrients

Compound[2]	Molecular weight	Concentration of stock solution, mM	Concentration of stock solution, g/liter	Volume of stock solution per liter of final solution, ml	Element	Final concentration of element, μM	Final concentration of element, ppm
KCl	74.55	50	3.728		Cl	50	1.77
H_3BO_3	61.84	25	1.546		B	25	0.27
$MnSO_4 \cdot H_2O$	169.01	2.0	0.338	1.0	Mn	2.0	0.11
$ZnSO_4 \cdot 7H_2O$	287.55	2.0	0.575		Zn	2.0	0.131
$CuSO_4 \cdot 5H_2O$	249.71	0.5	0.125		Cu	0.5	0.032
$H_2MoO_4 (85\% \ MoO_3)$	161.97	0.5	0.081		Mo	0.5	0.05
Fe–EDTA[3]	346.08	20	6.922	1.0	Fe	20	1.12

[1] After Johnson *et al.* (1957), modified.

[2] A combined stock solution is made up containing all micronutrients except iron.

[3] Ferrous dihydrogen ethylenediamine tetraacetic acid.

The second aspect of the pH of nutrient media has to do with the effect of hydrogen and hydroxyl ions on plant roots, especially on the ion transporting membranes of root cortical cells. These pH effects are discussed in Chapter 6 which deals with the process of active ion absorption.

The Osmotic Potential of Nutrient Solutions.   The higher the concentration of solutes the lower is the chemical potential of the water in which they are dissolved. This means that the tendency of water to diffuse and to react is diminished by the presence of solutes in the water. When pure water is separated from a solution by a membrane permeable to water but not to the solute, the pure water will be at a higher water potential than the water of the solution. As a result, water will tend to move across the membrane into the compartment containing the solution. This process, called "osmosis," can be reversed and water can be extruded from the solution across the membrane into the pure water by applying pressure to the solution so that its water potential becomes higher than that of the pure water on the other side of the membrane. This process is called "reverse osmosis." If such pressure is exerted on the solution as just to prevent net movement of water across the membrane, that pressure is the osmotic pressure of the solution. It has a positive sign. The osmotic potential of the solution is numerically equal to the osmotic pressure but its sign is negative.

The modern quantitative terminology of the water relations of plant cells is as follows (Kramer, 1969; Slatyer, 1967), where Ψ_{cell} is the potential of water in the cell, Ψ_s is the osmotic potential, and Ψ_p is the pressure potential:[1]

$$\Psi_{cell} = \Psi_s + \Psi_p$$

The water potential of pure water at atmospheric pressure equals zero. The osmotic potential, Ψ_s, is the difference between the chemical potential of water in a solution and that of pure water, and is therefore negative. The pressure potential, Ψ_p, is positive except in the rare situation when the cell wall pressure is negative. Under conditions of what Meyer and Anderson (1952) used to call a diffusion pressure deficit (DPD), the water potential of the cell water, Ψ_{cell}, is negative, because under these conditions the absolute value of Ψ_p, whose sign is positive, is less than the absolute value of Ψ_s, whose sign is negative. It is important to keep in

---

[1] Matric forces and their contribution to the water potential are disregarded here. They are forces by which water is bound by colloids and surfaces of cell components. For most plant cells they make but a small contribution to the water potential.

mind that the greater the water stress to which the cell is exposed, the lower will its water potential, Ψ_{cell}, tend to be. The water potential of the water of the fully turgid cell is zero because in this condition, the absolute values of Ψ_s and Ψ_p are equal. Since Ψ_s is negative and Ψ_p is positive their sum is then zero.

The units of water potential are in terms of energy per unit volume (erg cm^{-3}). This is dimensionally equivalent to pressure (dyne cm^{-2}) and in the context of plant cell water relations the convention is to express it in pressure units, either bars or atmospheres (1 bar $= 10^6$ dyne cm$^{-2} = 0.987$ atmosphere).

Typical osmotic potentials of the cells of plants growing in non-saline media are on the order of -10 to -20 bars, but there are large differences among species and even in plants of the same species grown in different media (Holley, 1966). The osmotic potential of most conventional nutrient solutions lies between -0.5 and -1 bars. For example, the nutrient solution of Table 3-1 has an osmotic potential of about -0.74 bar, depending somewhat on the temperature, pH, and other factors. The osmotic potentials of soil solutions in non-saline areas and under conditions of adequate water supply are also on this same order, seldom being less than -1 to -2 bars.

Certain aspects of plant–water relations are discussed in Chapter 13, dealing with ecological aspects.

MODIFIED SOLUTION CULTURES

Certain difficulties, both mechanical and chemical, encountered in work with solution culture set-ups like that shown in Figure 3-3 have led to modifications designed to circumvent or minimize these disadvantages. However, the modifications themselves introduce other undesirable features for at least some kinds of work, and for most purposes the classical, straightforward solution culture technique is to be preferred.

Mechanical Modifications.   In solution culture set-ups like that of Figure 3-3 the plants must be supported at the root-shoot transition zone. This is usually done by wrapping cotton or synthetic batting around the stem and then inserting the plant in a hole in the cover of the container. This being inconvenient and time-consuming, "sand culture" or "gravel culture" is sometimes substituted for solution culture proper. Washed sand or gravel is placed in containers and periodically irrigated by the culture solution, often by means of pumping devices which fill the containers with the solution at intervals and then let excess solution drain out (Eaton and Bernardin, 1962; Maynard *et al.,* 1970). Plants grow in these cultures, mechanically supported as in soil, though the solid phase is chemically inert and the solution alone is meant to supply nutrients.

There are two main disadvantages in this technique. First, if the experiment concerns micronutrients, even acid-washed, well-rinsed sand or gravel is likely to contribute contaminant elements. Secondly, it is often desirable to obtain the roots for analysis, and part of the root system is apt to be lost when the plants are removed from the solid rooting medium. These disadvantages are of no concern, however, when sand or gravel cultures are used not for experimental studies but for hydroponics, the commercial production of food and ornamental plants. The method is likely to become increasingly important, especially for high-priced crops like ornamental plants which can be grown in areas where the climate is favorable and good soil at a premium. Schwarz (1968) has given an account of commercial hydroponics, with emphasis on the extensive Israeli experience but applicable elsewhere. He gives a list of other publications on this subject.

Use of Ion Exchange Materials. Synthetic, solid ion exchange materials are available which bear positive or negative charges. A negatively charged exchange material, when immersed in a solution of a salt, will tend to attract and restrain positively charged ions from the solution, i.e., it will act as a cation exchanger, like negatively charged clay particles (see Figure 3-2). Positively charged exchange materials act as anion exchangers.

Ion exchange materials may be used to maintain culture solutions at a desired pH, without resort to salts of weak acids to buffer the solutions. For example, Hageman *et al.* (1961) added to a conventional culture solution a weak-acid cation exchange material bearing calcium and hydrogen ions in such proportions as to impart to the solution a pH of 4.2 to 4.5. The exchange resin maintained the pH of the solution within this range over a period of several weeks during which corn plants grew in the solution. Parr and Norman (1963, 1964) have used a similar exchange resin technique for controlling the pH of experimental solutions in short-term experiments with excised roots.

Ion exchange resins may be used not only for buffering otherwise conventional nutrient solutions but as sources of nutrient cations and anions (Converse *et al.,* 1943; Graham and Albrecht, 1943; Jenny, 1946; Schlenker, 1942). Plants may grow well in such "exchanger cultures." Nevertheless, this technique introduces a fairly complex medium in which the concentrations of various ions, including hydrogen and hydroxyl ions, are not uniform throughout but vary with the distance from the particles of exchange material. In other words, the technique reintroduces a measure of the complexity encountered in soils. As a result, it has not found much favor among those workers who wish to study the physiology and metabolism of nutrient absorption from media of strictly defined composition.

5. CULTURE SOLUTIONS COMPARED WITH SOIL SOLUTIONS

Conventional culture solutions like the one given in Table 3-1 are relatively concentrated compared with soil solutions. Reisenauer (1966) has surveyed and compiled numerous data on the chemical composition of soil solutions as indicated in Table 3-2. Inspection of the table shows that in many samples of soil solution the concentration of the macronutrient cations and anions is less than 100 ppm, and for phosphate, the great majority of samples showed concentrations less than 0.5 ppm.

A few comparisons between the concentrations of nutrient ions in the nutrient solution (Table 3-1) and soil solutions (Table 3-2) are revealing. Over 50 per cent of the soil solutions listed by Reisenauer contained potassium at concentrations not exceeding 50 ppm. The potassium concentration of the nutrient solution is almost 5 times as high. Nearly all soil solution samples contained phosphate phosphorus in the 0–0.50 ppm range while in the nutrient solution, the value is 62 ppm.

The reason for the high concentrations of the usual nutrient solutions is not that plants require such high concentrations. Williams (1961) grew barley plants, *Hordeum vulgare,* in culture solutions in which the potassium concentration was maintained at 0.01 ppm (0.25 μM). The plants grew normally, without any evidence of potassium deficiency, and contained normal concentrations of the element in their tissues. Asher and Ozanne (1967) have reported adequate rates of potassium supply to several species of Western Australian range plants grown in solutions maintained at 24 μM potassium. A phosphate concentration of 0.5 ppm PO_4 (5.3 μM) was found adequate for maximum growth of corn, *Zea mays,* sorghum, *Sorghum vulgare,* and tomato, *Lycopersicon esculentum,* in early experiments by Tidmore (1930). About the same concentration of phosphate (5 μM) proved adequate for optimal growth of several species in the experiments of Asher and Loneragan (1967).

It is apparent from these findings and from the results of the analyses of soil solutions that plants can grow well in solutions in which the concentrations of important nutrient elements are very low, compared with their

TABLE 3-2. Mineral Nutrients in Soil Solution[1]

Element (No. of samples)	Concentration, ppm	Fraction of samples, %
Potassium (155)	0–10	7.7
	11–20	11.0
	21–30	12.9
	31–40	12.9
	41–50	10.3
	51–60	7.7
	61–80	11.6
	81–100	10.3
	101–200	10.3
	>200	5.2
Calcium (979)	0–50	23.1
	51–100	54.6
	101–200	8.1
	201–300	2.4
	301–400	1.9
	401–500	3.8
	501–600	1.8
	601–700	1.5
	701–800	0.9
	801–1000	1.3
	>1000	0.4
Magnesium (337)	0–25	9.2
	26–50	21.4
	51–100	38.6
	101–200	25.2
	201–300	0.9
	301–500	0.6
	501–700	1.8
	701–1000	—
	>1000	2.4

Element (No. of samples)	Concentration, ppm	Fraction of samples, %
Nitrogen (as NO_3) (879)	0–25	4.9
	26–50	14.3
	51–100	28.8
	101–150	32.2
	151–200	10.5
	201–300	2.7
	301–400	4.9
	401–500	1.0
	501–1000	0.4
	>1000	0.4
Phosphorus (as PO_4) (149)	0–0.03	25.5
	0.031–0.06	18.8
	0.061–0.10	16.8
	0.101–0.15	12.1
	0.151–0.20	2.7
	0.201–0.25	2.0
	0.251–0.30	4.0
	0.301–0.40	6.0
	0.401–0.50	4.0
	>0.50	8.1
Sulfur (as SO_4) (693)	0–25	16.5
	26–50	40.1
	51–100	38.1
	101–200	3.2
	201–400	1.3
	401–600	0.1
	601–1000	0.1
	1001–2000	0.3
	>2000	0.3

[1] After Reisenauer (1966).

concentrations in conventional nutrient solutions. The reason for use of high concentrations in nutrient solutions lies in the experimental difficulty of maintaining a low concentration of an element in a nutrient solution in the face of constant withdrawal by the plant (Asher *et al.,* 1965; Reisenauer, 1969). It is the need for a large total supply of the elements if plants are to grow to maturity, not for high external concentrations of them, that has led to the conventional use of nutrient solutions which are unrealistically concentrated when compared with soil solutions. This difficulty has been overcome in only a few investigations like those referred to above, in which relatively few plants drew nutrients from large volumes of circulating solution which was replenished at regular intervals so as to maintain nutrient concentrations at the desired low levels. In soil, this function of constant replenishment of the soil solution is performed by the solid phase, as discussed above.

REFERENCES

Asher, C. J. and J. F. Loneragan. 1967. Response of plants to phosphate concentration in solution culture: I. Growth and phosphorus content. Soil Sci. 103:225–233.

Asher, C. J. and P. G. Ozanne. 1967. Growth and potassium content of plants in solution cultures maintained at constant potassium concentrations. Soil Sci. 103:155–161.

Asher, C. J., P. G. Ozanne and J. F. Loneragan. 1965. A method for controlling the ionic environment of plant roots. Soil Sci. 100:149–156.

Baker, K. F. and W. C. Snyder, eds. 1965. Ecology of Soil-Borne Plant Pathogens—Prelude to Biological Control. University of California Press, Berkeley, Los Angeles.

Barber, D. A. 1968. Microorganisms and the inorganic nutrition of higher plants. Ann. Rev. Plant Physiol. 19:71–88.

Barber, D. A. 1969. The influence of the microflora on the accumulation of ions by plants. In: Ecological Aspects of the Mineral Nutrition of Plants. I. H. Rorison, ed. Blackwell Scientific Publications, Oxford and Edinburgh. Pp. 191–200.

Bergersen, F. J. and E. H. Hipsley. 1970. Presence of nitrogen fixing bacteria in intestines of man and animals. J. Gen. Microbiol. 60:61–65.

Björkman, E. 1970. Forest tree mycorrhiza—the conditions for its formation and the significance for tree growth and afforestation. Plant and Soil 32:589–610.

Black, C. A. 1968. Soil-Plant Relationships. 2nd ed. John Wiley and Sons, Inc., New York.

Bould, C. and E. J. Hewitt. 1963. Mineral nutrition of plants in soils and in culture media. In: Plant Physiology. F. C. Steward, ed. Academic Press, New York and London. Vol. 3, pp. 15–133.

Burges, A. and F. Raw, eds. 1967. Soil Biology. Academic Press, London and New York.

Converse, C. D., N. Gammon and J. D. Sayre. 1943. The use of ion exchange materials in studies of corn nutrition. Plant Physiol. 18:114–121.

Dittmer, H. J. 1937. A quantitative study of the roots and root hairs of a winter rye plant (*Secale cereale*). Am. J. Bot. 24:417–420.

Eaton, F. M. and J. E. Bernardin. 1962. Soxhlet-type automatic sand cultures. Plant Physiol. 37:357.

Eyre, S. R. 1968. Vegetation and Soils—A World Picture. 2nd ed. Aldine Publishing Company, Chicago.

Fried, M. 1948. The absorption of sulfur dioxide by plants as shown by the use of radioactive sulfur. Soil Sci. Soc. Am. Proc. 13:135–138.

Fried, M. and H. Broeshart. 1967. The Soil-Plant System in Relation to Inorganic Nutrition. Academic Press, New York and London.

Fritsch, F. E. 1936. The rôle of the terrestrial alga in nature. In: Essays in Geobotany in Honor of William Albert Setchell. T. H. Goodspeed, ed. University of California Press, Berkeley. Pp. 195–217.

Gilmour, C. M. and O. N. Allen, eds. 1965. Microbiology and Soil Fertility. Oregon State University Press, Corvallis.

Gorham, E. 1961. Factors influencing supply of major ions to inland waters, with special reference to the atmosphere. Geol. Soc. Am. Bull. 72:795–840.

Graham, E. R. and W. A. Albrecht. 1943. Nitrate absorption by plants as an anion exchange phenomenon. Am. J. Bot. 30:195–198.

Gray, T. R. G. 1967. Stereoscan electron microscopy of soil microorganisms. Science 155:1668–1670.

Hageman, R. H., D. Flesher, J. J. Wabol and D. H. Storck. 1961. An improved nutrient culture technique for growing corn under greenhouse conditions. Agron. J. 53:175–180.

Hall, N. S., W. F. Chandler, C. H. M. van Bavel, P. H. Reid and J. H. Anderson. 1953. A tracer technique to measure growth and activity of plant root systems. North Carolina Agric. Expt. Sta. Tech. Bull. 101:1–40.

Harley, J. L. 1956. The mycorrhiza of forest trees. Endeavour 15:43–48.

Harley, J. L. 1963. Mycorrhiza. In: Vistas in Botany. W. B. Turrill, ed. Pergamon Press, Ltd., London. Vol. 3, pp. 79–103.

Harley, J. L. 1969. The Biology of Mycorrhiza. 2nd ed. Leonard Hill Books, London.

Hewitt, E. J. 1966. Sand and Water Culture Methods Used in the Study of Plant Nutrition. Revised 2nd ed. Commonwealth Bureau of Horticulture and Plantation Crops, East Malling. Tech. Communication No. 22.

Hoagland, D. R. and D. I. Arnon. 1950. The Water-Culture Method for Growing Plants without Soil. Calif. Agric. Expt. Sta. Circ. 347.

Hodgson, J. F. 1963. Chemistry of the micronutrient elements in soils. Adv. Agron. 15:119–159.

Holley, K. T. 1966. Freezing point depression, osmotic pressure, and conductivity of plant sap: angiosperms. In: Environmental Biology. P. L. Altman and D. S. Dittmer, eds. Federation of American Societies for Experimental Biology, Bethesda. Pp. 540–541.

Homès, M. V. L. 1961. L'Alimentation Minérale Equilibrée des Végétaux. Théorie, Méthodologie, Applications. Vol. I. L'Alimentation sur Milieux Dépourvus de Fertilité Naturelle. Universa, Wetteren, Belgium.

Homès, M. V. 1963. The method of systematic variations. Soil Sci. 96:380–386.

Jenny, H. 1941. Factors of Soil Formation. A System of Quantitative Pedology. McGraw-Hill Book Company, Inc., New York and London.

Jenny, H. 1946. Adsorbed nitrate ions in relation to plant growth. J. Colloid Sci. 1:33–47.

Johnson, C. M., P. R. Stout, T. C. Broyer and A. B. Carlton. 1957. Comparative chlorine requirements of different plant species. Plant and Soil 8:337–353.

Kellog, C. E. and A. C. Orvedal. 1969. Potentially arable soils of the world and critical measures for their use. Adv. Agron. 21:109–170.

Kramer, P. J. 1969. Plant and Soil Water Relationships: A Modern Synthesis. McGraw-Hill Book Company, Inc., New York.

Larsen, S. 1967. Soil phosphorus. Adv. Agron. 19:151–210.

Maynard, D. N., A. V. Barker and H. F. Vernell. 1970. A semiautomatic sand culture system for greenhouse plant nutrition research. Agron. J. 62:304–306.

McLaren, A. D. and G. H. Peterson, eds. 1967. Soil Biochemistry. Marcel Dekker, Inc., New York.

Meyer, B. S. and D. B. Anderson. 1952. Plant Physiology. 2nd ed. D. Van Nostrand Company, Inc., Princeton.

Monteith, J. L. 1963. Dew: facts and fallacies. In: The Water Relations of Plants. A. J. Rutter and F. H. Whitehead, eds. John Wiley and Sons, Inc., New York. Pp. 37–56.

National Academy of Sciences—National Research Council. 1962. Basic

Problems and Techniques in Range Research. N.A.S.–N.R.C. Publication No. 890, Washington, D. C. Pp. 85–108.

Norman, A. G. 1961. The biological environment of roots. In: Growth in Living Systems. M. X. Zarrow, ed. Basic Books, Inc., New York. Pp. 653–664.

Olsen, R. A. 1957. Absorption of sulfur dioxide from the atmosphere by cotton plants. Soil Sci. 84:107–111.

Parkinson, D. 1967. Soil micro-organisms and plant roots. In: Soil Biology. A. Burges and F. Raw, eds. Academic Press, London and New York. Pp. 449–478.

Parr, J. F. and A. G. Norman. 1963. A procedure for control of pH in cation uptake studies with excised barley roots. Soil Sci. Soc. Am. Proc. 27:531–534.

Parr, J. F. and A. G. Norman. 1964. pH control in nitrate uptake studies with excised roots. Plant and Soil 21:185–190.

Ponnamperuma, F. N. 1955. The Chemistry of Submerged Soils in Relation to the Growth and Yield of Rice. Ph.D. Thesis, Cornell University.

Reisenauer, H. M. 1966. Mineral nutrients in soil solution. In: Environmental Biology. P. L. Altman and D. S. Dittmer, eds. Federation of American Societies for Experimental Biology, Bethesda. Pp. 507–508.

Reisenauer, H. M. 1969. A technique for growing plants at controlled levels of all nutrients. Soil Sci. 108:350–353.

Rovira, A. D. 1965. Interactions between plant roots and soil microorganisms. Ann. Rev. Microbiol. 19:241–266.

Rovira, A. D. 1969. Plant root exudates. Bot. Rev. 35:35–57.

Russell, E. W. 1961. Soil Conditions and Plant Growth. 9th ed. John Wiley and Sons, Inc., New York.

Schlenker, F. S. 1942. Availability of adsorbed ions to plants growing in quartz sand substrate. Soil Sci. 54:247–251.

Schwarz, M. 1968. Guide to Commercial Hydroponics. Israel Universities Press, Jerusalem.

Shields, L. M. and L. W. Durrell. 1964. Algae in relation to soil fertility. Bot. Rev. 30:92–128.

Shive, J. W. 1915. A three-salt nutrient solution for plants. Am. J. Bot. 2:157–160.

Shive, J. W. and W. H. Martin. 1918. A comparative study of salt requirements for young and for mature buckwheat plants in solution cultures. J. Agric. Res. 14:151–175.

Simonson, R. W. 1968. Concept of soil. Adv. Agron. 20:1–47.

Slatyer, R. O. 1967. Plant-Water Relationships. Academic Press, London and New York.

Starkey, R. L. 1958. Interrelations between microorganisms and plant roots in the rhizosphere. Bact. Rev. 22:154–172.

Steiner, A. A. 1961. A universal method for preparing nutrient solutions of a certain desired composition. Plant and Soil 15:134–154.

Stone, E. C. 1957. Dew as an ecological factor. II. The effect of artificial dew on the survival of *Pinus ponderosa* and associated species. Ecology 38:414–422.

Stout, P. R. and R. Overstreet. 1950. Soil chemistry in relation to inorganic nutrition of plants. Ann. Rev. Plant Physiol. 1:305–342.

Taylor, H. M. 1969. The Rhizotron at Auburn, Alabama—A Plant Root Observation Laboratory. Auburn University Agric. Expt. Sta. Circ. 171. Pp. 1–9.

Thomas, M. D., R. H. Hendricks and G. R. Hill. 1950. Sulfur metabolism of plants—effects of sulfur dioxide on vegetation. Ind. Engin. Chem. 42:2231–2235.

Tidmore, J. W. 1930. Phosphate studies in solution cultures. Soil Sci. 30:13–31.

Trench, R. K., R. W. Greene and B. G. Bystrom. 1969. Chloroplasts as functional organelles in animal tissues. J. Cell. Biol. 42:404–417.

Walker, R. B., E. T. Wherry and W. F. Bennett. 1964. Soil pH: spermatophytes. In: Biology Data Book. P. L. Altman and D. S. Dittmer, eds. Federation of American Societies for Experimental Biology, Washington. P. 442.

Weaver, J. E. 1968. Prairie Plants and their Environment. University of Nebraska Press, Lincoln.

Weaver, J. E. and F. E. Clements. 1938. Plant Ecology. 2nd ed. McGraw-Hill Book Company, New York.

Weaver, J. E. and R. W. Darland. 1949. Soil-root relationships of certain native grasses in various soil types. Ecol. Monographs 19:303–338.

Went, F. W. and N. Stark. 1968a. The biological and mechanical role of soil fungi. Proc. Nat. Acad. Sci. 60:497–504.

Went, F. W. and N. Stark. 1968b. Mycorrhiza. BioScience 18:1035–1039.

Williams, D. E. 1961. The absorption of potassium as influenced by its concentration in the nutrient medium. Plant and Soil 15:387–399.

4

THE INORGANIC
COMPONENTS
OF PLANTS

1. WATER

Life first evolved in the ocean or in tidal pools, and water remains the universal medium in which the activities of living cells take place. We distinguish between terrestrial organisms and aquatic ones, on the basis of their habitats, but physiologically, all living beings are aquatic, for their cells can only function when bathed by water and permeated by it. Where water is in short supply, as in deserts, the population density of living things, animals as well as plants, is very low. Plant life is most luxuriant in the tropical rain forests where a thousand centimeters of rain or more (several hundred inches) fall each year.

Aquatic plants live immersed in water, but in land plants only the roots normally have constant access to water. In the course of the evolution of terrestrial plants complex structures and mechanisms have been elaborated which supply even the topmost leaves of tall trees with water.

Fresh plant matter is about 80–95 per cent water. Inactive or dead tissues, like the heartwood of trees, contain much less water, and seeds may contain as little as a few per cent. Such seeds, however, must imbibe water if they are to germinate. Kramer (1969) lists the water contents of various plant materials. Because of the prominence of water, hydrogen and oxygen atoms combined outnumber all other elements as constituents of living plants.

Why is it that water, of all substances, is the medium and the matrix of living things? The unique properties of water which make for its fitness as the ambiance of life have been discussed by Blum (1968) and by Needham (1965), among others. We shall here draw attention to only a few which are particularly relevant.

First is its abundance on earth—about 1.25×10^{24} grams, enough for a layer 2.5 kilometers (1.5 miles) deep if it were spread evenly over the face of the globe. Second is the fact that at the temperatures prevailing over most of the surface of the earth water is a liquid. Gases are so dynamic and structureless as to prevent the generation of organized units, and solids too rigid for the degree of dynamism and chemical activity essential for life processes.

The great heat capacity of water causes it to be a temperature buffer—relatively large quantities of heat must be gained or lost by a given quantity of water to effect a given change in temperature. Living things are therefore protected against violent changes in temperature which might upset their chemical homeostasis or steady state. Also important for temperature regulation is the high latent heat of vaporization of water: a large input of heat is required for the evaporation of a given amount of water, i.e., 500–600 calories per gram, depending on the temperature. This value is higher than that of any other liquid. This property has two important advantages for living things: it minimizes the loss of water, and it provides maximal cooling per unit water lost, since the heat which goes into the evaporation of the water is withdrawn from the immediate environment.

The extremely high surface tension of water is still another important property because it functions to maintain discrete boundaries where moist cells abut on air and where aqueous phases border on lipid membranes. This same property also promotes the retention of films of water on soil particles when much of the water filling the interstices among the particles after a rain has drained away. The cohesiveness of water responsible for these features of its behavior is also important in its movement through the plant body, a subject discussed in Chapter 7.

Water is the best solvent known. Few substances are not at least somewhat soluble in water, and most substances of biological significance are extremely soluble in it. In the context of mineral plant nutrition the high solubility of inorganic salts in water is especially significant. The growth of plants depends on a constant traffic of mineral elements into and through the plant body, and no known solvent other than water could serve the function of a carrier of these mineral nutrients. The viscosity of water is fairly low, a factor which permits rapid diffusion of solutes including inorganic ions, and makes possible the rapid flow of water through the narrow conduits of the vascular system.

Many of the unique properties of water are due to the hydrogen bond—
a low-energy bond in which hydrogen atoms form a "bridge" between two
other atoms which "share" it. As a result of hydrogen bonding, molecules
of water are loosely knit together:

$$\begin{array}{ccc} H & & H \\ | & & | \\ H\!-\!O & \cdots & H\!-\!O \end{array}$$

The dots represent the hydrogen bond between two water molecules. The
energy needed to break these bonds contributes to the stability of water
reflected in such properties as its high latent heat of vaporization and high
surface tension.

DRY MATTER

When fresh plant material is dried at 70°C for 24 to 48 hours the dry
matter remaining will be roughly 10–20 per cent of the initial fresh or wet
weight. The results of chemical analyses of plant matter are usually expressed on the basis of the dry weight rather than the fresh weight because
the latter is variable, changing with the time of day, the amount of moisture
available in the soil, the temperature, wind velocity, and other factors (see,
however, section 9 in this chapter).

Over 90 per cent of the dry weight of most plant matter consists of
the three elements carbon, oxygen, and hydrogen. Reference to Table 4-1
will show that a crop species, corn, contains these elements in about the
same proportions in which they are combined in carbohydrate, i.e.,
$C(H_2O)$. This is a reflection of the fact that the bulk of the dry weight
of plants is due to the cell walls, which consist mainly of cellulose, i.e.,
a carbohydrate polymer. Surrounded by the cell wall is the cytoplasm—the
sum total of the proteins and other chemical entities which make up the
living machinery of the cell. On a weight basis, however, the cytoplasm
and its non-living inclusions such as the vacuoles make up but a small percentage of the dry matter of plants.

MINERAL COMPOSITION

If only 10–20 per cent, say 15 per cent, of the fresh weight of a plant
is dry matter, and all but 10 per cent of the dry matter is represented
by carbon, oxygen, and hydrogen, it follows that all the many elements
that make up the plant, except carbon and the elements of water, together

TABLE 4-1. Elemental Composition of some Organisms and Substances[1]

Element	Corn plant, Zea mays	Man, Homo sapiens	Per cent of dry weight		
			Carbohydrate	Fat	Protein
O	44.43	14.62	51.42	11.33	24
C	43.57	55.99	42.10	76.54	52
H	6.24	7.46	6.48	12.13	7
N	1.46	9.33	—	—	16
Si	1.17	0.005	—	—	—
K	0.92	1.09	—	—	—
Ca	0.23	4.67	—	—	—
P	0.20	3.11	—	—	—
Mg	0.18	0.16	—	—	—
S	0.17	0.78	—	—	1
Cl	0.14	0.47	—	—	—
Al	0.11	—	—	—	—
Fe	0.08	0.012	—	—	—
Mn	0.04	—	—	—	—
Na	—	0.47	—	—	—
Zn	—	0.010	—	—	—
Rb	—	0.005	—	—	—

[1] Corn plant, from Miller (1938); man, recalculated after Oser (1965); carbohydrate, on the basis of sucrose; fat, on the basis of oleic acid; protein, after various sources.

account for roughly $0.15 \times 0.10 = 0.015$, or a mere 1.5 per cent of the fresh weight of the plant. These elements represent the mineral composition of the plant, its content of phosphorus, sulfur, potassium, calcium, magnesium, silicon, and many other elements taken up from the surroundings, principally the soil.

To determine the mineral elements the investigator ashes the dry plant material, thereby removing the organic carbon compounds, and then analyzes the ash for the elements in which he is interested. Special procedures and precautions must be followed if analyses are to be done of nitrogen, sulfur, or other elements that might be lost by volatilization during the ashing. Chapman and Pratt (1961), among many, have brought together methods of chemical analysis that have been successfully used for the analysis of plant materials.

4. ESSENTIAL AND OTHER MINERAL ELEMENTS

The finding that a certain element is present in a plant does not in itself signify that this element plays an essential role in the life of that plant. Soil contains numerous chemical elements; with sufficiently sensitive methods, a majority of the elements in the periodic table of elements could probably be detected in any sample of soil taken at random. A plant growing on soil might therefore contain at least traces of most of the elements, those essential for its growth as well as others that are taken up because the absorption mechanisms make no absolute selection between essential and other elements.

At this point we should clarify by what criteria an element is classified as an essential nutrient element or one that is not essential. There are two criteria of essentiality, each self-sufficient (Epstein, 1965). An element is essential (a) if, without it, the plant cannot complete its life cycle, and (b) if it is part of the molecule of an essential plant constituent or metabolite. To give first an example of the second criterion, plant life as we know it is inconceivable without water; water is one of the chief metabolites of plants and all other living things. By the second criterion, then, hydrogen and oxygen are essential elements. Another example is magnesium, a constituent of the chlorophyll molecule.

Historically, however, it has been the first criterion by which the essentiality of nutrient elements has been established. This criterion was laid down by Arnon and Stout (1939a) who also included the specification that indirect beneficial effects of an element, such as the correction of some unfavorable microbial or chemical condition of the nutrient medium, do not qualify an element for "essential" status. An "essential" element must be directly involved in the metabolism of the plant. To apply this test, plants are grown with the element of interest omitted from the culture medium, or present in inadequate amount or concentration. Until about the middle of the 19th century, plant scientists did not have an experimental technique that would allow them to grow plants in well defined media from which elements could be purged at will. This situation changed with the development of the method of solution culture, which has enabled experimenters to establish the essentiality of nutrient elements (Hoagland, 1948; Stout, 1956).

5. MACRONUTRIENT ELEMENTS

The composition of nutrient solutions can be varied by the investigator, and it soon was found that plants would only grow in solutions containing at least the following seven mineral elements: nitrogen, phosphorus, sulfur, potassium, calcium, magnesium, and iron. Carbon by then was known to be taken up from the air, so that, counting those of water, ten elements were recognized as being essential plant nutrients. No generally accepted additions were made to this classical list of essential elements during the remainder of the 19th century. With the exception of iron, these elements are needed in relatively large amounts; they are "macronutrient elements" or simply, "macronutrients."

6. MICRONUTRIENT ELEMENTS

In 1905 the French investigator, G. Bertrand, concluded that manganese is an essential nutrient element for plants, and subsequent research has confirmed this conclusion. The amounts of manganese required were much smaller than those of most of the other elements then known to be essential. Only one element in the classical list of ten, namely iron, is needed in comparably small amounts. Such elements are called "micronutrients," to distinguish them from the nine macronutrient elements.

Beginning in 1914, the Frenchman P. Mazé published papers in which he claimed essentiality for several other elements, basing his conclusions on the results of experiments with solution cultures. Some of these claims have stood the test of later, more refined experimentation; others have not been confirmed. Mazé's work was useful, in any event, because it drew to the attention of plant physiologists the possibility that additional elements might be shown to be essential nutrients.

Katherine Warington of Oxford, England, showed in the 1920's that boron is an essential micronutrient for several species of the legume family, and A. L. Sommer and C. B. Lipman of the University of California proved boron to be essential for non-leguminous plants as well. J. S. McHargue in Kentucky conclusively demonstrated the need for manganese, and Lipman and his associates soon added zinc and copper to the list of essential micronutrients. By 1931, then, in addition to the classical ten essential nutrient elements, four more were known to be required: boron, manganese, zinc, and copper.

In the late 1930's P. R. Stout and D. I. Arnon, working in the University of California laboratory of plant nutrition headed by D. R. Hoagland, perfected techniques for removing contaminating elements from nutrient solutions and in 1939 were able to publish evidence proving conclusively that molybdenum is an essential micronutrient for the tomato plant. Molybdenum is now recognized as essential for plants generally (Stout, 1956;

Stout and Johnson, 1956). Molybdenum present in nutrient solutions at a concentration of 0.1 μmole/liter (10 parts per billion) completely prevents the symptoms of molybdenum deficiency disease from appearing. This gives some indication of the need for rigorous exclusion of adventitious contamination in experiments on the micronutrient requirements of plants (Arnon and Stout, 1939b; Hoagland, 1944; Stout and Arnon, 1939).

No further additional micronutrient requirements were uncovered till 1954 when a team at the University of California headed by T. C. Broyer proved the need for chlorine, first for the tomato plant (Broyer *et al.,* 1954) and subsequently for many other species as well (Johnson *et al.,* 1957). Oddly enough, the quantitative requirements for chlorine are considerably higher than those for any of the other micronutrients, and one might reason that the need for chlorine should have become manifest earlier than the essentiality of the other micronutrient elements. Part of the answer to this puzzle seems to be the fact that plants may acquire considerable chlorine from the atmosphere, especially in locations near the ocean. Chlorine is the latest addition to the roster of elements known to be required by green plants generally. For an excellent historical account of the discovery and significance of the micronutrient elements by one of the major contributors to this research the reader is referred to the paper by Stout (1956). For a recent discussion by specialists, see the symposium volume edited by Mortvedt *et al.* (1971).

7. COMPARATIVE ELEMENTAL REQUIREMENTS OF HIGHER PLANTS

Our discussion of essential elements has shown that at the present time, sixteen elements are known to be generally required by higher green plants: carbon, hydrogen, oxygen, nitrogen, phosphorus, sulfur, potassium, calcium, magnesium, iron, manganese, zinc, copper, boron, molybdenum, and chlorine. There are, however, certain plants which are known to require additional elements. Other plants may not require one or more of these elements, and in still others certain elements may to a greater or lesser degree substitute for elements listed. In this and the following section we shall explore these variations on the standard theme.

Sodium.   Sodium is not known to be generally required by green plants. However, certain halophytes, plants indigenous to saline soils, not only tolerate the high concentrations of salt in such soils but actually require sodium. *Atriplex vesicaria* is a perennial pasture species of inland Australia. Although not a typical halophyte (it is not usually found on highly saline soils) it is very tolerant of high concentrations of salt. Brownell and Wood (1957) and Brownell (1965) showed that this species requires sodium as a micronutrient element. Several other species of *Atriplex* also have a sodium requirement (Brownell, 1968).

Another such species is *Halogeton glomeratus,* a poisonous weed which has spread over many millions of acres of rangeland in the western United States. This halophytic species has such a high salt requirement that sodium may be considered a macronutrient element for it (Williams, 1960).

Selenium.   Selenium is generally toxic to plants. However, certain plants indigenous on soils rich in selenium not only accumulate and tolerate high concentrations of this element but may even have a requirement for it (Rosenfeld and Beath, 1964; Shrift, 1969). Such plants, for example *Astragalus racemosus* and other species of this genus in the legume family, serve as selenium indicator plants.

Cobalt.   A general requirement by higher green plants for cobalt has not been established. However, plants which are hosts to nitrogen-fixing bacteria contained in root nodules, for example, the legumes, represent a special case. Except as noted below, when such plants have fixed nitrogen such as ammonium or nitrate available, a requirement for cobalt has not been shown. On the other hand, when these plants depend for their nitrogen supply on fixation of atmospheric nitrogen by their bacterial symbionts, cobalt is essential for their growth (Ahmed and Evans, 1959, 1961; Evans and Kliewer, 1964; Reisenauer, 1960). It appears that cobalt is required for nitrogen fixation by the symbiotic root-nodule bacteria. According to a report by Wilson and Nicholas (1967), the legume, subterranean clover, *Trifolium subterraneum,* grown with a supply of fixed nitrogen and not depending on symbiotic nitrogen fixation, and the nonlegume, wheat, *Triticum durum,* require cobalt. If these findings are repeated and extended cobalt may join the ranks of generally essential micronutrients for higher plants.

Silicon.   Plants grown in soil invariably contain appreciable amounts of silicon (Fidanovski, 1968; Lewin and Reimann, 1969), but experiments with plants grown in nutrient solutions from which this element was omitted have in general failed to show that it is essential. However, silicon plays a role in the normal development of at least certain plants which, when growing in soil, accumulate large amounts of this element. Yoshida *et al.* (1959) and Mitsui and Takatoh (1960, 1963a,

b) found that rice failed to grow normally in solutions lacking silicon. Its essentiality, however, has not been demonstrated for rice. On the other hand, Chen and Lewin (1969) showed silicon to be essential for *Equisetum arvense,* one of the "scouring rushes" whose ash was used in bygone days to scour pots and pans, because of the gritty crystals of silica that it contains.

Smaller amounts of numerous other elements can readily be demonstrated in all soil-grown plants but apparently can be dispensed with as shown by experiments with purified solutions. Just the same, it is improper to call such elements "nonessential." "Apparently nonessential" or "not known to be essential" are better expressions. Even highly purified solutions may still contain undetectable traces of the element under consideration, and the plants growing in these solutions may benefit by scavenging these small amounts from the solutions.

Woolley (1957) conducted extremely careful experiments with tomato plants grown in solutions from which silicon had been as carefully removed as available techniques permitted. When the plants were analyzed they were found to contain 0.15 μmole silicon per gram dry weight of tissue. Plants which were deliberately given silicon did not grow better than the controls. The conclusion from this experiment was not that silicon is "nonessential," but that if it is essential, the requirement is less than about 0.2 μmole silicon per gram of dry tissue.

Other Elements.   Many other elements, though not apparently essential, are nevertheless important in the mineral economy of plant life. Some of them are discussed in Chapter 13 which deals with ecological aspects of mineral plant nutrition. Bollard and Butler (1966) have discussed many of these elements and their putative significance.

COMPARATIVE ELEMENTAL REQUIREMENTS OF ALGAE, FUNGI, AND BACTERIA

Algae.   With the exception of certain groups, the green algae, *Chlorophyta,* require the same macronutrient elements that are essential for higher green plants. Calcium, however, is for most algae a micronutrient element rather than a macronutrient (Gerloff and Fishbeck, 1969), and in some strains of some species strontium can substitute for calcium (Walker, 1956). Strains also have been obtained in which rubidium can substitute for potassium (Pirson, 1939; Kellner, 1955).

Many marine and brackish water algae have a requirement for sodium, and chlorine is often a macronutrient rather than a micronutrient for them (Guillard, 1962). In at least some of these algae, bromide can substitute for chloride (McLachlan and Craigie, 1967). Some fresh water algae, for example *Anabaena cylindrica* and other blue-green algae, *Cyanophyta*, also have a sodium requirement (Allen and Arnon, 1955; Kratz and Myers, 1955).

The diatoms, *Bacillariophyceae*, have an unusual macronutrient requirement. These organisms are yellow-green algae with beautifully sculptured walls or "frustules" composed of almost pure hydrated silica ($SiO_2 \cdot nH_2O$) (Reimann *et al.*, 1965). Experimental studies have shown silicon to be essential for these algae; without silicon, the cells fail to divide (Darley and Volcani, 1969; Lewin, 1962). Boron does not seem to be required by at least some green algae (Dear and Aronoff, 1968), but it is essential for diatoms (Lewin, 1966). Several blue-green algae have been shown to respond to it (Gerloff, 1968).

One species of green algae, *Scendesmus obliquus,* has been reported to require vanadium, an element not known to be essential for any other plants (Arnon and Wessel, 1953). Several blue-green algae, *Cyanophyta,* require cobalt as a micronutrient element (Holm–Hansen *et al.*, 1954). Algae from several groups have been shown to require vitamin B_{12} (Droop, 1962; Evans and Kliewer, 1964; Pedersén, 1969b; Provasoli, 1963). Since the vitamin B_{12} molecule contains an atom of cobalt, the requirement for this vitamin shows the essentiality of cobalt for these algae. Other algae which do not need an external supply of vitamin B_{12} may nevertheless require it and meet the requirement by synthesizing the vitamin themselves. For such algae cobalt would be an essential micronutrient. A general discussion of the micronutrient requirements of algae has been given by Wiessner (1962).

Certain algae, especially in the group of brown algae, *Phaeophyta,* accumulate high concentrations of iodide; in some marine species, internal concentrations of iodide may reach levels 30,000 times higher than the iodide concentration of sea water. Whether the iodide-accumulating algae generally require this element is not known but for some of these species iodine is essential (Fries, 1966; Pedersén, 1969a; von Stosch, 1962, 1963). For a general discussion of the role of halogens in algae see Shaw (1962).

Fungi.   The true fungi, *Eumycota,* require the same macronutrient elements as do higher green plants (Cochrane, 1958) except that potassium is a micronutrient for some fungi (Hickman, 1969), and not all fungi can be shown to need calcium. For those that do, the requirement is small, making calcium a micronutrient element for these organisms. The other known micronutrient elements of fungi are iron, manganese, zinc, copper, and molybdenum. Boron has not been shown to be essential for fungi (Bowen and Gauch, 1966), and neither have chlorine, sodium, and cobalt (Lilly, 1965).

Bacteria. The macronutrient elements of bacteria, *Schizomycophyta,* are the following, in addition to carbon, nitrogen, and the elements of water: potassium, magnesium, phosphorus, and sulfur. In some species of bacteria, rubidium can largely substitute for potassium (MacLeod and Snell, 1950; Wyatt, 1963). Calcium has not been shown to be generally required. For those bacteria that do need it, calcium is a micronutrient element (Nicholas, 1963).

The micronutrient element requirements of bacteria are not well established. Iron and manganese are known to be needed, but such well known micronutrients for higher plants as zinc, copper, and molybdenum have been shown to be essential for only a few species of bacteria (Guirard and Snell, 1962). Free-living, nitrogen-fixing bacteria such as *Azotobacter chroococcum* need cobalt, as do the symbiotic species already discussed (Iswaran and Sundara Rao, 1964).

There are several groups of bacteria which are halophilic (Larsen, 1962). Some are extreme halophiles which grow best in solutions with concentrations of sodium chloride on the order of 4 M, or about 25 per cent salt. Most if not all of these bacteria are obligate halophiles—they cannot grow except in the presence of high concentrations of NaCl. The same is true of most moderate halophiles. They require a lower concentration of salt for optimal growth than do the extreme halophiles; however, this requirement is also absolute and specific. Cations other than sodium and anions other than chloride either cannot substitute, or can substitute to only a limited extent. Marine bacteria are moderate halophiles. Most of them seem to have a specific requirement for sodium which cannot be met by other cations (MacLeod and Onofrey, 1963). The requirement for chlorine seems to be somewhat less specific, but information on this point is meager. As is true of marine algae, the salt requirement of marine bacteria is partly osmotic and partly nutritional (Larsen, 1962; Pratt and Austin, 1963).

Information on the comparative mineral requirements of plants is summarized in Table 4-2. Not included are elements which in some species can substitute for the normally functional essential element, for example rubidium (for potassium), strontium (for calcium), and bromine (for chlorine).

TABLE 4-2.   Essentiality of Mineral Elements for Plants[1]

Elements	Higher plants	Algae	Fungi	Bacteria
N, P, S, K, Mg Fe, Mn, Zn, Cu	+	+	+	+
Ca	+	+	±	±
B	+	±	−	−
Cl	+	+	−	±
Na	±	±	−	±
Mo	+	+	+	±
Se	±	−	−	−
Si	±	±	−	−
Co	−[2]	±	−	±
I	−	±	−	−
V	−	±	−	−

[1] Essential for the group: +; not known to be essential for the group: −. Essential for some but not known to be generally essential for members of the group: ±.
[2] See page 58 for discussion.

9.   QUANTITATIVE CONSIDERATIONS

Returning now to higher green plants, let us examine the relative amounts in which the essential nutrient elements occur. It must be kept in mind that in different plant species the concentrations and ratios of the different elements will vary, and the same is true even for plants of the same geno-type growing under different conditions. Nevertheless, an approximate value can be given for each element, based on numerous analyses of many plant materials. Such values are given in Table 4-3 (Epstein, 1965), modified after one compiled by P. R. Stout (1961) of the University of California at Davis. In this table, the elements are arranged in order of increasing relative amounts commonly found in plant matter, beginning with molybdenum which is the nutrient required in least amount.

In this table, the concentration of the elements is expressed on the basis of the dry weight of the tissue, as is the almost universal convention. However, many nutrient and other elements are highly water soluble and are present in the living plant largely as ions in solution. Potassium, sodium, and chloride are examples of such elements. Cassidy (1966, 1970) has

TABLE 4-3. Concentrations of Nutrient Elements in Plant
Material at Levels Considered Adequate[1]

| Element | Chemical symbol | Atomic weight | Concentration in dry matter | | Relative number of atoms with respect to molybdenum |
			μmole/g	ppm or %	
				ppm	
Molybdenum	Mo	95.95	0.001	0.1	1
Copper	Cu	63.54	0.10	6	100
Zinc	Zn	65.38	0.30	20	300
Manganese	Mn	54.94	1.0	50	1,000
Iron	Fe	55.85	2.0	100	2,000
Boron	B	10.82	2.0	20	2,000
Chlorine	Cl	35.46	3.0	100	3,000
				%	
Sulfur	S	32.07	30	0.1	30,000
Phosphorus	P	30.98	60	0.2	60,000
Magnesium	Mg	24.32	80	0.2	80,000
Calcium	Ca	40.08	125	0.5	125,000
Potassium	K	39.10	250	1.0	250,000
Nitrogen	N	14.01	1,000	1.5	1,000,000
Oxygen	O	16.00	30,000	45	30,000,000
Carbon	C	12.01	40,000	45	40,000,000
Hydrogen	H	1.01	60,000	6	60,000,000

[1] From Epstein (1965), after Stout (1961). Copyright 1965 by Academic Press,
New York and London.

shown that it is more meaningful to express the concentration of these
elements in tissue on the basis of the tissue water (sap). For example,
there is a progressive accumulation of chloride in leaves of citrus trees
as the season progresses. The reason for this is that the leaves, as they
mature, have a progressively lesser water content. The continued import
of chloride into the leaf, coupled with a diminished moisture content, may
result in a steep rise in the concentration of chloride in the leaf sap. This
may escape attention or appear minimal when the chloride content is expressed on the basis of the dry weight of the tissue. Although it is more

rational to express the concentrations of such ions on the basis of the tissue water, plant scientists will no doubt continue to use the customary dry weight basis on account of its convenience and conventionality. It is important, however, to keep in mind the implications of this procedure.

A generalized summary of plant composition like that given in Table 4-3 implies that the biosynthetic reactions of plant growth are subject to the same rules of stoichiometry that govern the well-known reactions of the inorganic analytical laboratory. This means that the important metabolites are synthesized in more or less fixed ratios to each other, in accordance with the genetic constitution of the plant, depending, of course, on the availability of all needed elements in the environment. Certain elements, however, if readily available, may be absorbed in amounts well in excess of metabolic requirements. For example, Tables 4-1 and 4-3 suggest that an adequate concentration of potassium is about 1 per cent of the dry matter. But when roots are exposed to high concentrations of potassium, as may happen right after a heavy application of potassium fertilizer

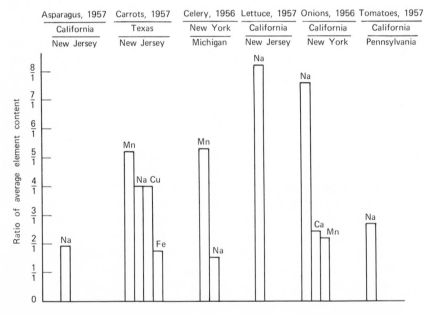

Figure 4-1.   Ratio of average content of elements in six vegetables from one state to their content in the vegetables from another state. The vegetables were obtained in the Washington, D. C., wholesale market in the years indicated. After Hopkins and Eisen (1959). Copyright 1959 by the American Chemical Society.

to a field, the potassium content of the plants may reach several per cent of the dry matter. Absorption of such large amounts—larger than required for optimal growth—is called "luxury consumption."

Hopkins and Eisen (1959) compared the mineral composition of vegetables delivered to the Washington, D. C., wholesale market from different geographic areas of the United States. Figure 4-1 illustrates the extent of the variations in mineral content they found in six vegetables. Examples of such variability could be multiplied at will, for both crops and natural vegetation. McIlrath (1964) has compiled a large listing of values of the mineral composition of plants. Inasmuch as forest trees grow on soils with low levels of available nutrients, compared with agricultural soils, we might be led to the assumption that trees do not require as high internal percentages of nutrients as do agricultural crops. However, analyses for nutrients in forest trees give values similar to those of Table 4-3 (Baule and Fricker, 1967; Bengtson *et al.*, 1968; Tamm, 1964).

DEFICIENCIES AND TISSUE ANALYSIS

If the concentration of an essential nutrient element in plant tissue drops below a level necessary for optimal growth, the plant is said to be deficient in that element. A deficiency may develop if the concentration of the element in the soil or other nutrient substrate is low, or if the element is present in chemical forms that render it unavailable for absorption. Sometimes excessive concentrations of some other element may so reduce the rate of absorption of a nutrient that the plant becomes deficient in that nutrient. Such an "induced" deficiency resulting from the "antagonistic" action of another element may develop even when the nutrient is present in the substrate at a concentration that would be more than adequate were it not for the presence of the antagonistic element at high concentration.

When tissue is deficient in an essential element, far-reaching changes in metabolism and growth are brought about. First those metabolic processes in which the element normally participates are slowed down. Since each metabolic reaction is part of the entire intricate, interconnected network of metabolism, processes other than those immediately involved are in turn affected until, under conditions of severe deficiency, the entire metabolic pattern becomes deranged. These biochemical and metabolic aspects of nutrient deficiencies are discussed in Chapter 11.

Until early in the present century, the only mineral deficiencies recognized were those of the macronutrients nitrogen, phosphorus, and potassium, and chemical fertilization usually took into account only the elements

TABLE 4-4. Extent of Micronutrient Deficiencies in the United States[1,2]

Crop	Number of states	Crop	Number of states	Crop	Number of states
Boron		Molybdenum		Manganese	
Alfalfa	38	Alfalfa	13	Bean	13
Beet	12	Clover	6	Corn	5
Celery	10	Cruciferae	9	Fruit trees	9
Clover	13	Soybean	3	Small grains	10
Cruciferae	25			Spinach	8
Fruit trees	21				
Iron		Copper		Zinc	
Bean	5	Corn	3	Bean	7
Corn	3	Fruit trees	3	Corn	20
Fruit trees	11	Grasses	3	Fruit trees	12
Grasses	7	Onion	7	Nut trees	10
Shade trees	7	Small grains	4	Onion	4
Shrubs	11			Potato	3

[1] After Berger (1962).
[2] Deficiencies occur in many other crops but with less frequency.

of this triad, N–P–K. Since that time, the importance and extent of other macronutrient and especially, of micronutrient deficiencies have become dramatically apparent. Table 4-4 gives the number of states in the United States in which micronutrient deficiencies occur in important crops.

Deficiencies, if sufficiently severe, will manifest themselves through the development of more or less distinct symptoms. Familiarity with these symptoms aids farmers and agricultural specialists in identifying nutrient deficiencies in the field. Symptoms of various deficiencies are discussed in section 11 of this chapter. The disturbance of normal metabolism caused by deficiencies may slow down the growth of plants and decrease yields. Such reduction in growth may happen without development of any obvious deficiency symptoms, or well before such symptoms make their appearance. To growers, this naturally is a matter of concern, and the technique of "tissue analysis" has been developed as a means of obtaining from the plant itself information about its nutrient status.

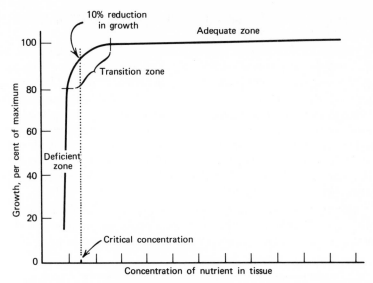

Figure 4-2. Generalized plot of growth as a function of the concentration of a nutrient in plant tissue. See text for explanation. From Ulrich and Hills (1967) in: Soil Testing and Plant Analysis, Part II—Plant Analysis. Soil Science Society of America, Inc., Madison.

The basic concept in assaying the nutrient status of crops by tissue analysis is the "critical concentration." The critical concentration is that concentration of a nutrient in the tissue just below the level giving optimal growth. The concept is illustrated in Figure 4-2. In the deficient zone, where levels are below the critical concentration, the yield increases sharply as the supply of the nutrient in the soil or the culture solution is raised. For each additional increment of nutrient absorbed by the plant, there is a corresponding increment of additional growth, so that the amount of nutrient per unit mass of tissue (the concentration) remains approximately constant. Beyond the transition zone, absorption of the element is further increased by additional supplies of the nutrient in the growth medium, but the nutrient no longer is the factor that limits growth, which remains constant. As a result, the additional amounts absorbed raise the concentration of the nutrient in the tissue to levels of "luxury consumption." A detailed discussion of yields as affected by nutrient supply has been given by Steenbjerg and Jakobsen (1963). Interactions among ions and their bearing on the interpretation of tissue analyses have been discussed by Bates (1971), Clements (1964), Emmert (1961), Shear *et al.* (1953), and Smith (1962).

Curves like the one shown in Figure 4-2 are obtained by growing plants in soil (or nutrient solution) amply supplied with all nutrients except the one whose critical concentration is to be determined. Graded additions of this nutrient are made to a series of containers, starting from a low value and going stepwise to high levels that are expected to give tissue concentrations well above the critical concentration. The plants are grown until those in the low range of supply show deficiency symptoms. The plants are harvested, yields are measured, and the tissue is analyzed. A calibration plot like that shown in Figure 4-2 is then made and the critical concentration established (Ulrich, 1961; Ulrich and Hills, 1967).

For elements like potassium or nitrogen which are mobile in the phloem and readily redistributed within the plant, the concentration in the leaf is usually a good index of the potassium or nitrogen status of the plant

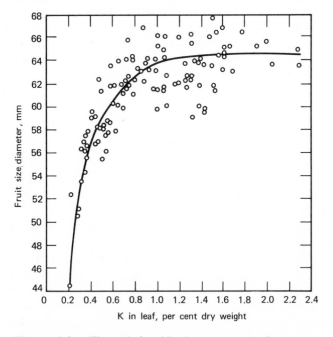

Figure 4-3. The relationship between potassium concentration in leaves of 118 "Halford" peach trees, *Prunus persica,* and fruit size at harvest. The data indicate a critical potassium concentration of about 1.0 per cent. After Lilleland *et al.* (1962).

as a whole and correlates well with indices of physiological performance and growth elsewhere. Figure 4-3 from Lilleland *et al.* (1962) shows the relationship between potassium levels in the leaves of peach trees, *Prunus persica,* and the diameter of fruits at harvest, and Figure 4-4 from Mitchell and Chandler (1939) shows the curve relating the nitrogen concentration of leaves of white oak, *Quercus alba,* and chestnut oak, *Q. montana,* to the annual radial increment of the trunks. But for elements which are not mobile in the phloem, analysis of leaves may be unrevealing of the nutritional status of other organs, as discussed later in connection with calcium in the last section of Chapter 11.

For each crop careful research is needed to determine which part of the plant is most indicative of the nutrient status of the plant, what field sampling procedure is best, and how often during the season analyses should be made. The precision and cost of the analytical survey are also

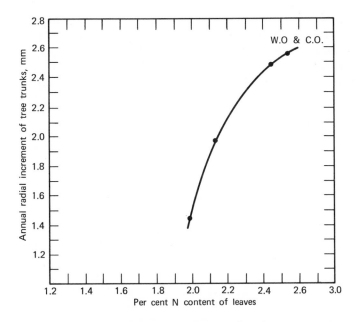

Figure 4-4. The relationship between the nitrogen concentration in leaves of white oak, *Quercus alba,* and chestnut oak, *Q. montana,* plotted together, and the annual radial increment of the trunks. The data indicate that the critical nitrogen concentration was not reached. After Mitchell and Chandler (1939).

factors to be considered. If the results show that the concentration of a given nutrient, say nitrate, has fallen to the critical concentration or below, supplementary fertilization, if done in time, can keep yields from being reduced or quality from being impaired.

Periodically, international colloquia on plant analysis and fertilizer problems are held by workers in this field. The published Proceedings of these meetings (Bould *et al.*, 1964; Reuther, 1961) contain much information on the methods, findings, and interpretations of tissue analysis. Bates (1971) has given a critical review of the concept of critical concentration. The compilation edited by Chapman (1966) of the University of California at Riverside is an indispensable reference work on diagnostic criteria in plant nutrition. For the mineral nutrition of forest trees, the book by Baule and Fricker (1967) and the symposium volume edited by Bengtson *et al.* (1968) provide a wealth of quantitative information. The International Potash Institute (1967) has published the proceedings of a Colloquium on Forest Fertilization.

Understandably, most work on critical concentrations has been done with economically useful crop plants, but Figure 4-5 shows the same typical relationship between the phosphorus content and growth of an aquatic higher plant, *Vallisneria americana*, and Figure 4-6 shows it for magnesium and the growth of the filamentous green alga, *Stigeoclonium tenue*.

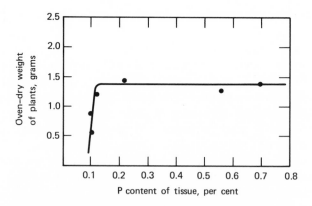

Figure 4-5. The relationship between the phosphorus concentration of *Vallisneria americana*, an aquatic higher plant, and its growth in terms of dry weight. The data indicate a critical phosphorus concentration of about 0.13 per cent. After Gerloff and Krombholz (1966).

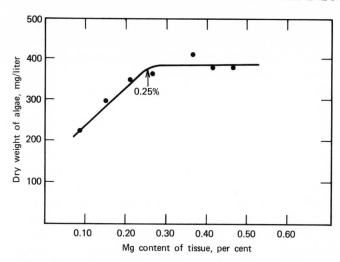

Figure 4-6. The relationship between the magnesium concentration of *Stigeoclonium tenue,* a filamentous green alga, and its growth in terms of dry weight. The data indicate a critical magnesium concentration of 0.25 per cent. After Gerloff and Fishbeck (1969).

. DEFICIENCY SYMPTOMS: GENERAL DISCUSSION

The metabolic derangements brought about by deficiencies of essential nutrients (Chapter 11 on Mineral Metabolism) eventually manifest themselves in visible abnormalities. The overall growth and development of the plant may be affected, there may be characteristic macroscopic symptoms, and there may be changes in the appearance of cellular structures that can be studied with the microscope or electron microscope.

The overall appearance of the plant and localized specific symptoms are an important aid to knowledgeable observers in identifying deficiencies. The symptoms of iron deficiency, for example, are fairly similar in many crop species, and sometimes are all that is needed to pinpoint a deficiency of this element. With this in mind, several authors have published extensive descriptions and pictures of plants afflicted with nutrient deficiencies (Baule and Fricker, 1967; Malavolta *et al.,* 1962; Sprague, 1964; Wallace, 1961). Nevertheless, visible symptoms should be regarded as just one of several kinds of evidence of deficiency of a given element. Several factors combine to introduce considerable uncertainty in diagnoses based on symptomology only. The more important of these factors are discussed below.

Symptoms of a deficiency of a certain element may differ so greatly in different crops that knowledge of the deficiency syndrome in one species affords little aid in identifying the same deficiency in another species. For example, zinc deficiency in many kinds of fruit trees and some annual crops as well results in a condition aptly called "little leaf." Leaves of severely affected plants may be a mere fraction of the size of normal leaves. In corn, on the other hand, a general blanching or chlorosis, most pronounced in the young emerging leaves, is the most characteristic symptom ("white bud"). A person familiar with white bud in corn would not be able, on that basis, to identify little leaf of peach or citrus trees as being due to zinc deficiency.

But not only different symptoms of the deficiency of a single element complicate diagnosis; similar or identical symptoms may result from deficiencies of different elements. For example, both nitrogen and sulfur deficiencies cause a general chlorosis, and even experienced observers may often be unable to distinguish between deficiencies of these elements on the basis of observation alone.

A third difficulty arises from the fact that deficiencies occurring in the field may be severe enough to reduce yields and impair the quality of the crop while the visible symptoms are still too slight to be readily spotted. The extremely severe deficiencies that are commonly pictured and described in publications on this subject are often of little avail to the man called upon to deal with the less spectacular manifestations of deficiencies encountered in the field.

Multiple deficiencies add other complications. A plant may have symptoms of deficiencies of two or more elements, making identification of the deficiencies more difficult than when they occur singly. And finally, conditions that are not deficiencies at all may engender symptoms which might be mistaken for deficiency symptoms. Such "simulated" deficiency symptoms may be caused by many factors ranging from elemental toxicities to bacterial diseases and virus infections.

Because of the complexities outlined above, visual diagnosis of nutrient deficiency conditions is practiced successfully only by experts—specialists in plant nutrition, farm advisors, knowledgeable growers, and others who by dint of long observation and experience have learned to recognize those deficiency symptoms that are encountered in the area where they work. In making diagnoses they also draw on their knowledge of soil conditions in the area, and verify their conclusions by means of tissue analysis and experimental applications of the nutrient suspected to be in short supply.

2. DEFICIENCY SYMPTOMS: INDIVIDUAL ELEMENTS

Nitrogen. Excepting only drought, no other deficiency is as dramatic in its effects as is nitrogen deficiency. General chlorosis and an etiolated habit are the most characteristic symptoms. Growth is retarded and slow, and the plants have an unthrifty, spindly appearance. Fruit is often exceptionally well colored. The more mature parts of the plant are first affected because nitrogen is translocated from older to young, actively growing regions.

Sulfur. Sulfur deficiency symptoms resemble those of nitrogen deficiency. Plants are chlorotic, spindly, and grow poorly.

Phosphorus. Dark green or blue-green foliage is one of the first symptoms of phosphorus deficiency in many species. Often red, purple, or brown pigments develop in the leaves, especially along the veins. Growth is reduced, and under conditions of severe deficiency the plants become stunted.

Potassium. Potassium deficiency in many species causes leaves to be dark green or blue-green, as in phosphorus deficiency. Necrotic spots often develop on the leaves. There may also be marginal necrosis or leaf scorch. Growth is subnormal and under severe conditions terminal and lateral buds may die ("dieback").

Calcium. Symptoms of calcium deficiency appear earliest and most severely in meristematic regions and young leaves. Calcium requirements seem to be high in such tissues, and calcium contained in older, mature tissues tends to become immobilized there and is not appreciably retranslocated to the young, actively growing regions (Chapters 8 and 11). Growing points are damaged or die ("dieback"); in flowers and developing fruit the symptoms are known as "blossom-end rot." The growth of roots is severely affected. Many calcium deficient soils are acid, so that calcium deficiency proper may be accompanied by toxic levels of hydrogen ions, and of ions of heavy metals such as aluminum, manganese and others which tend to go into solution at low pH values. The damaged roots become prone to infection by bacteria and fungi.

Magnesium. Unlike calcium, magnesium is readily translocated from mature to young, actively growing regions of the plant. As a result it is in mature leaves that deficiency symptoms first make their appearance.

Marginal chlorosis is common, often accompanied by development of a variety of pigments. Chlorosis may also begin in patches or blotches which later merge and spread to the leaf margins and tips. The variety of symptoms in different species is so great that a generalized description of magnesium deficiency symptoms is even more difficult than it is for those of other deficiencies.

Iron. A general chlorosis of young leaves is the most telling symptom of iron deficiency. At first the veins may remain green, but in most species in which the deficiency has been observed the veins also become chlorotic eventually. The deficiency is common in fruit trees. Where it is associated with high levels of calcium carbonate in the soil it is called "lime-induced" chlorosis.

Manganese. The symptoms of manganese deficiency vary greatly from one species to the next. Leaves often show an interveinal chlorosis, the veins making a green pattern on a yellow background, much as in early stages of iron deficiency. There may be necrotic spots or streaks on leaves ("gray speck" of oats), and in legume seeds, necrosis may appear in the embryo or the adjacent inner surfaces of the cotyledons. Leaves of some species become malformed ("mouse ear" of pecans). In severe cases, plants are badly stunted.

Zinc. "Little leaf" and "rosette" are the classical symptoms of zinc deficiency in fruit trees. Both these syndromes result from failure of tissues to grow normally. Failure of leaves to expand causes little leaf, and failure of internodes to elongate causes leaves at successive nodes to be so closely telescoped as to give rise to the rosette symptom. In some species leaves become chlorotic, but in others leaves may be dark green or blue-green. Leaves may become twisted and necrotic. Flowering and fruiting are much reduced under conditions of severe zinc deficiency, and the entire plant may be stunted and misshapen.

Copper. Symptoms vary greatly depending on the species. Leaves may be chlorotic or deep blue-green with margins rolled up. The bark of trees is often rough and blistered, and gum may exude from fissures in the bark ("exanthema"). Young shoots often die back, whereupon new shoots emerge from multiple buds further back, making for a bushy appearance. Flowering and fruiting are curtailed; annual plants may fail to develop and may die in the seedling stage.

Chlorine. The essentiality of chlorine was discovered in experiments with tomato plants grown in purified nutrient solutions (Broyer *et al.,* 1954). The symptoms of chlorine deficiency in this species are, at first, a blue-green and shiny appearance of young leaves. In the heat of the day the tips of the young leaves wilt and dangle down, though they may recover at night or on cool, cloudy days. As the deficiency progresses a characteristic "bronzing" shows up on the leaves, followed by chlorosis and necrosis. Under severe deficiency conditions plants are spindly and stunted. Wilting, discoloration (bronzing), and necrosis are also observed in other species (Johnson *et al.,* 1957; Ozanne *et al.,* 1957). This deficiency is the only one that is nowhere of economic importance. No deficiency of chlorine has yet been observed in any plants grown under the open sky.

Boron. Growing tips often are damaged by boron deficiency and may be killed. Tissues of plants with this deficiency appear hard, dry, and brittle. Leaves may become distorted and stems rough and cracked, often with corky ridges or spots. Flowering is severely affected. If fruit sets, it often shows symptoms similar to those encountered in stems. Roots suffer greatly, and bacterial infections often are a secondary consequence of boron deficiency both in roots and shoots.

Molybdenum. Molybdenum deficiency, first identified in tomato plants, causes in this and other species an interveinal chlorosis. The veins remain pale green, so that the chlorosis gives the leaf a mottled appearance, somewhat like manganese deficiency. Leaf margins tend to curl or roll. In severe cases necrosis follows, and the entire plant is stunted. In brassicas the leaf blades become necrotic and disintegrate, leaving only a much reduced strip along the midrib ("whiptail").

When plants are given nitrogen in the form of ammonium, molybdenum requirements are considerably smaller than when nitrate is the source of nitrogen. The reason is that one of the important functions of molybdenum is in the reduction of nitrate, a sequence of reactions that is obviated when nitrogen is offered in reduced form, as ammonium ions. Plants which are given ammonium may show no molybdenum deficiency symptoms or only mild symptoms at molybdenum concentrations low enough to result in severe symptoms when nitrate is the source of nitrogen. Molybdenum is essential also for symbiotic nitrogen fixation in plants that bear root nodules containing bacterial symbionts. When such plants depend for nitrogen on symbiotic fixation of atmospheric nitrogen, molybdenum deficiency renders them nitrogen deficient, with symptoms characteristic of this deficiency.

Hewitt (1963) has given a comprehensive account of the essential elements, their interactions, and the effects of deficiencies. The functions of essential nutrients are discussed in Chapter 11.

REFERENCES

Ahmed, S. and H. J. Evans. 1959. Effect of cobalt on the growth of soybeans in the absence of supplied nitrogen. Biochem. Biophys. Res. Comm. 1:271–275.

Ahmed, S. and H. J. Evans. 1961. The essentiality of cobalt for soybean plants grown under symbiotic conditions. Proc. Nat. Acad. Sci. 47:24–36.

Allen, M. B. and D. I. Arnon. 1955. Studies on nitrogen-fixing blue-green algae. II. Sodium requirement of *Anabaena cylindrica*. Physiol. Plantarum 8:653–660.

Arnon, D. I. and P. R. Stout. 1939a. The essentiality of certain elements in minute quantity for plants with special reference to copper. Plant Physiol. 14:371–375.

Arnon, D. I. and P. R. Stout. 1939b. Molybdenum as an essential element for higher plants. Plant Physiol. 14:599–602.

Arnon, D. I. and G. Wessel. 1953. Vanadium as an essential element for green plants. Nature 172:1039–1040.

Bates, T. E. 1971. Factors affecting critical nutrient concentrations in plants and their evaluation: a review. Soil Sci. In press.

Baule, H. and C. Fricker. 1967. Die Düngung von Waldbäumen. Bayerischer Landwirtschaftsverlag GmbH, München.

Bengtson, G. W., R. H. Brendemuehl, W. L. Pritchett and W. H. Smith, eds. 1968. Forest Fertilization—Theory and Practice. Tennessee Valley Authority, Muscle Shoals.

Berger, K. C. 1962. Micronutrient deficiencies in the United States. Agric. Food Chem. 10:178–181.

Blum, H. F. 1968. Time's Arrow and Evolution. 3rd ed. Princeton University Press, Princeton.

Bollard, E. G. and G. W. Butler. 1966. Mineral nutrition of plants. Ann. Rev. Plant Physiol. 17:77–112.

Bould, C., P. Prevot and J. R. Magness, eds. 1964. Plant Analysis and Fertilizer Problems. IV. American Society for Horticultural Science.

Bowen, J. E. and H. G. Gauch. 1966. Nonessentiality of boron in fungi and the nature of its toxicity. Plant Physiol. 41:319–324.

Brownell, P. F. 1965. Sodium as an essential micronutrient element for a higher plant (*Atriplex vesicaria*). Plant Physiol. 40:460–468.

Brownell, P. F. 1968. Sodium as an essential micronutrient element for some higher plants. Plant and Soil 28:161–164.

Brownell, P. F. and J. G. Wood. 1957. Sodium as an essential micronutrient element for *Atriplex vesicaria,* Heward. Nature 179:635–636.

Broyer, T. C., A. B. Carlton, C. M. Johnson and P. R. Stout. 1954. Chlorine—a micronutrient element for higher plants. Plant Physiol. 29:526–532.

Cassidy, N. G. 1966. A rational method for recording and comparing concentrations of plant constituents that are water soluble, with particular reference to chloride and potassium. Plant and Soil 25:372–384.

Cassidy, N. G. 1970. The distribution of potassium in plants. Plant and Soil 32:263–267.

Chapman, H. D., ed. 1966. Diagnostic Criteria for Plants and Soils. University of California, Division of Agricultural Sciences.

Chapman, H. D. and P. F. Pratt. 1961. Methods of Analysis for Soils, Plants, and Waters. University of California, Division of Agricultural Sciences.

Chen, C. H. and J. Lewin. 1969. Silicon as a nutrient element for *Equisetum arvense.* Can. J. Bot. 47:125–131.

Clements, H. F. 1964. Interaction of factors affecting yield. Ann. Rev. Plant Physiol. 15:409–442.

Cochrane, V. W. 1958. Physiology of Fungi. John Wiley and Sons, Inc., New York.

Darley, W. M. and B. E. Volcani. 1969. Role of silicon in diatom metabolism. A silicon requirement for deoxyribonucleic acid synthesis in the diatom *Cylindrotheca fusiformis* Reimann and Lewin. Expt. Cell Res. 58:334–342.

Dear, J. M. and S. Aronoff. 1968. The non-essentiality of boron for *Scenedesmus.* Plant Physiol. 43:997–998.

Droop, M. R. 1962. Organic micronutrients. In: Physiology and Biochemistry of the Algae. R. A. Lewin, ed. Academic Press, New York. Pp. 141–159.

Emmert, F. H. 1961. The bearing of ion interactions on tissue analysis results. In: Plant Analysis and Fertilizer Problems. W. Reuther, ed. American Institute of Biological Sciences, Washington. Pp. 231–243.

Epstein, E. 1965. Mineral metabolism. In: Plant Biochemistry. J. Bonner and J. E. Varner, eds. Academic Press, New York and London. Pp. 438–466.

Evans, H. J. and M. Kliewer. 1964. Vitamin B_{12} compounds in relation to the requirements of cobalt for higher plants and nitrogen-fixing organisms. Ann. New York Acad. Sci. 112, Art. 2:735–755.

Fidanovski, F. 1968. Silicium, ein für Pflanzen "nützliches" Element. Z. Pflanzenern. Bodenk. 120:191–207.

Fries, L. 1966. Influence of iodine and bromine on growth of some red algae in axenic culture. Physiol. Plantarum 19:800–808.

Gerloff, G. C. 1968. The comparative boron nutrition of several green and blue-green algae. Physiol. Plantarum 21:369–377.

Gerloff, G. C. and K. A. Fishbeck. 1969. Quantitative cation requirements of several green and blue-green algae. J. Phycol. 5:109–114.

Gerloff, G. C. and P. H. Krombholz. 1966. Tissue analysis as a measure of nutrient availability for the growth of angiosperm aquatic plants. Limnol. and Oceanography 11:529–537.

Guillard, R. R. L. 1962. Salt and osmotic balance. In: Physiology and Biochemistry of the Algae. R. A. Lewin, ed. Academic Press, New York. Pp. 529–540.

Guirard, B. M. and E. E. Snell. 1962. Nutritional requirements of microorganisms. In: The Bacteria. I. C. Gunsalus and R. Y. Stanier, eds. Academic Press, New York. Vol. 4, pp. 33–93.

Hewitt, E. J. 1963. The essential nutrient elements: requirements and interactions in plants. In: Plant Physiology. F. C. Steward, ed. Academic Press, New York. Vol. 3, pp. 137–360.

Hickman, D. W. 1969. Potassium as a micronutrient for aquatic *Hyphomycetes*. Plant Physiol. 44 (Supplement):21.

Hoagland, D. R. 1944. Lectures on the Inorganic Nutrition of Plants. Chronica Botanica Company, Waltham.

Holm–Hansen, O., G. C. Gerloff and F. Skoog. 1954. Cobalt as an essential element for blue-green algae. Physiol. Plantarum 17:665–675.

Hopkins, H. and J. Eisen. 1959. Mineral elements in fresh vegetables from different geographic areas. Agric. Food Chem. 7:633–638.

International Potash Institute. 1967. Colloquium on Forest Fertilization. International Potash Institute, Berne.

Iswaran, V. and W. V. B. Sundara Rao. 1964. Role of cobalt in nitrogen fixation by *Azotobacter chroococcum*. Nature 203:549.

Johnson, C. M., P. R. Stout, T. C. Broyer and A. B. Carlton. 1957. Comparative chlorine requirements of different plant species. Plant and Soil 8:337–353.

Kellner, K. 1955. Die Adaption von *Ankistrodesmus braunii* an Rubidium und Kupfer. Biol. Zentralblatt 74:662–691.

Kramer, P. J. 1969. Plant and Soil Water Relationships: A Modern Synthesis. McGraw-Hill Book Company, Inc., New York.

Kratz, W. and J. Myers. 1955. Nutrition and growth of several blue-green algae. Am. J. Bot. 42:282–287.

Larsen, H. 1962. Halophilism. In: The Bacteria. I. C. Gunsalus and R. Y. Stanier, eds. Academic Press, New York. Vol. 4, pp. 297–342.

Lewin, J. C. 1962. Silicification. In: Physiology and Biochemistry of the Algae. R. A. Lewin, ed. Academic Press, New York. Pp. 445–455.

Lewin, J. 1966. Boron as a growth requirement for diatoms. J. Phycol. 2:160–163.

Lewin, J. and B. E. F. Reimann. 1969. Silicon and plant growth. Ann. Rev. Plant Physiol. 20:289–304.

Lilleland, O., K. Uriu, T. Muraoka and J. Pearson. 1962. The relationship of potassium in the peach leaf to fruit growth and size at harvest. Proc. Am. Soc. Hort. Sci. 81:162–167.

Lilly, V. G. 1965. Chemical constituents of the fungal cell. I. Elemental constituents and their roles. In: The Fungi—An Advanced Treatise. G. C. Ainsworth and A. S. Sussman, eds. Academic Press, New York. Vol. 1, pp. 163–177.

MacLeod, R. A. and E. E. Snell. 1950. Ion antagonism in bacteria as related to antimetabolites. Ann. New York Acad. Sci. 52, Art. 8:1249–1259.

MacLeod, R. A. and E. Onofrey. 1963. Studies on the stability of the Na^+ requirement of marine bacteria. In: Symposium on Marine Microbiology. C. H. Oppenheimer, ed. Charles C. Thomas, Publ., Springfield. Pp. 481–489.

Malavolta, E., H. P. Haag, F. A. F. Mello and M. O. C. Brasil Sobr°. 1962. On the Mineral Nutrition of some Tropical Crops. International Potash Institute, Berne.

McIlrath, W. J. 1964. Plant tissues and organs: mineral composition. In: Biology Data Book. P. L. Altman and D. S. Dittmer, eds. Federation of American Societies for Experimental Biology, Washington. Pp. 405–416.

McLachlan, J. and J. S. Craigie. 1967. Bromide, a substitute for chloride in a marine algal medium. Nature 214:604–605.

Miller, E. C. 1938. Plant Physiology. 2nd ed. McGraw-Hill Book Company, Inc., New York and London.

Mitchell, H. L. and R. F. Chandler, Jr. 1939. The Nitrogen Nutrition and Growth of Certain Deciduous Trees of Northeastern United States. The Black Rock Forest Bull. No. 11. Pp. 1–94.

Mitsui, S. and H. Takatoh. 1960. Nutritional study of silica in graminaceous crops. I. Growth of rice plants without SiO_2 and symptoms of SiO_2 deficiency. J. Sci. Soil Manure 30: 335–339.

Mitsui, S. and H. Takatoh. 1963a. Nutritional study of silicon in graminaceous crops. Part 1. Soil Sci. and Plant Nutr. (Tokyo) 9:49–53.

Mitsui, S. and H. Takatoh. 1963b. Nutritional study of silicon in graminaceous crops. Part 2. Soil Sci. and Plant Nutr. (Tokyo) 9:54–58.

Mortvedt, J. J., P. M. Giordano and W. L. Lindsay, eds. 1971. Micronutrients in Agriculture. Soil Science Society of America, Madison.

Needham, A. E. 1965. The Uniqueness of Biological Materials. Pergamon Press, Oxford.

Nicholas, D. J. D. 1963. Inorganic nutrient nutrition of microorganisms. In: Plant Physiology. F. C. Steward, ed. Academic Press, New York. Vol. 3, pp. 363–447.

Oser, B. L., ed. 1965. Hawk's Physiological Chemistry. 14th ed. McGraw-Hill Book Company, Inc., New York.

Ozanne, P. G., J. T. Woolley and T. C. Broyer. 1957. Chlorine and bromine in the nutrition of higher plants. Austral. J. Biol. Sci. 10:66–79.

Pedersén, M. 1969a. The demand for iodine and bromine of three marine brown algae grown in bacteria-free cultures. Physiol. Plantarum 22:680–685.

Pedersén, M. 1969b. Marine brown algae requiring vitamin B_{12}. Physiol. Plantarum 22:977–983.

Pirson, A. 1939. Über die Wirkung von Alkaliionen auf Wachstum und Stoffwechsel von Chlorella. Planta 29:231–261.

Pratt, D. and M. Austin. 1963. Osmotic regulation of the growth rate of four species of marine bacteria. In: Symposium on Marine Microbiology. C. H. Oppenheimer, ed. Charles C. Thomas, Publ., Springfield. Pp. 629–637.

Provasoli, L. 1963. Organic regulation of phytoplankton fertility. In: The Sea. M. N. Hill, ed. Interscience Publishers, John Wiley and Sons, New York. Vol. 2, pp. 165–219.

Reimann, B. E. F., J. C. Lewin and B. E. Volcani. 1965. Studies on the biochemistry and fine structure of silica shell formation in diatoms. I. The structure of the cell wall of Cylindrotheca fusiformis Reimann and Lewin. J. Cell Biol. 24:39–55.

Reisenauer, H. M. 1960. Cobalt in nitrogen fixation by a legume. Nature 186:375–376.

Reuther, W., ed. 1960. Plant Analysis and Fertilizer Problems. American Institute of Biological Sciences, Washington. Publication No. 8.

Rosenfeld, I. and O. A. Beath. 1964. Selenium. Academic Press, New York and London.

Shaw, T. I. 1962. Halogens. In: Physiology and Biochemistry of the Algae. R. A. Lewin, ed. Academic Press, New York. Pp. 247–253.

Shear, C. B., H. L. Crane and A. T. Myers. 1953. Nutrient element balance: response of Tung trees grown in sand culture to potassium, magnesium, calcium, and their interactions. U. S. Department of Agriculture Tech. Bull. No. 1085.

Shrift, A. 1969. Aspects of selenium metabolism in higher plants. Ann. Rev. Plant Physiol. 20:475–494.

Smith, P. F. 1962. Mineral analysis of plant tissues. Ann. Rev. Plant Physiol. 13:81–108.

Sprague, H. B., ed. 1964. Hunger Signs in Crops. 3rd ed. David McKay Company, New York.

Steenbjerg, F. and S. T. Jakobsen. 1963. Plant nutrition and yield curves. Soil Sci. 95:69–88.

Stout, P. R. 1956. Micronutrients in crop vigor. Agric. Food Chem. 4:1000–1006.

Stout, P. R. 1961. Micronutrient needs for plant growth. Proc. Ninth Ann. Calif. Fertilizer Conf. Pp. 21–23.

Stout, P. R. and D. I. Arnon. 1939. Experimental methods for the study of the role of copper, manganese, and zinc in the nutrition of higher plants. Am. J. Bot. 26:144–149.

Stout, P. R. and C. M. Johnson. 1956. Molybdenum deficiency in horticultural and field crops. Soil Sci. 81:183–190.

Tamm, C. O. 1964. Determination of nutrient requirements of forest stands. Internat. Rev. Forestry Res. 1:115–170.

Ulrich, A. 1960. Plant analysis in sugar beet nutrition. In: Plant Analysis and Fertilizer Problems. W. Reuther, ed. American Institute of Biological Sciences, Washington. Publication No. 8. Pp. 190–211.

Ulrich, A. and F. J. Hills. 1967. Principles and practices of plant analysis. In: Soil Testing and Plant Analysis. Part II. Plant Analysis. Soil Science Society of America, Madison, Special Publication Series, No. 2. Pp. 11–24.

von Stosch, H. A. 1962. Jodbedarf bei Meeresalgen. Naturwiss. 49:42–43.

von Stosch, H. A. 1963. Wirkungen von Jod und Arsenit auf Meeresalgen in Kultur. Proc. Fourth Internat. Seaweed Symp. Pp. 142–150.

Walker, J. B. 1956. Strontium inhibition of calcium utilization by a green alga. Arch. Biochem. Biophys. 60:264–265.

Wallace, T. 1961. The Diagnosis of Mineral Deficiencies in Plants (A Colour Atlas and Guide). 3rd ed. H. M. Stationary Office, London.

Wiessner, W. 1962. Inorganic micronutrients. In: Physiology and Biochemistry of the Algae. R. A. Lewin, ed. Academic Press, New York. Pp. 267–286.

Williams, M. C. 1960. Effect of sodium and potassium salts on growth and oxalate content of *Halogeton*. Plant Physiol. 35:500–505.

Wilson, S. B. and D. J. D. Nicholas. 1967. A cobalt requirement for non-nodulated legumes and for wheat. Phytochem. 6:1057–1066.

Woolley, J. T. 1957. Sodium and silicon as nutrients for the tomato plant. Plant Physiol. 32:317–321.

Wyatt, H. V. 1963. The effect of alkali metals on the growth of *Staphylococcus pyogenes*. Expt. Cell Res. 30:56–73.

Yoshida, S., Y. Onishi and K. Kitagishi. 1959. Role of silicon in rice nutrition. Soil Plant Food (Tokyo) 5:127–133.

PART II

TRANSPORT

5

"OUTER" OR "FREE" SPACES

THE NEED FOR A MEMBRANE

In the preceding two chapters evidence was presented concerning the elemental composition of plants and of the media in which they grow. If this evidence is considered jointly it becomes clear that there is startlingly little correlation between the relative concentrations of various elements in the substrates and in the plants. For example, potassium is present in most plants at a level of about 1 per cent or more of the dry matter, which makes it the quantitatively most prominent inorganic cation of plant material. Yet in soil solutions (see Table 3-2) the concentration of potassium is usually less than 50 ppm, and may be exceeded by the concentrations of calcium and magnesium. In the soils of arid and semiarid regions the concentration of sodium is often vastly in excess of the potassium concentration, yet analysis of crop plants growing on these soils nearly always reveals a much higher concentration of potassium than of sodium in the tissues.

This difference between the composition of plants and their media is most striking in the case of many algae and bacteria, for here the internal solution (the cell sap or intracellular fluid) of unicellular organisms is found to differ markedly from the composition of the medium bathing the cells, the distance between the internal and external solutions equaling the thickness of the outer cell membrane—a mere fraction of a micron.

Now if two solutions, say soil solutions, are separated by considerable distances, no particular barrier needs to exist to keep the solutions from mixing. The slowness of the processes of convection and diffusion suffices to keep the solutions apart and maintain their respective compositions for long periods. On the other hand, if two solutions of greatly different chemical composition are separated from each other by a distance measured in fractions of a micron there must exist a barrier to the diffusion of the solutes of the two solutions. Without such a barrier, the thermal agitation of both solvent and solutes, causing diffusion and convection, would result in speedy mixing of the two solutions. The finding that external and intracellular solutions may differ markedly in chemical composition is therefore evidence for the existence of a barrier to free diffusion. The plasma membrane of cells is this essential barrier.

It is this membrane that represents the line of demarcation between the external solution with its particular composition and the cell interior with its characteristically different composition. There are intracellular organelles which in turn are separated by membranes from the cytoplasm and maintain distinctive concentrations and ratios of various solutes within the volume enclosed by their membranes. But the entire cytoplasmic system is separated by the plasma membrane from the external medium. Without such a barrier, the cell would be swamped by the external chemical environment, a situation which is incompatible with the intricate and chemically delicate machinery of life.

Plant life first evolved in sea water, and sea water represents therefore the primeval nutrient solution. Yet the high concentration of salts found in sea water, especially the concentration of sodium chloride which is nearly 0.5 M (Table 2-2), is inimical to cellular metabolism, and the elaboration of a cell membrane which maintains an intracellular ionic environment differing from the external one was therefore a prerequisite for the evolution of the complex metabolic systems of plant cells.

2. THE STRUCTURE OF PLANT CELLS

A typical young plant cell (see Figure 5-1) consists of cytoplasm, the cell wall outside it, and numerous vacuoles within. As the cell matures the vacuoles coalesce to form a single, large central vacuole. The cytoplasm is an aqueous phase containing organelles which specialize in various metabolic functions. The plasma membrane or plasmalemma is the outer cytoplasmic membrane which lies appressed against the cell wall. The inner cytoplasmic membranes, the tonoplasts, form the boundary between the cytoplasm and the vacuoles. In the mature cell, a single, continuous tonoplast separates the cytoplasm and the central vacuole. The aqueous solu-

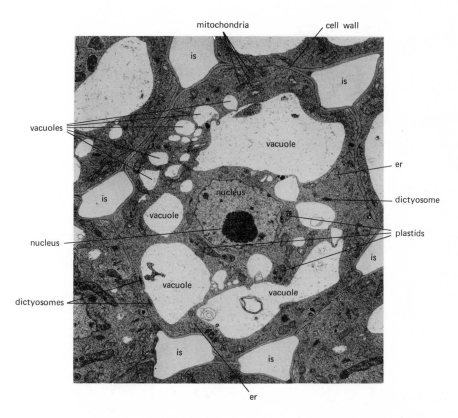

Figure 5-1. Electron micrograph of a cortical cell from the root tip of bean, *Phaseolus vulgaris,* prepared by Katherine Esau, University of California at Santa Barbara. Key: is, intercellular space; er, endoplasmic reticulum. Magnification, ×4,500.

tion in the vacuoles contains mainly inorganic ions and organic solutes of low molecular weight such as sugars and amino acids, but also some enzymes (Matile, 1966). Excellent books on the structure of plant cells are those by Frey-Wyssling and Mühlethaler (1965), Clowes and Juniper (1968), Ledbetter and Porter (1970), and the symposium volume edited by Pridham (1968).

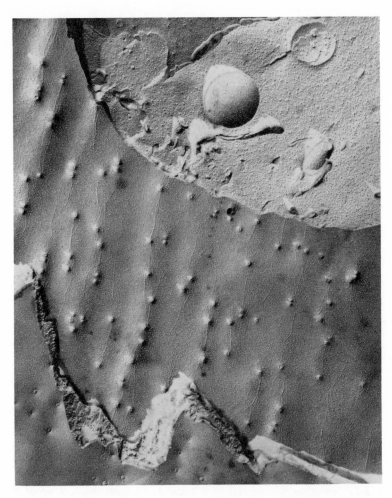

Figure 5-2. Freeze-etch electron micrograph of a cortical cell of onion root tip, *Allium cepa,* prepared by Daniel Branton, University of California at Berkeley. In the lower left corner is shown some of the cell wall, with rough broken edges where it has been fractured. Next is shown a surface view of the plasmalemma with plasmodesmata (large dots) and membrane particles (small dots) which are interpreted as membrane protein extending from one surface of the membrane to the other. The cytoplasm is shown in the upper right part, including a vacuole (the large, dome-like structure) and endoplasmic reticulum (along the upper edge, center and middle left). Magnification, ×21,000.

The architecture of a plant cell dictates that an ion from the external solution, if it is to reach the central vacuole, must negotiate several regions in turn. This is brought out vividly in the freeze-etch electron micrograph prepared by Daniel Branton of the University of California at Berkeley (Figure 5-2). First the ion has to traverse the cell wall, a fractured portion of which is shown in the lower left-hand corner of the picture. Next it must pass through the outer cytoplasmic membrane, the plasmalemma, which underlies the wall and a large expanse of which is shown in surface view where the wall has been stripped away. Having moved through this membrane it has reached the cytoplasm, shown in the upper right quarter of the picture where it has been laid bare by removal of the plasmalemma. Finally, it moves across the inner cytoplasmic membrane, the tonoplast, into the vacuole, shown as a dome-like structure within the cytoplasm. (The picture is of a meristematic cell. Such cells have many small vacuoles. A mature cell would have a single, large, central vacuole).

8. DIFFUSION

Much of the evidence concerning the movement of ions from the external medium into the cell comes from investigations with entire tissues, such as excised roots or slices of tuberous tissue. Nevertheless, in our initial discussions of these ionic movements we shall consider them in terms of the structure of individual cells. In other words, we shall look upon tissue as though it were merely an aggregate of individual cells, disregarding for the moment the organization and architecture of the tissue and the fact that it contains more than a single type of cell. We shall see later that for the present purpose this is a justifiable simplifying assumption.

The point was made earlier that it is the plasmalemma which in a functional sense forms the outer boundary of the cell where it abuts on the extracellular fluid. The implication of this statement is that the cell wall, which is external to the plasmalemma, is not a functional barrier between cell and environment. Physiological and chemical observations bear out this conclusion.

When plant tissue is immersed in a solution of a salt, the external solution almost immediately permeates a certain fraction of the volume of the tissue. In an investigation with barley roots, *Hordeum vulgare,* this was demonstrated as follows (Epstein, 1955). Roots weighing 1.00 g fresh weight were immersed in a solution of K_2SO_4 whose concentration was 20 μmoles/ml. The roots were then taken from the solution, blotted to remove solution adhering to the surface, and were then transferred to a known volume of water. They were found to release sulfate to this water, the total amount lost being 4.45 μmoles sulfate.

The interpretation of this simple experiment is as follows. It is assumed that some fraction of the volume of the tissue is not separated by any diffusion barrier or membrane from the external medium. When the tissue is immersed in the solution the solution permeates this accessible volume of the tissue so that the concentration of the solution within this volume becomes equal to the external concentration. In the experiment discussed above, the concentration of sulfate in the accessible volume of the tissue would be 20 μmoles/ml, equal to that of the external solution. When now the tissue is transferred, after blotting, to a volume of water which is large compared with the volume of the tissue, virtually all the sulfate diffuses out of the accessible volume, since the solution in this internal volume tends always to equilibrate with the external solution.

In this experiment, the amount of sulfate lost from one gram of roots was 4.45 μmoles. If this amount of sulfate, while still in the tissue, was indeed at the same concentration as the external solution (20 μmoles/ml), it was occupying an internal volume of 4.45/20 = 0.22 ml. In other words, on this evidence 0.22 ml per gram fresh weight of these roots constitute a volume which stands in free and reversible communication with the external solution. Following earlier work by Conway and Downey (1950) with yeast, this volume within the tissue was called the "outer" space of the tissue. The general equation is (Epstein, 1955):

$$\frac{\text{[Diffusible ions]}}{\text{[External concentration]}} = \text{"Outer" space,}$$

where the diffusible ions are expressed in μmole/g fresh weight of tissue, the external concentration is given as μmole/ml, and the "outer" space in ml/g fresh weight.

Theoretical considerations (Levitt, 1957) and experimental work (Bernstein and Nieman, 1960; Ingelsten and Hylmö, 1961) have since shown that a thin film of the initial solution remains on the tissue despite careful blotting, so that the "outer" space measured in this way tends to be too high. For most tissues of higher plants, this space seems to occupy a volume of about 0.10–0.15 milliliter per gram fresh weight.

4. CATION EXCHANGE

When tissue is immersed for some minutes in a solution of a salt, say $CaCl_2$, and then taken out and rinsed with water, that amount of the salt that had diffused into the "outer" space of the tissue would be expected to diffuse out again during the water rinse and be removed from the tissue. However, upon subsequently immersing the tissue in a solution of another salt, say $MgSO_4$, it is found that an additional amount of calcium (but not of chloride) is lost from the tissue. Evidently, this fraction of calcium had failed to diffuse out into the rinse water used before. In other words, the tissue acts as a cation exchange material, in the manner discussed earlier in connection with clay particles (see Figure 3-2).

Cations held by the negatively charged cation exchange matrix of the

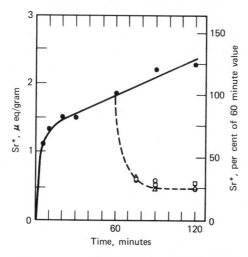

Figure 5-3. Uptake and loss of strontium labeled with ^{89}Sr by barley roots. Upon immersion of the roots in the solution of labeled strontium, the cation is rapidly taken up at first, and more slowly thereafter. On transfer to a solution of unlabeled strontium at 60 minutes, or to solutions of calcium or magnesium, much of the labeled strontium taken up initially is displaced by cation exchange with these ions. The labeled strontium taken up by the slow, steady process evident after 30 minutes is non-exchangeable—it has moved across the plasmalemma by the processes described in Chapter 6. ●, ^{89}Sr; ○, Sr; △, Ca; □, Mg. Sr* indicates labeled strontium. After Epstein and Leggett (1954).

tissue are not free to diffuse out into water because of the electrostatic forces between the negative charges of the tissue and the positively charged ions. These adsorbed cations can, however, be displaced by cations of a chemically or isotopically different kind (Figure 5-3). The negative sites of the tissue which make up its cation exchange capacity are called electro-negative sites (Epstein and Leggett, 1954) or fixed negative charges (Dainty and Hope, 1961), and the volume in which cations held by them are confined has been termed the "Donnan free space" (D.F.S.) of the tissue (Briggs, Hope and Pitman, 1958). If one disregards the distinction between freely diffusing ions in the "outer" space and those restrained by the negative charges of the cell wall exchange sites, the totality of ions in the cell wall diffusion and exchange system is said to be in the "apparent free space" (A.F.S.) of the tissue (Briggs, 1957; Briggs, Hope and Robertson, 1961; Briggs and Robertson, 1957; Hope and Stevens, 1952).

5. THE LOCALE OF THE "OUTER" AND "DONNAN FREE" SPACE OF ROOTS

Some of the early explorers of botanical "outer" space, or "free" space, thought that it included at least part of the cytoplasm as well as the cell wall (Briggs, 1957; Briggs and Robertson, 1957). Though recently revived by Laties of the University of California at Los Angeles, as explained in the next chapter, this concept no longer commands much support, and most students of ion transport in plants have concluded that the space accessible by diffusion and ion exchange lies in the cell wall and the film of moisture lining the intercellular spaces. In other words, these spaces are extracellular, outside the outermost membrane, the plasmalemma. Evidence for this conclusion is as follows:

(1) Bernstein and Nieman (1960) measured the "outer" space of bean roots, *Phaseolus vulgaris,* by a method similar to that described in section 3 of this chapter. As solutes they used mannitol and NaCl. For each of these solutes a series of concentrations was set up, ranging in osmotic potential from −1 to −8 atmospheres. The measured volume of the "outer" space was independent of the concentration of the solutions in the range −1 to −4.1 atmospheres, but at concentrations beyond −4.1 atmospheres it sharply increased. Independent measurements showed this concentration to be the one at which the cells plasmolyzed. The interpretation of these findings is that the plasmalemma is the inner boundary of the "outer" space; when the membrane pulls away from the cell wall during plasmoly-sis, the "outer" space into which the external solute is free to diffuse is

thereby increased. If NaCl had entered the cytoplasm by diffusion the measurements of the "outer" space using NaCl should have given consistently greater values than those made with mannitol, known to be largely excluded from the cytoplasm of plant cells, or taken up very slowly. No such differences were observed.

(2) Other evidence for the plasmalemma as the inner boundary of the "outer" space comes from experiments, to be discussed in detail in the next chapter, which lead to the conclusion that the plasmalemma is highly impermeable to ions, and is traversed by them only via mechanisms of metabolically active transport. It is at this membrane that selective, metabolically active mechanisms come into play—mechanisms for which there is no evidence from the passive (non-metabolic) diffusion and exchange phenomena discussed above. The locale of the latter phenomena must therefore be external to the plasma membrane, in the cell wall.

(3) Like the "outer" space, the negative sites giving rise to the observed cation exchange properties of plant tissues are located within the cell walls. Jansen et al. (1960) determined the total cation exchange capacity of Avena coleoptile and found it to agree very closely with the cation exchange capacity of the cell wall pectic substances. Specifically, the cation exchange capacity of the cell walls was found to be due to pectic carboxyl groups which act as fixed negative charges: $R \cdot COO^-$. That these findings have general validity is apparent from data of Knight et al. (1961) who examined the cation exchange properties of roots of different plants and found them to be related to the content of pectic substances of the tissues. Chapters 6 and 7 contain further evidence for the conclusion that the space in which ions are free to diffuse and exchange is external to the plasmalemma, and that the latter represents a barrier to diffusion and exchange of ions.

Up to this point, the discussion of diffusion and cation exchange phenomena has been entirely in terms of the structure of plant cells, without regard to the organization of tissues and organs. The organ for which a discussion of this subject in terms of supracellular organization is most important is the root. The root is directly exposed to the ionic environment which is the principal source of nutrient and other ions, and is involved in the initial ionic adjustments between the plant and its source of mineral ions. How far does the "outer" space extend radially into the root?

The answer to this question almost certainly is: to the endodermis, and not beyond. In other words, the "outer" space is located in the walls of the epidermal and cortical cells of the root. Figure 5-4 shows the cross-section of the root of a barley seedling. Since the "outer" space occupies an appreciable percentage of the total volume of the root it is evident that

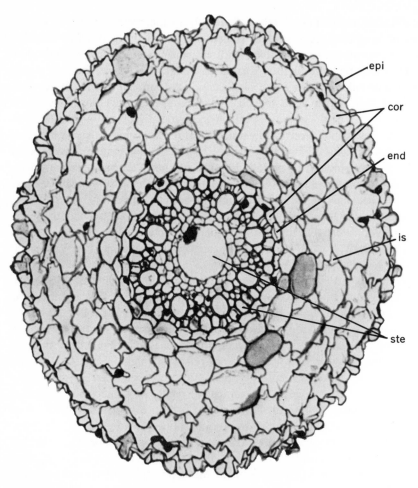

Figure 5-4. Cross section of the primary root of barley, *Hordeum vulgare:* epi, epidermis; cor, cortex; end, endodermis; is, intercellular space; ste, stele.

the epidermis does not constitute a barrier to diffusion. The walls of the cells of the cortex must be the principal locale of the "outer" space.

On the other hand, it is unlikely that the "outer" space extends beyond the endodermis, the innermost layer of cells of the cortex. It has often been pointed out that the Casparian strip, embedded in the radial and transverse walls of the endodermis, constitutes a barrier to the diffusion of ions (Van Fleet, 1961). Furthermore, if ions were free to diffuse into the stele beyond the endodermis, the ionic content of the cells of the stele should tend to be similar to that of the cells of the cortex. However, discussion of long-distance transport (Chapter 7) will bring out the fact that the concentration of ions in the stele frequently exceeds their concentration in the cortex by large factors. Free diffusive movement into the stele would seem to be ruled out for any solutes for which there is evidence of accumulation in the stele to levels much in excess of their concentrations in the cortex.

More direct evidence for the role of the endodermis as the inner limit of the "outer" space has been provided by Lüttge and Weigl (1962, 1964), using the technique of tissue autoradiography. After exposing roots of corn, *Zea mays,* and pea, *Pisum sativum,* to radioactivity labeled sulfate or calcium they caused these ions to be precipitated *in situ* in the form of insoluble salts ($BaSO_4$ and CaC_2O_4, respectively) and found that diffusive movement was largely restricted to the walls of the cells of the cortex; the endodermis evidently blocked diffusion into the stele.

So far, we have mainly considered ion movements in seedling roots showing only primary growth. However, Kramer (1969) has drawn attention to the fact that at least in perennial plants, much absorption of water and possibly, of mineral ions, may occur in older, partially or wholly suberized roots. It is an important point, especially in respect to the mineral nutrition of trees, but direct evidence is scant (O'Leary, 1965; O'Leary and Kramer, 1964).

"OUTER" SPACE IN LEAVES

Like the cells of the cortex of the root, the mesophyll cells of the leaf are bathed by a solution which pervades their walls. This solution is delivered into the extracellular leaf space by the xylem (Chapter 7). Before ions in this solution can participate in essential metabolic and biochemical processes within the cells of the leaf they must be absorbed by these cells from the solution bathing them (Smith and Epstein, 1964). The solution piped by the xylem into the extracellular space of the leaf becomes the immediate medium for absorption across the plasmalemmas of the mesophyll cells, much as the soil solution penetrating the "outer" space of the root cortex serves as the mineral substrate for the cells of the cortex.

Recognition of the existence of extracellular spaces in leaves where ions are free to diffuse and exchange, without the direct intervention of metabolism, is important in several connections. In experiments on the absorption of solutes by leaf cells, discs cut from the leaves with cork borers have often been used. The discs are then floated on the experimental solutions.

Smith and Epstein (1964) showed that in experiments with leaf discs of corn, *Zea mays,* only those cells near the cut edges of the discs were reached by the external solutions and absorbed ions from them. The cells in the center of the discs were too far removed from the cut edge where the external solution entered the tissue. When progressively narrower slices of tissue were cut, instead of discs, it was found that slices with a width of 300 microns were sufficiently narrow so that this diffusion limitation was avoided. The evidence was that in such narrow slices, corresponding in width to the diameters of a few cells, the ions from the external solution quickly diffused throughout the "outer" space of the tissue and became available to all the mesophyll cells for metabolically active absorption. The validity of these conclusions has been confirmed (Jacoby and Dagan, 1967; Osmond, 1968).

Normally, by far the greatest portion of mineral nutrients reaches the extracellular spaces of the mesophyll via the xylem, from the roots. However, mineral ions descending onto leaves in rain (see Chapters 2 and 3) may slowly penetrate through the stomates and the cuticle and reach the interior of the leaf, i.e., its extracellular spaces, and thereby become available for absorption by the mesophyll cells. The same process occurs when nutrient ions in solution are directly sprayed onto the leaves of crop plants. This is an increasingly important practice, called "foliar application," useful especially for the correction of deficiencies of micronutrient elements which tend to be immobilized in soil (Krantz *et al.,* 1962). The details of the penetration of extraneous substances through the leaf surface into the interior have been examined and reviewed, especially by members of the Michigan group in plant nutrition (Wittwer *et al.,* 1965) and other investigators (Franke, 1967; Hull, 1970), the latter emphasizing the penetration of pesticides and other esoteric compounds rather than that of inorganic nutrients.

Since these penetrations of extraneous substances into the cell wall spaces are passive movements not directly depending on metabolism, they are completely reversible and movement of solutes from the cell wall spaces of the leaf onto its surface also occurs, provided that the cell surface is rinsed and an outward gradient for the diffusion of solutes is thereby maintained. Such leaching happens in rain, during sprinkler irrigation, and when artificial mists are applied to horticultural plants for cooling and to reduce transpiration (Tukey, 1970).

The most direct evidence for the extracellular location of the leaf "outer" or "free" space comes from experiments in which lead in solution (complexed by ethylenediamine tetraacetic acid) was administered to wheat seedlings, *Triticum* sp. After 12 hours, the leaves were placed in an atmosphere of hydrogen sulfide gas for 1 minute. This precipitated the lead in the leaves as lead sulfide, and microscopic examination showed the precipitate to be confined to the walls (Crowdy and Tanton, 1970).

The most significant difference between the "outer" spaces of roots and leaves is the cuticle covering the leaves which tremendously reduces the rate of equilibration of an external solution with the solution in the cell walls of the interior of the leaves. Martin and Juniper (1970) have furnished an excellent account of the cuticles of plants. If the cells within the leaves of terrestrial plants are to be exposed, without significant diffusion lag, to experimental solutions for studies on the processes of ion transport by these cells, the only available technique is that of Smith and Epstein (1964) in which the leaves are sliced into strips so narrow that all cells within the tissue are almost instantly reached by the experimental solution which diffuses into the "outer" space via the cut edges.

DIFFUSION AND CATION EXCHANGE
VS. METABOLIC ION TRANSPORT

The hallmarks of the ionic movements discussed above—diffusion and cation exchange in the cell wall—are, first, that they are independent of concurrent metabolic activity of the tissue, second, that they are readily reversible, and third, that they are not selective.

Diffusion and cation exchange occur as readily in dead as in living, metabolizing tissue. Ions that have diffused into the "outer" space may diffuse out again, and cations restrained by electronegative sites of cell wall carboxyl groups can readily be displaced by other cations. This process of cation exchange in plant cell walls differs in no way from cation exchange on other carboxylic cation exchangers such as the synthetic cation exchange resins of the chemical laboratory or manufacturing plant. Divalent cations are adsorbed more strongly than are monovalent cations, and within each group adsorption of the ions follows the usual lyotropic series. However, these differences in strength of adsorption are not such as to impart to the system a high degree of selectivity. On the contrary, any cation readily displaces any other (Epstein and Leggett, 1954; Mengel, 1963). If absorption of ions from the external solution depended on the cation exchange properties of the cell wall, the ionic composition of the cells could be readily predicted on the basis of the known properties of carboxylic cation exchangers.

As shown in Chapters 6 and 9, the mineral composition of the plant depends on processes which are, first, dependent on metabolic processes, second, largely unidirectional rather than readily reversible, and third, highly selective. These statements apply to the absorption of cations and anions alike. The processes of ion absorption or ion transport, then, which are responsible for the mineral composition of the plant in no wise parallel the known properties of the cell wall diffusion and cation exchange space in which ions move in a manner strictly predictable on the basis of what is known about diffusion and cation exchange in non-living systems. Other mechanisms must be invoked to account for the remarkable selectivity which is one of the outstanding characteristics of biological ion transport. These processes of metabolically active, selective transport are discussed in the next chapter.

8. SIGNIFICANCE OF DIFFUSION AND CATION EXCHANGE FOR SOIL-PLANT RELATIONSHIPS

The outer cytoplasmic membrane or plasmalemma lies behind the cell wall. Ions in the soil solution, before impinging upon this membrane, must therefore pass through the interstices of the cell wall or along the cation exchange surfaces of cell wall fibrils. During this passage through the cell walls, the ions are in a sense still part of the external solution: the solution in the "outer" space is an extension of the soil solution. Diffusive movement through the cell wall is fast so that in slender, growing rootlets, this process is not normally the one that limits the rate of absorption (see Chapter 6).

As for cation exchange on the root cation exchange sites, there is a school of thought which holds that this process determines relative rates of absorption of monovalent and divalent cations (Drake, 1964). However, the measurements of root "cation exchange capacity" used as evidence often involve drastic treatment of the roots under unphysiological conditions, and the values obtained depend much on the procedure used (Bartlett, 1964). The "cation exchange capacity" measured in many of these experiments is not only that of the cell wall but includes all other acid-titrable groups of the tissue. Furthermore, as already mentioned in the preceding section, the selectivity of the absorption process reflected in the ultimate composition of the plant does not correspond to the exchange properties of carboxylic cation exchangers. A high cation exchange capacity of the root is said to favor absorption of divalent cations over that of monovalent ones, but this explanation fails to account for the sharp selection plants effect between cations within each of these groups, for example, between potassium and sodium. In short, neither specific investigations (Epstein and Leggett, 1954; Lider and Sanderson, 1960) nor general surveys of the available evidence (Black, 1968; Fried and Broeshart, 1967) support the view that cation exchange on electronegative sites of the cell walls of roots determines the selectivity of cation absorption by roots and plants.

Nevertheless, the absorption of ions present in soil at extremely low concentrations may be significantly influenced by the cation exchange properties of root cell walls. Evidence for such effects has been discussed since the 1930's by Jenny (1961) and his collaborators. Exchangeable cations present on a cation exchange material, for example, clay particles, may diffuse from there to another exchange material, for example, root cell walls, with extreme slowness if an appreciable distance intervenes between the two solid phases, although eventually the ions will reach an equilibrium throughout the system. When, however, the two exchange materials are brought in close proximity of each other so that the cation swarms in the vicinity of their negative sites intermingle, the process of equilibration between the two phases is greatly speeded up ("contact exchange").

While such exchange-migration of cations doubtless may occur between solid negatively charged surfaces in intimate contact (Glauser and Jenny, 1960), its significance for the nutrition of soil-grown plants is difficult to assess. It may be of importance for the absorption of ions which are present at vanishingly low concentrations, for example, ferric iron in a well-aerated, calcareous soil. At the high pH of such a soil (above pH 8) the very low solubility of ferric hydroxide will keep the concentration of iron in the soil solution at an extremely low value. Surface migration and contact exchange may provide a mechanism whereby ferric ions may migrate from the soil into the cell wall spaces of the root (Jenny, 1961).

On the other hand, highly soluble monovalent cations, mainly potassium, probably invade the cell wall space without becoming involved in cation exchange on the cell wall cation exchange spots. With the exception of hydrogen ions, monovalent cations are less strongly adsorbed on carboxylic exchange spots than are divalent and trivalent ones. The predominant cations on the electronegative sites of the cell walls of roots in soil are therefore calcium, magnesium, iron, manganese, aluminum, and hydrogen ions, to the virtual exclusion of monovalent cations, especially if they are present at low concentrations. Monovalent cations under these conditions behave precisely like anions: they are not exchangeably adsorbed on the electronegative sites of roots to a measurable extent but exist in free solution (Epstein, Rains, and Schmid, 1962).

By the processes of diffusion and ion exchange ions from the external medium migrate into the cell wall space. The movement of water toward and into the root, caused ultimately by transpirational loss of water to the atmosphere, also may hasten the process. Once the ions impinge upon the outer cytoplasmic membrane, the plasmalemma, their onward movement by these same processes is impeded because the membrane is not highly permeable to polar solutes such as inorganic ions. They nevertheless may cross this membrane and enter the cell. The processes whereby ions are transported across the cellular membranes are metabolically active: they require the participation of metabolic events and are in fact a manifestation of metabolic activity. These processes of "active transport" form the subject of the next chapter.

REFERENCES

Bartlett, R. J. 1964. Measurement of cation- and anion-exchange capacities of roots using NaCl exchange. Soil Sci. 98:351–357.

Bernstein, L. and R. H. Nieman. 1960. Apparent free space of plant roots. Plant Physiol. 35:589–598.

Black, C. A. 1968. Soil-Plant Relationships. 2nd ed. John Wiley and Sons, Inc., New York.

Briggs, G. E. 1957. Some aspects of free space in plant tissues. New Phytol. 56:305–324.

Briggs, G. E., A. B. Hope and M. G. Pitman. 1958. Exchangeable ions in beet disks at low temperature. J. Expt. Bot. 9:128–141.

Briggs, G. E., A. B. Hope and R. N. Robertson. 1961. Electrolytes and Plant Cells. Blackwell Scientific Publications, Oxford.

Briggs, G. E. and R. N. Robertson. 1957. Apparent free space. Ann. Rev. Plant Physiol. 8:11–30.

Clowes, F. A. L. and B. E. Juniper. 1968. Plant Cells. Blackwell Scientific Publications, Oxford and Edinburgh.

Conway, E. J. and M. Downey. 1950. An outer metabolic region of the yeast cell. Biochem. J. 47:347–355.

Crowdy, S. H. and T. W. Tanton. 1970. Water pathways in higher plants. I. Free space in wheat leaves. J. Expt. Bot. 21:102–111.

Dainty, J. and A. B. Hope. 1961. The electric double layer and the Donnan equilibrium in relation to plant cell walls. Austral. J. Biol. Sci. 14:541–551.

Drake, M. 1964. Soil chemistry and plant nutrition. In: Chemistry of the Soil. F. E. Bear, ed. Reinhold Publishing Corporation, New York.

Epstein, E. 1955. Passive permeation and active transport of ions in plant roots. Plant Physiol. 30:529–535.

Epstein, E. and J. E. Leggett. 1954. The absorption of alkaline earth cations by barley roots: kinetics and mechanism. Am. J. Bot. 41:785–791.

Epstein, E., D. W. Rains and W. E. Schmid. 1962. Course of cation absorption by plant tissue. Science 136:1051–1052.

Franke, W. 1967. Mechanisms of foliar penetration of solutions. Ann. Rev. Plant Physiol. 18:281–300.

Frey–Wyssling, A. and K. Mühlethaler. 1965. Ultrastructural Plant Cytology, with an Introduction to Molecular Biology. Elsevier Publishing Company, Amsterdam.

Fried, M. and H. Broeshart. 1967. The Soil-Plant System. Academic Press, New York and London.

Glauser, R. and H. Jenny. 1960. Two-phase studies on availability of iron in calcareous soils. III. Contact, and exchange diffusion in ionic membranes. Agrochimica 5:1–9.

Hope, A. B. and P. G. Stevens. 1952. Electric potential differences in bean roots and their relation to salt uptake. Austral. J. Sci. Res. B 5:335–343.

Hull, H. M. 1970. Leaf structure as related to absorption of pesticides and other compounds. Residue Rev. 31:1–155.

Ingelsten, B. and B. Hylmö. 1961. Apparent free space and surface film determined by a centrifugation method. Physiol. Plantarum 14:157–170.

Jacoby, B. and J. Dagan. 1967. A comparison of two methods of investigating sodium uptake by bean-leaf cells and the vitality of isolated leaf-cells. Protoplasma 64:325–329.

Jansen, E. F., R. Jang, P. Albersheim and J. Bonner. 1960. Pectic metabolism of growing cell walls. Plant Physiol. 35:87–97.

Jenny, H. 1961. Two-phase studies on availability of iron in calcareous soils. V. Kinetics of iron transfer as conditioned by ion exchange capacity and structure of roots. Agrochimica 5:281–289.

Knight, A. H., W. M. Crooke and R. H. E. Inkson. 1961. Cation-exchange capacities of tissues of higher and lower plants and their related uronic acid contents. Nature 192:142–143.

Kramer, P. J. 1969. Plant and Soil Water Relationships—A Modern Synthesis. McGraw-Hill Book Company, Inc., New York.

Krantz, B. A., A. L. Brown, B. B. Fischer, W. E. Pendery and V. W. Brown. 1962. Foliage sprays correct iron chlorosis in grain sorghum. Calif. Agric. 16(5):5–6.

Ledbetter, M. C. and K. R. Porter. 1970. Introduction to the Fine Structure of Plant Cells. Springer-Verlag, Berlin, Heidelberg, New York.

Levitt, J. 1957. The significance of "apparent free space" (A.F.S.) in ion absorption. Physiol. Plantarum 10:882–888.

Lider, L. A. and G. W. Sanderson. 1960. Some measurements of cation-exchange capacity of grape roots. Am. J. Enol. and Vit. 11:174–178.

Lüttge, U. and J. Weigl. 1962. Mikroautoradiographische Untersuchungen der Aufnahme und des Transportes von $^{35}SO_4^{--}$ und $^{45}Ca^{++}$ in Keimwurzeln von *Zea mays* L. und *Pisum sativum* L. Planta 58:113–126.

Lüttge, U. and J. Weigl. 1964. Der Ionentransport in intakten und entrindeten Wurzeln. Ber. Deutsch. Bot. Ges. 77:63–70.

Martin, J. T. and B. E. Juniper. 1970. The Cuticles of Plants. Edward Arnold, London.

Matile, P. 1966. Enzyme der Vakuolen aus Wurzelzellen von Maiskeimlingen. Ein Beitrag zur funktuellen Bedeutung der Vakuole bei der intrazellulären Verdauung. Z. Naturforsch. 21b:871–878.

Mengel, K. 1963. Die Bedeutung von Kationenkonkurrenzen im free space der Pflanzenwurzel für die aktive Kationenaufnahme. Agrochimica 7:236–257.

O'Leary, J. W. 1965. Root-pressure exudation in woody plants. Bot. Gaz. 126:108–115.

O'Leary, J. W. and P. J. Kramer. 1964. Root pressure in conifers. Science 145:284–285.

Osmond, C. B. 1968. Ion absorption in *Atriplex* leaf tissue. I. Absorption by mesophyll cells. Austral. J. Biol. Sci. 21:1119–1130.

Pridham, J. B., ed. 1968. Plant Cell Organelles. Academic Press, London and New York.

Smith, R. C. and E. Epstein. 1964. Ion absorption by shoot tissue: technique and first findings with excised leaf tissue of corn. Plant Physiol. 39:338–341.

Tukey, H. B., Jr. 1970. The leaching of substances from plants. Ann. Rev. Plant Physiol. 21:305–324.

Van Fleet, D. S. 1961. Histochemistry and function of the endodermis. Bot. Rev. 27:165–220.

Wittwer, S. H., W. H. Jyung, Y. Yamada, M. J. Bukovac, R. De, S. Kannan, H. P. Rasmussen and S. N. Haile Mariam. 1965. Pathways and mechanisms for foliar absorption of mineral nutrients as revealed by radioisotopes. In: Isotopes and Radiation in Soil-Plant Nutrition Studies. International Atomic Energy Agency, Vienna. Pp. 387–403.

6

ACTIVE ION
TRANSPORT IN
CELLS AND TISSUES

1. PLANT MATERIALS

The processes of diffusion and ion exchange in the cell wall described in the preceding chapter do not differ from those that occur in non-living systems, for example in clays or other ion exchange materials. But when sap from the interior of cells is analyzed its composition almost invariably is found to differ from that of the external medium in such a way as to prove that processes other than the familiar ones of diffusion and ion exchange must have been at work.

In classical experiments, Professor D. R. Hoagland of the University of California and his collaborators analyzed the internal sap of cells of the fresh water alga, *Nitella clavata,* which had been growing in pond water (Hoagland, 1944). This alga has very large cells, up to several centimeters long, and from their vacuoles nearly uncontaminated sap can be obtained for analysis. It was found that the principal ions present in the pond water were also present within the cell sap, and all of them, cations and anions, at concentrations much higher than their concentrations in the pond water. The evidence was, furthermore, that these intracellular ions were in free solution in the cell sap, not bound to cell constituents. Together, this evidence meant that salts had been transferred from a region where their concentration was low (the pond water) to a region where their concentration was high (the cell sap).

Such "uphill" transfer of salt does not occur spontaneously; the spontaneous (diffusive) movement of salt is in the opposite direction, from regions of high to those of low concentration. These simple experiments therefore proved that the cells expend energy in causing salt to move against the diffusion gradient. The necessary energy is derived from the active metabolism of the tissue, and the terms "active absorption" and "active transport" were used to refer to such movement of salt dependent on metabolic activity.

Hoagland and his co-workers soon extended this type of research to other plant materials. In Hoagland's laboratory, roots of young barley seedlings, *Hordeum vulgare,* became a favorite experimental object, and many workers have subsequently used excised barley roots for experiments on the absorption of ions. In the original, classical technique of Hoagland and Broyer (1936), barley seedlings were grown for three weeks in a dilute nutrient solution. The solution was not renewed and slowly became depleted of nutrients, through absorption by the roots. The roots in turn became depleted of ions because a large fraction of the ions they absorbed was translocated to the shoots. The "low-salt" roots obtained by this method were then excised and used in experiments on the absorption of various ions. Most experiments lasted from several to 24 hours. The initial and final content of the element in the roots was determined by chemical analysis and its uptake calculated as the difference. The roots had a high capacity for salt accumulation: "Significant salt accumulation by highly active root systems can occur within 2 hours."

In a modern version of this technique (Epstein, 1961), barley seedlings are grown in a dilute solution of calcium sulfate (0.2 mM $CaSO_4$) for five days (Figure 6-1). The roots are excised and 1.00-gram samples enclosed in open-weave cheesecloth "teabags" (Epstein *et al.,* 1963). The samples can then be handled with ease by means of a thread attached to each "teabag" and transferred into and out of experimental solutions with precise timing (Figure 6-2). The ion under investigation is usually radioactively labeled, and routine experimental periods are 10 or 20 minutes. At the end of the absorption period, the sample is normally transferred to an unlabeled solution of the same ion to remove the diffusible radioions from the "outer" space and those held exchangeably by fixed negative charges, leaving for final radioassay only that fraction that has been transferred by metabolically active processes into the "inner" spaces. This fraction is not subject to rapid exchange. With this technique, significant accumulation can be shown within a minute (Epstein *et al.,* 1962).

At the same time that Hoagland and Broyer developed the excised root technique, F. C. Steward initiated a series of investigations in which thin

Figure 6-1. Barley seedlings grown according to the method of Epstein (1961) for experiments with excised roots.

discs of potato tuber tissue, *Solanum tuberosum,* were exposed to experimental solutions and absorbed ions from them (Steward and Harrison, 1939). Discs of other storage tissue (carrot, *Daucus carota,* beet, *Beta vulgaris*) have since been used extensively, notably by R. N. Robertson and his co-workers in Australia (Robertson and Wilkins, 1948). Discs freshly cut from storage tissue do not have the capacity for rapid accumulation of ions. The discs acquire this capacity only after a period of incuba-

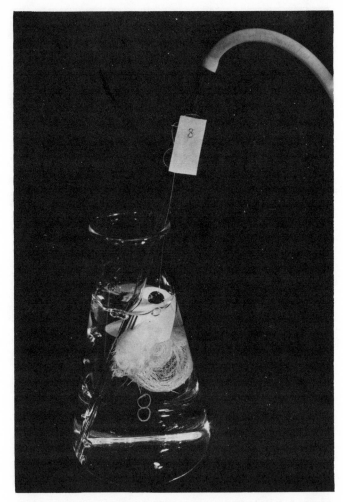

Figure 6-2. A 1.00-g sample of excised barley roots enclosed in an open-weave cheesecloth "teabag" and immersed
in aerated experimental solution. In actual experiments,
arrays of flasks are kept in temperature-controlled water
baths.

tion in aerated water or solution of a calcium salt. During this period of "aging" profound alterations occur in the metabolism of the tissue, due ultimately to the derepression of gene activity and consequent initiation of protein synthesis (Ellis and MacDonald, 1967; Kahl *et al.,* 1969). The connection between these events and the initiation of the capacity for ion transport is as yet poorly understood (Polya, 1968; Pitman *et al.,* 1971). Later, Steward and his co-workers switched from discs cut from storage tissue to the technique of tissue culture of carrot explants. This work has been exhaustively reviewed by Steward and Mott (1970).

W. H. Arisz and his collaborators in Holland have done much work on absorption and translocation in leaf tissue of the submerged aquatic plant, *Vallisneria spiralis,* tape grass (Arisz, 1964). Recently leaf tissue of terrestrial plants and cells separated from such tissue have been used (Jacoby and Dagan, 1967; Jyung *et al.,* 1965; Smith and Epstein, 1964a). Practically everything we know about the mechanisms of ion absorption by plant cells has been discovered through experiments with a few species of algae, roots of barley and some other *Gramineae,* storage tissue (potato, carrot, beet), and leaf tissue of a few species. Like those who conduct public opinion polls, biologists base far-reaching conclusions on knowledge gained from a very meager sampling of sources of information. But unlike those of the pollsters, the chief conclusions that have been drawn from these experiments appear to be quite generally valid. We shall now examine the principal findings leading to the conclusion that the transport of ions into plant cells is a metabolically active process.

EVIDENCE FOR ACTIVE ION TRANSPORT

Accumulation Ratios. The most compelling piece of evidence for active transport is the one already mentioned in the preceding section: the fact that salt may reach concentrations within cells far higher than its concentration outside. Such movement against the diffusion gradient is often called "accumulation," and the ratio—concentration within the cell divided by the concentration of the same solute in the solution bathing the cell—is called the "accumulation ratio." Steward and Sutcliffe (1959) have listed accumulation ratios for various ions in cells of algae and tissues of higher plants. For potassium and chloride, values on the order of 1000/1 or higher are common. Accumulation ratios are usually higher the lower the external concentration. MacDonald *et al.* (1960) exposed discs of rutabaga, *Brassica napus,* to flowing tap water containing potassium at a concentration of 0.01 mM. After several days, the accumulation ratio was over 10,000/1.

Temperature Effects.   The rate of physiological processes is markedly influenced by the temperature. Within the range of about 10°C to 30°C an increase of 10°, say from 15 to 25°, usually causes the rate of metabolic processes to go up by a factor of 2 or more. Purely physical processes like ion exchange on inert exchangers are much less temperature sensitive, with a temperature coefficient (Q_{10}) of about 1.1 to 1.2. The temperature coefficient for the process of ion absorption is 2 or higher, as expected for a metabolically active process, and at temperatures near

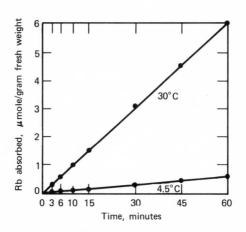

Figure 6-3. Absorption of rubidium by excised barley roots as a function of time and temperature. Concentration of RbCl, 0.1 mM, and of CaCl₂ 0.5 mM. After Epstein *et al.* (1962).

freezing absorption is severely inhibited (Hoagland, 1944). Figure 6-3 shows the effect of temperature on the absorption of rubidium by excised barley roots.

Oxygen.   Hoagland and his collaborators, and many investigators since, observed that under anaerobic conditions or at low concentrations of oxygen in the solution bathing the tissue, absorption of many ions is inhibited. Although some metabolic activity persists even under anaerobic conditions, steady absorption of the ions in question evidently depends on aerobic metabolism. Figure 6-4 shows the effect of the oxygen tension in the solution upon the rate of phosphate absorption by excised barley

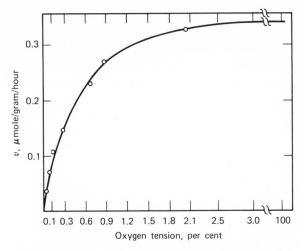

Figure 6-4.   Rate, v, of phosphate absorption by excised barley roots as a function of the oxygen tension of the solution. After Hopkins (1956).

roots. At about 2 per cent oxygen, the rate was almost maximal, and at 0.3 per cent, it was half-maximal. The absorption of many ions has been found to be affected by the oxygen tension in the medium.

Poisons.   There are many substances which, at appropriate concentrations, interfere with the normal operations of metabolism. For at least some of these poisons or inhibitors the specific metabolic steps which they block are known. In the hands of discerning investigators, such poisons are therefore important tools in the study of metabolism. They can be used to block specific metabolic processes or enzymic reactions, and the effects of these metabolic blocks can then be determined.

Ordin and Jacobson (1955) investigated the effects of ten widely used metabolic inhibitors on the absorption of potassium, bromide, and on oxygen uptake by barley roots, *Hordeum vulgare.* All ten under appropriate conditions inhibited the absorption of potassium and of bromide, but the degree of inhibition was often not the same for the two ions. For example, a 2-hour pretreatment with 40 mM fluoride inhibited the subsequent absorption of potassium 97 per cent, while that of bromide was inhibited only 60 per cent. Oxygen uptake (i.e., aerobic respiration) was inhibited to the same extent as was bromide absorption. On the other hand, dinitrophenol at 1.1×10^{-6} M inhibited potassium absorption 90 per cent, bromide absorption 67.3 per cent, and was without effect on oxygen uptake.

Experiments like these lead to the conclusion that ion absorption is one of the manifestations of metabolism; derangements in metabolism produced by poisons are reflected in abnormalities, usually in inhibition, of ion transport. The experiments also show that the cations and anions of a salt are not necessarily teamed up in their transport, since a given poison may affect the transport of the two ions of a salt quite differently.

Carbohydrate. The energy required for the active transport of ions must be derived from metabolism, more specifically, from the oxidation of substrate metabolites. The most important substrates are carbohydrates derived ultimately from photosynthesis. It is therefore logical to look for connections between the utilization of sugars and the accumulation of ions.

Hoagland and his collaborators, in their pioneering work on ion absorption as an aspect of metabolism (Hoagland, 1944), observed that barley roots low in reserves of sugar absorbed more potassium from a solution to which sugar had been added than from control solutions lacking sugar. Later, more detailed experiments of this kind have confirmed their finding (Helder, 1952; Mengel, 1962).

Light. The synthesis of carbohydrates and other respiratory substrates depends on the energy of light trapped in the process of photosynthesis (Chapter 9, pp. 226–230). If the absorption of ions requires metabolic energy, light may therefore be expected to promote ion absorption in photosynthetic cells, and this is found to be the case. The role of light in ion absorption was first studied in plants in which the same cells perform the functions of both photosynthesis and the primary acquisition of mineral nutrients. This is so in algae and in submerged vascular plants such as *Vallisneria* (tape grass) and *Elodea* (waterweed).

Hoagland pioneered in this, as he did in other aspects of research dealing with the connection between metabolism and ion transport. He and Davis (1923) reported that the accumulation of chloride, bromide, and nitrate by cells of the green alga, *Nitella clavata,* depends on light as a source of energy. Much work on ion transport in green algae has confirmed the role of photosynthetic energy in this process (MacRobbie, 1962, 1966; Scott and Hayward, 1953).

As for aquatic vascular plants, van Lookeren Campagne (1957), working in the laboratory of W. H. Arisz in the Netherlands and following up earlier leads obtained by Arisz, found in a thorough investigation that light promoted chloride absorption by leaf tissue of *Vallisneria*. The effect was mediated by the photosynthetic apparatus of the cell. More recent work leads to similar conclusions for absorption of phosphate and sulfate (Jeschke and Simonis, 1965) and of chloride (Jeschke, 1967) by leaves of waterweed, *Elodea densa.*

Most of the work done on absorption of ions by leaves of terrestrial plants has been concerned with the mechanisms by which ions in an externally applied solution penetrate into the "outer" space of the tissue, as discussed in the preceding chapter. However, using the technique of Smith and Epstein (1964a) by means of which the mesophyll cells can be directly exposed to experimental solutions (Chapter 5, pp. 96–97), Rains (1967, 1968) has shown that leaf tissue of corn, *Zea mays,* accumulated potassium in the light at twice the rate it did in the dark, and that the energy for this enhancement was derived from photosynthetic mechanisms. Nobel (1969) presented similar evidence for leaf tissue of pea, *Pisum sativum.*

Many lines of evidence, then, point to the conclusion that the transport of inorganic ions by plant cells is effected through mechanisms whose functioning depends on energy made available through cellular metabolism. It was through the discovery of this relationship that Hoagland initiated the modern era of research in plant nutrition. Important as this realization is, it still leaves two crucial questions unanswered. One concerns the nature of the transport mechanism itself—the agency whereby the ion is moved from one compartment into another one across the intervening membrane. And the second question has to do with the linkage between metabolism, the source of energy, and the energy-requiring transport mechanism. Leaving this second question for detailed consideration in Chapter 9, we shall consider in this chapter the central hypothesis concerning the transport mechanism and the evidence upon which it rests. First, however, we must discuss the membranes across which transport takes place.

MEMBRANES

Structure.   Transport of ions into the cell takes place across the outer cell membrane, the plasmalemma. Delivery into the chief cellular repository of salts, the vacuole, entails transport across the vacuolar membrane or tonoplast. There is also ionic traffic across the membranous envelopes of nuclei, mitochondria, and other organelles (see Figure 5-1). We shall therefore briefly consider the structure of biological membranes and those of plant cells in particular. Brevity is dictated only in part by considerations of space. There are two additional reasons for it. The first is that despite an immense amount of research we do not yet have unequivocal, definitive evidence on the structure of any biological membrane. The second reason for not expanding on the structure of membranes is that as yet, what we have learned about membranes has helped remarkably little in understanding transport. The reverse is also true: studies on transport have been of little help to either biochemists or electron microscopists attempting to clarify the structure of membranes. Evidently, our knowledge is as yet insufficient for studies of membrane transport and of membrane structure to provide mutual intellectual stimulation, although this situation will surely change soon.

The classical view of the structure of biological membranes is that first proposed by Danielli and Davson (1935). The model consists essentially of two layers of phospholipid, with their hydrophobic "tails" pointing inwards, toward the middle of the membrane, and their polar hydrophilic "heads'" facing outward, toward the surfaces of the membrane, where they interact weakly with layers of protein (Figure 6-5). This model emphasizes the lipoidal nature of the membrane and hence, its function as a permeability barrier to polar solutes including inorganic ions. It makes no provision, however, for the exceedingly active transport function of the membrane for which an immense amount of evidence has accumulated since the model was first proposed. The proteins, which must be involved in the

Exterior

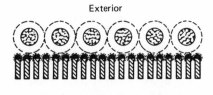

Lipoid

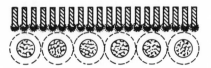

Interior

Figure 6-5. The Danielli–Davson scheme of a membrane. The hydrophobic fatty acid "tails" of the molecules of lipid point toward the center of the membrane and their polar hydrophilic "heads" face the surfaces. A layer of protein is adsorbed on each surface. From Danielli and Davson (1935).

dynamic function of transport, are relegated in this model to the outer surfaces of the membrane, associated with it by weak ionic bonds.

More recent views of the plasma membrane all stress the importance of lipoprotein subunits as both structural and functional entities (Benson, 1964; Branton, 1969). These protein units are visualized as extending through the membrane, an arrangement compatible with their presumed function as carriers of solutes from one surface of the membrane to the other (Korn, 1968). Figure 6-6 is a recent concept of a two-dimensional

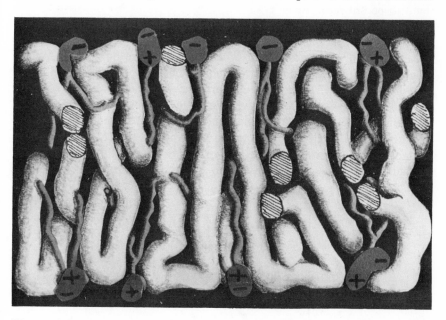

Figure 6-6. The Benson scheme of a membrane, showing the molecular arrangement of lipids and protein within a section of membrane lipoprotein. The polar "heads" of the lipids are at the surfaces of the membrane. Protein extends through the membrane. From Benson (1968).

molecular assembly of lipids and protein within a section of membrane lipoprotein (Benson, 1968). Further development of our view of the structure of cell membranes is likely to retain the basic model of a lipid bilayer, with modifications to accommodate transport and other functions (Hendler, 1971).

Impermeability of Membranes to Ions. The point was made in the previous chapter that the interior of the cell is separated from the medium bathing the cell by a membrane which acts as a barrier to the free diffusion of inorganic ions and many other solutes. Only that part of the cell external to the plasmalemma is in diffusion and exchange equilibrium with the external solution. Beyond the plasmalemma lie the "inner" spaces. The kinds and concentrations of ions in these spaces are under the active, metabolic control of the cell. Ions within cannot readily diffuse out, even when their internal concentration is high and the external concentration of ions of the same kind is very low, so that diffusion would tend to be outward.

Failure of a given species of ion to diffuse out of the cell might be due not to the impermeability of the membrane but to the existence within the cell of ions of opposite sign which cannot diffuse out. These fixed or non-diffusible ions would electrostatically hold the potentially diffusible ions within the cell. However, numerous experiments with radioactively labeled ions have revealed that even exchange of one kind of ion within the cell for others of the same kind outside is greatly impeded by the cellular membranes. Such exchange would be expected to occur readily if the membranes were permeable.

Figure 6-7 shows the results of an experiment on the absorption and

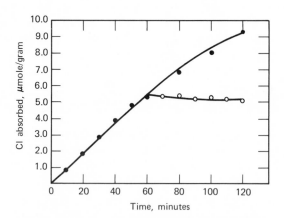

Figure 6-7. Absorption and retention of labeled chloride by barley roots. Dark dots, roots in solution of labeled chloride. Light dots, roots in solution of unlabeled chloride. Concentration of KCl, both labeled and unlabeled, 0.10 mM. Concentration of CaSO₄, 0.5 mM. Temperature, 30°C throughout. After Elzam and Epstein (1965).

retention of radiochloride by barley roots. In this experiment, samples of roots absorbed chloride labeled with ^{36}Cl. When after one hour samples were transferred to a chemically identical solution in which the chloride was not labeled the roots lost very little of the previously absorbed radiochloride, evidence that isotopic exchange (labeled chloride $\rightleftarrows$ unlabeled chloride) across the cell membrane was exceedingly slow. (Even what little loss of radiochloride took place was not via a back movement across the plasmalemma but represents xylem exudate from the cut end of the roots, as explained in the next chapter.)

When the membranes responsible for this retention are injured, there is a rapid migration of labeled chloride out of the tissue, and the interior of the cell becomes indistinguishable from the external solution, so far as the distribution of both labeled and unlabeled chloride ions is concerned. But in normal, healthy tissue, the membrane acts as a barrier to the free diffusion and exchange of chloride. This experiment was done with barley roots and chloride ions, but similar experiments have been done, with similar results, using diverse ions and plant materials. A few examples are sodium, potassium and chloride in roots of two species of *Agropyron* (Elzam, 1966), phosphate in roots of barley and corn seedlings (Carter and Lathwell, 1967; Crossett and Loughman, 1966), rubidium in roots of barley (Helder, 1958), of soybean, *Glycine max* (Hanson, 1960) and of corn (Venrick and Smith, 1967), chloride in roots of barley (Helder, 1964) and in leaves of *Vallisneria* (Arisz, 1964), potassium in leaf tissue of corn (Rains, 1968), of the mangrove, *Avicennia marina* (Rains and Epstein, 1967c) and in roots of barley seedlings (Johansen *et al.,* 1970), and zinc in barley roots (Schmid *et al.,* 1965).

It has sometimes been argued that there is a very appreciable ionic traffic into and out of plant tissue through diffusion and exchange across the plasmalemma. Much of this work takes its inspiration from experiments on efflux of labeled ions from algal cells (MacRobbie and Dainty, 1958a, b). The cells are first immersed in a solution containing a radioactively labeled ion which they absorb. They are then transferred to an unlabeled solution of the same ion and the loss of radioactivity from the cells is followed as a function of time (Figure 6-8). The initial, very rapid loss is considered to be due to diffusion and exchange of radioions from the cell walls or "free" space. The next fraction, which is lost more slowly, is interpreted as representing the ions from the cytoplasm which pass out across the plasmalemma, considered permeable. The final phase represents the slow loss of ions from the vacuole. On this view, the tonoplast is less permeable than the plasmalemma, and therefore exit of ions from the vacuole is the slowest of the effluxes observed.

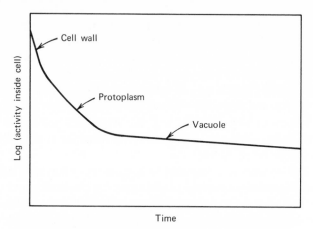

Figure 6-8. "Efflux curve" according to MacRobbie and Dainty. The plasmalemma is considered much more permeable than the tonoplast. After Dainty (1969) in The Physiology of Plant Growth and Development. M. B. Wilkins, ed. © 1969 McGraw-Hill Publishing Company, Ltd., Maidenhead.

This "efflux analysis" was first applied by Pitman (1963) to tissue from a higher plant, discs of beet, *Beta vulgaris*. He interpreted his results as MacRobbie and Dainty did—in terms of diffusion of the ions across the plasmalemma and tonoplast. A few others have used the same approach, among them Pitman and Saddler (1967) with barley roots, Osmond and Laties (1968) with beet discs, Cram (1968) with carrot discs, and Pierce and Higinbotham (1970) with segments of *Avena* coleoptiles.

Whatever the validity of this method as applied to algal cells (see on this Barber, 1968; MacRobbie, 1968), its application to tissues and organs of higher plants is fallacious. Algal cells are isotropic systems delimited by the plasmalemma on all sides. This is not true of tissues of higher plants. Roots do not merely absorb ions, they forward them through the xylem to the shoots (Chapter 7, pp. 165–179). When excised roots containing labeled ions are immersed in unlabeled solutions for appreciable periods (hours), exudate will redeliver labeled ions into the solution—quite a different process from back diffusion across the plasmalemma of an individual algal cell (Greenway, 1967). This effect becomes magnified when the roots are cut into short pieces and then kept in the eluting solution for many hours (Pallaghy *et al.,* 1970; Pitman *et al.,* 1971). Even discs of storage tissue contain vascular elements. Furthermore, they are bordered on all sides by cut and injured cells which face the external solution and are connected by plasmodesmata with the cells in the interior of the discs. Finally, efflux may be carrier-mediated (metabolic) rather than diffusive (Poole, 1969).

Aspects of experimentation are also questionable. The tissue was often kept in solutions for many hours, long enough to become depleted of substrate and to suffer other impairment. In some of these and similar experiments, the tissue was kept in water or experimental solutions containing high concentrations of monovalent salt in the absence of calcium. Calcium is so important for normal functioning of plant cell membranes that we shall devote a separate discussion to it.

The Role of Calcium. The older literature contains many references to the toxicity of monovalent salts to plants and the beneficial effect of calcium which was said to "counteract" or "neutralize" the .harmful effects of alkali salts (Kearney and Cameron, 1902). But for many years it was not realized that even brief exposure of plant tissue to solutions lacking calcium will cause injury, and many experiments were done with excised tissues in the absence of calcium, or at least, without its deliberate inclusion in the experimental solutions.

In 1960 J. B. Hanson of the University of Illinois showed that roots of soybean, *Glycine max,* and of corn, *Zea mays,* suffered drastic impairment of their ability to absorb and retain solutes when treated with ethylenediamine tetraacetic acid (Hanson, 1960). Hanson attributed the damage to the removal of calcium from the tissue, especially its membranes, by the chelator. These findings, extended by Foote and Hanson (1964), showed that the permeability of the tissue to ions was much increased by depleting it of part of its calcium. At the same time, Jacobson et al. (1960) observed that the disastrous effects of pH values below 4 on potassium absorption and retention by barley roots were counteracted if not abolished by inclusion of calcium in the experimental solutions.

Treatment of plant tissue with ethylenediamine tetraacetic acid and with mineral acid solutions is drastic. Epstein (1961) showed the essentiality of calcium for unimpaired ion transport in experiments done under much milder conditions. He simply rinsed roots of barley seedlings, grown for five days in 0.2 mM solutions of $CaSO_4$, with water and then placed them in solutions of monovalent salts, with and without the addition of $CaCl_2$. One of the experiments is shown in Figure 6-9. The experiment was done with radioactively labeled rubidium, an element which is a close analog of potassium. Without calcium in the solutions (bottom curve) the rate of absorption declined at once and was nil after one hour. With calcium

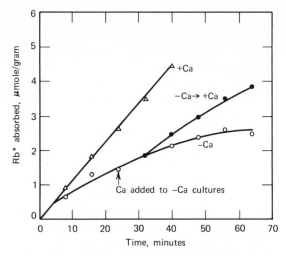

Figure 6-9. Absorption of rubidium by barley roots as a function of time and the absence and presence of calcium in the solution. Concentration of RbCl, 1 mM, of NaCl, 20 mM, and of CaCl₂ when present, 0.1 mM. After Epstein (1961).

present (top curve) the rate of rubidium absorption was steady, the difference between the two treatments becoming evident within minutes of the beginning of the experiment. When calcium was added to solutions lacking it (middle curve) there was an immediate, if not complete, reversal of the impairment of ion transport caused by the absence of calcium.

There have since been many experiments of a similar kind, confirming the essentiality of calcium for normal transport and retention of ions (Carter and Lathwell, 1967; Elzam and Epstein, 1965, 1969; Rains *et al.,* 1964). When Maas and Leggett (1968) exposed corn roots to solutions containing low concentrations of potassium and labeled rubidium, but no calcium, potassium was lost from the tissue where its concentration was high compared with its concentration outside, and labeled rubidium entered the tissue where initially its concentration was nil. In other words, the tissue behaved as though permeable to potassium and rubidium, and each ion followed its diffusion gradient, potassium outward and labeled rubidium inward.

Läuchli and Epstein (1970) found that no such two-way traffic occurred when calcium was included in the experimental solutions. Potassium and labeled rubidium were absorbed in parallel fashion; there was no measur-

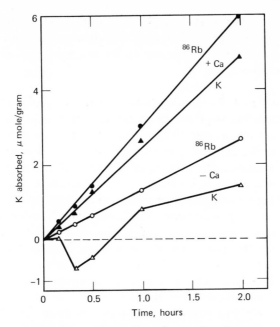

Figure 6-10. Absorption of potassium by corn roots as a function of time and the absence and presence of calcium in the solution. Concentration of KCl, 0.1 mM, and of CaSO₄ when present, 0.5 mM. Top two curves, calcium present. Bottom two curves, calcium omitted. "⁸⁶Rb," uptake of K calculated on the basis that ⁸⁶Rb radioactive tracer in the solution serves to label the K; "K," uptake of K determined by chemical analysis of roots. After Läuchli and Epstein (1970).

able loss of potassium from the tissue (top two curves of Figure 6-10). When calcium was left out (bottom two curves) chemical analysis for potassium revealed a loss of potassium from the tissue while rubidium entered it. In other words, as in Maas and Leggett's experiments, the tissue was "leaky" in the absence of calcium, and each ion followed its own diffusion gradient.

Healthy, whole plant tissue, then, in the presence of calcium absorbs ions essentially in a unidirectional fashion; there is little if any measurable efflux, and the membrane separating the cell interior from the external medium, the plasmalemma, is highly impermeable to diffusive permeation by inorganic ions. We must share the view of Collander (1959) who was impressed by the competence of plant cell membranes, both "their resistance to penetration and their capacity for transport," as the title of his excellent review has it.

Evidence from Electropotentials.   Another line of evidence bearing on the permeability of plant cell membranes is based largely on work done by investigators using cells of algae and of animals. A microelectrode is inserted into the cell and a reference electrode is kept in the external solution. The potential difference between the two phases is then measured. If the inside is negative in respect to the medium there exists an electrochemical driving force exerted on cations in the medium. This needs to be taken into account in addition to the chemical concentration (activity) gradient in determining the direction of the movement of the cations to be expected on the basis that the membrane is sufficiently permeable to the cations to be traversed by them at appreciable rates under the influence of these gradients. Similar considerations apply to anions and their movements as calculated on this basis.

There are great experimental difficulties connected with this method, but the cells of giant algae have several features which make them suitable objects for such measurements. The main one is that the cells are large enough so that the microelectrode can be inserted into a more or less well-defined region, either the vacuole or the cytoplasm. The relatively large size of these compartments in cells of giant algae also makes it rather easy to obtain samples for analysis. Finally, it is necessary to measure ion fluxes into and out of these compartments; this is difficult even with algae. With higher plants, the difficulties are much greater, mainly because the cells are so small that the position of the microelectrode cannot be determined, samples of sap in the various compartments cannot be obtained for analysis, and flux analysis so far has not been successful.

The method was first used with tissue of higher plants by Etherton and Higinbotham (1960), who worked with cells of *Avena sativa, Pisum sativum,* and *Zea mays.* Later investigations by Higinbotham and his group have emphasized the difficulties of the approach (Higinbotham *et al.,* 1967; Macklon and Higinbotham, 1970). However, the general conclusion has emerged that the major cations and anions of plant cells are not in electrochemical equilibrium across the plasmalemma. Like the work discussed under the previous heading, the electrochemical research therefore emphasizes the paramount importance of active ion transport and the means whereby it is brought about. It is therefore to this topic that we now turn.

CARRIERS

Concept. As we have seen, the outer cell membrane is highly impermeable to ions, whether they are driven by activity or electrochemical gradients. Nevertheless, they enter the cell. We are faced with the paradox of ions moving across membranes impermeable to them. This ion movement depends on the machinery of cellular metabolism, and the mechanism, as we shall see, is selective: it discriminates among ions of different elements.

To account for these and other features of ion transport the "carrier" hypothesis has been elaborated. Subunits of the membrane called carriers bind the ion at the external face of the membrane, forming a carrier-ion complex (Figure 6-11). This complex then traverses the membrane, rotates within the membrane, or undergoes some other spatial rearrangement within the membrane as a result of which the ion is brought to the far (inner) side of the membrane. Along with or after this change, the carrier-ion complex breaks down as a result of a change in molecular configuration of the carrier so that the ion is released into the "inner" space at the far side of the membrane.

This scheme resolves the paradox discussed above. The membrane is not permeable to the free ion; it is not the free ion that penetrates across the membrane. Rather, within the membrane the ion is a constituent of a molecular subunit of the membrane. It becomes a free ion again only upon its release from the carrier to the "inner" space. The ion cannot diffuse out because of the impermeability of the membrane to free ions and the configuration of the carrier facing the inner surface of the membrane which is such as to release the ion rather than to bind it. The absorbed ion is trapped within the cell.

Many observations on ion transport lend themselves to analysis in terms of the carrier hypothesis, and support the hypothesis. We shall now examine the main lines of evidence and their interpretation. It will be seen later that there are at least two different mechanisms of transport for a given kind of ion. At very low concentrations, a "mechanism 1" operates; at high concentrations, "mechanism 2" with different properties comes into play. Our initial discussion of experiments on carrier-mediated transport will deal only with mechanisms of the first type.

The Kinetics of Ion Transport. When the process of active ion absorption by plant tissue is followed as a function of time, care being taken that the ions superficially associated with the "outer" and "Donnan free" spaces of the tissue are not included in the measurements, absorption

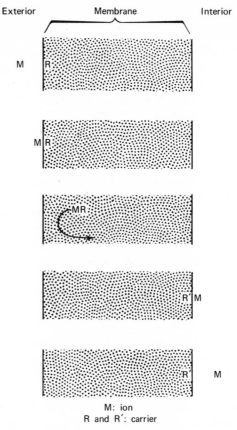

Exterior   Membrane   Interior

M: ion
R and R': carrier

Figure 6-11. Carrier-mediated ion transport according to the enzyme-kinetic concept: M, the ion; R, the carrier in its ion binding configuration at the outer surface of the membrane; MR, the carrier-ion complex; R', the carrier in its ion releasing configuration at the inner surface of the membrane. After Epstein (1953).

is usually found to proceed at a steady rate for at least an hour (Figure 6-7) and often for much longer. There is thus no difficulty in obtaining accurate values for rates of ion absorption under different conditions.

Figure 6-12 shows the results of an experiment on absorption of potassium by barley roots. The rate, v, of potassium absorption is plotted as a function of the potassium concentration, over the range 0.002 to 0.20

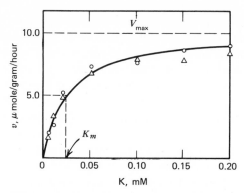

Figure 6-12. Rate, v, of absorption of potassium by barley roots as a function of the concentration of KCl ($\bigcirc$) or K_2SO_4 ($\triangle$) in the solution. Concentration of $CaCl_2$ or $CaSO_4$, 0.5 mM. The curve is a plot of the Michaelis–Menten equation. $K_m = 0.023$ mM; $V_{max} = 10.0$ μmole/g fresh weight/hour.

mM KCl or K_2SO_4. At low concentrations, the rate of absorption increases sharply with increasing external concentration of potassium, but at progressively higher concentrations, each added increment of concentration adds less and less of an increment in absorption rate, until, at the highest concentrations used, absorption of potassium is virtually independent of its concentration in the solution.

Such "saturation kinetics" have been found for the absorption of numerous ions by many different kinds of plant tissue. The Dutch plant physiologist T. H. van den Honert (1937) observed such a relationship between the concentration of phosphate and its absorption by sugar cane, *Saccharum officinarum*. He reasoned that there was some similarity between the operation of the phosphate absorbing mechanism and a steadily rotating belt conveyor. The more densely the conveyor is loaded, the more material it transports in unit time, but its capacity is limited and as its loading is increased a density of loading is reached which cannot be exceeded. Under these conditions, transport of load in unit time is maximal. Van den Honert's analogy between the cellular mechanism of ion transport and a mechanical conveyor represents the first intimation of the carrier concept in plant physiology.

Epstein and Hagen (1952) introduced a treatment of ion transport kinetics in terms familiar from enzymology. Enzymes catalyze the transformation of substrate into product through the transitory formation of a complex between enzyme and substrate. With graded concentrations of substrate, saturation kinetics are observed; at high substrate concentrations, the rate of reaction becomes independent of the concentration of substrate. That is, the kinetics of carrier-mediated transport and of enzymic catalysis are similar, and this similarity is thought to reflect a similarity in the mechanisms. In each case, a substrate (the ion or the organic substrate, respectively) attaches to an active agent (carrier or enzyme), and following the process promoted by the agent (transport or catalysis), the agent is free again to repeat the performance. The difference between carrier-mediated transport and enzyme-mediated catalysis is in the processes the two kinds of agents promote: the carrier effects the transport of the substrate ion from one side of a membrane to the other, whereas the enzyme promotes the chemical transformation of the substrate molecule into another. However, this difference in definition of the processes promoted by carriers and enzymes need not denote a fundamental difference in molecular events. Many enzymic reactions are transfer reactions, and the event promoted by, say, a phosphate carrier may be fundamentally similar to that of an enzyme which transfers a phosphate group from one compound to another.

The rate of carrier-mediated transport, like the rate of enzymic catalysis, can be shown to be dependent on two factors. One is a capacity factor, denoting the maximal rate of transport that can be achieved when all available carrier sites are loaded. The other is an intensity factor reflecting the fraction of the carrier actually occupied at a given concentration of the ion.

The capacity factor, $V_{\max}$, can be calculated from experimental data. It is represented by the value of the maximal rate of absorption asymptotically approached at high concentrations of the ion (cf. Figure 6-12). This rate is the product of the total concentration of carrier sites present and the rate of turn-over of the carriers within the membrane.

The intensity factor, the fraction θ of sites occupied at ionic concentration $[S]$, is given by the Langmuir adsorption isotherm,

$$\theta = \frac{[S]}{K_m + [S]}$$

where K_m is the dissociation constant of the carrier-ion complex. The rate of absorption, v, assuming there is no counterflow in the opposite direction, is given by the product of the capacity and intensity factors,

$$v = \frac{V_{\max} \cdot [S]}{K_m + [S]}$$

This is the Michaelis–Menten equation relating the rate of enzymic catalysis to the concentration of substrate, and equally applicable to the rate of ion transport as a function of the concentration of the ion, [S]. The line depicting this relationship for potassium absorption in Figure 6-12 is in fact a plot of the Michaelis–Menten equation, with the parameters (V_{max} and K_m) indicated in the legend.

Let us consider the case where the concentration [S] of the ion is such as to result in an observed rate of absorption, v, which is half the theoretical maximal rate, V_{max}, i.e., $v = \frac{1}{2}V_{max}$. Then

$$\frac{V_{max}}{2} = \frac{V_{max} \cdot [S]}{K_m + [S]}$$

$$K_m + [S] = 2[S]$$

and

$$K_m = [S]$$

That is, K_m, the "Michaelis constant," equals that concentration of the substrate ion giving half the maximal rate of absorption. The lower this value, that is, the lower the concentration of the ion resulting in half the maximal rate of its absorption, the higher is the affinity of the carrier sites for the ion.

The Michaelis–Menten equation can be rearranged in three ways which yield straight-line plots. They are useful for testing the fit of experimental data with the equation, and for calculating the parameters, V_{max} and K_m. The two most common linear transformations and the plots they yield are shown in Figure 6-13.

Many experiments on the relation between the concentration of an ion and the rates of its absorption have yielded results which conform to Michaelis–Menten kinetics. Table 6-1 gives examples, mainly from fairly recent work, for many ions and diverse plant materials.

Selectivity. According to the carrier hypothesis, and more specifically, the enzyme-kinetic formulation of it that we have outlined, the ion is bound in a transitory manner to the carrier. This implies that just as enzymes have "active sites" (molecular subgroupings which bind the substrate), so carriers possess active sites which bind the ions. The active sites of enzymes are often highly specific for the substrate; even closely related compounds may not be bound by the sites which bind the substrate, and may therefore not be acted upon by the enzyme. There is much evidence for a similar specificity or selectivity in carrier-mediated ion transport.

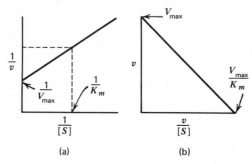

(a)   (b)

Figure 6-13. The two most common linear transformations of the Michaelis–Menten equation. (a) $1/v$ is plotted against $1/[S]$. This is often called the double-reciprocal or Lineweaver-Burk plot. (b) v is plotted against $v/[S]$. This is often called the Hofstee plot. In the presence of a competing ion at a given concentration, the slopes but not the y-axis intercepts are changed.

Figure 6-14 shows that absorption of potassium by barley roots, over the concentration range 0.005–0.20 mM, is quite insensitive to the presence of 0.50 mM sodium, even at the lowest potassium concentration where the sodium concentration is 100 times that of potassium. In terms of the enzyme-kinetic formulation of the carrier hypothesis this finding is inter-

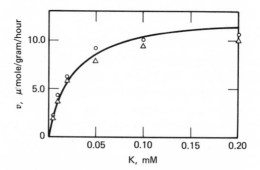

Figure 6-14. Rate, v, of absorption of potassium by barley roots as a function of the concentration of KCl in the solution, in the absence of sodium and in the presence of 0.50 mM NaCl: ○, control; △, +0.50 mM Na.

preted as evidence that the active sites of the carrier which bind and transport potassium have little affinity for sodium. Sodium, therefore, fails to compete with potassium. Many such cases of specificity or selectivity in ion absorption have been recorded. This might readily be expected for the transport of chemically very dissimilar ions, but it is very impressive when such discrimination is made between ions of elements belonging to the same group in the periodic table of elements, as is the case for potassium and sodium. In a similar manner, absorption of chloride by barley roots was found to be unaffected by the presence of two other halides, fluoride and iodide (Elzam and Epstein, 1965).

For less closely related ions, instances of even more spectacular specificity can be given. For example, it is commonly found that at low concentrations of potassium, its absorption is not interfered with by the presence of calcium in large excess. Absorption of halides is indifferent to the presence of sulfate or phosphate, and conversely, halides fail to inhibit the absorption of sulfate and phosphate. In Elzam's (1966) experiments with roots of two species of wheatgrass, *Agropyron,* absorption of chloride from a 0.1 mM solution was totally unaffected by a 500-fold excess of sulfate.

Nevertheless, certain ions so closely resemble each other in their chemical properties that the cellular transport mechanisms fail to discriminate between them. Such ions compete with each other in the process of absorption. Examples are the pairs potassium-rubidium, calcium-strontium, chloride-bromide, and sulfate-selenate (Epstein, 1962). This competition indicates that both ions can fit into a common carrier site. The frequent references to rubidium in the literature of ion absorption are not due to any particular interest in this rare element but rather to the fact that it acts as an analog or "stand-in" for potassium, and that findings concerning its transport can usually be taken to apply to potassium also. For experimental purposes, rubidium is often more convenient than potassium because it has a radioisotope,[86]Rb, with an 18.77 day half-life. The half-life of the potassium radioisotope, [42]K, is only 12.4 hours, and that makes [42]K a demanding task master in the laboratory!

Counter-Ion Effects. When cells accumulate potassium from a solution of KCl via a mechanism specific for potassium there is no a priori reason for assuming that this transport needs to be coupled to a simultaneous transport of chloride. It has, in fact, been shown that the transport of potassium may be quite independent of the transport of the anion of the salt. In experiments with barley roots (Epstein *et al.,* 1963) and corn roots (Lüttge and Laties, 1966), rates of absorption of potassium or rubidium were viturally the same whether the anion was

TABLE 6-1. Kinetics of Ion Absorption from Dilute Solutions: Selected Experiments (Concentration Range of Type 1 Mechanisms)[1]

Substrate ion	Plant species	Plant organ	Highest concentration of substrate ion, mM	Michaelis constant, mM	Competing ions	Non-competing or weakly competing ions	Remarks	Reference
K^+	Barley, *Hordeum vulgare*	Roots	0.2	0.021		Na^+		Epstein *et al.* (1963)
K^+	Corn, *Zea mays*	Leaf tissue	0.2	0.038 (mean)	Rb^+	Na^+		Smith and Epstein (1964b)
K^+	Corn	Leaf tissue	0.2	0.035 (light) 0.030 (dark)			^{86}Rb-labeled K	Rains (1968)
K^+	Tall wheatgrass, *Agropyron elongatum*	Roots	0.2	0.008		Na^+	^{86}Rb-labeled K	Elzam and Epstein (1969)
K^+	Waterweed, *Elodea densa*	Leaf	0.2	0.099 (light)	Rb^+			Jeschke (1970)
K^+	*Avicennia marina* (a tropical mangrove)	Leaf tissue	1.5	0.20			^{86}Rb-labeled K	Rains and Epstein (1967c)
Rb^+	Barley	Roots	0.2	0.016	H^+			Rains *et al.* (1964)
Rb^+	Barley	Roots	0.5	0.017 (mean)				Jackman (1965)
Rb^+	Corn	Leaf tissue	0.2	0.015 (mean)	K^+	Na^+, Li^+, Cs^+, NH_4^+		Smith and Epstein (1964b)
Rb^+	Ryegrass, *Lolium perenne*	Roots	0.4	0.012 (mean)				Jackman (1965)
Rb^+	Mung beans, *Phaseolus aureus*	Roots	0.3	0.012 (mean)				Jackman (1965)
Rb^+	Subterranean clover, *Trifolium subterraneum*	Roots	0.4	0.008 (mean)				Jackman (1965)

Ion	Plant	Tissue	Conc.	K_m	Interfering ions		Remarks	Reference
Na^+	Barley	Roots	0.2	0.32	K^+		See footnote[2]	Rains and Epstein (1967a)
Mn^{2+}	Sugarcane, *Saccharum officinarum*	Leaf tissue	5.0	0.0161	H^+	Zn^{2+}, Cu^{2+}		Bowen (1969)
Zn^{2+}	Sugarcane	Leaf tissue	0.5	0.0111	Cu^{2+}, H^+	Mn^{2+}	Intact plants	Bowen (1969)
Zn^{2+}	Wheat, *Triticum aestivum*	Roots	0.01	0.007	Cu^{2+}	Ca^{2+}, Mg^{2+}, Mn^{2+}, Fe^{2+}, Co^{2+}, H^+		Chaudhry (1971)
Cu^{2+}	Sugarcane	Leaf tissue	0.5	0.0145	Zn^{2+}, H^+	Mn^{2+}		Bowen (1969)
Cl^-	Barley	Roots	0.2	0.014	Br^-	F^-, I^-		Elzam and Epstein (1965)
Cl^-	Tall wheatgrass	Roots	0.2	0.013				Elzam and Epstein (1969)
Br^-	Barley	Roots	0.2	0.024	Cl^-			Böszörményi (1966)
NO_3^-	Corn	Roots	0.24	0.021			Intact plants	van den Honert and Hooymans (1955)
$H_2PO_4^-$	Corn	Roots	0.03	0.00609			At 30°C but differing with temperature	Carter and Lathwell (1967)
$H_2PO_4^-$	Waterweed	Leaf	0.1	0.015				Jeschke and Simonis (1965)
SO_4^{2-}	Waterweed	Leaf	0.1	0.006				Jeschke and Simonis (1965)
$H_2BO_3^-$	Sugar cane	Leaf tissue	0.2	0.0286			At 30°C but differing with temperature	Bowen (1968)

[1] No experiments are listed in which calcium was not included in the experimental solutions.

[2] Sodium is absorbed over the low (mechanism 1) range of concentrations via mechanism 1 of alkali cation transport, but with low affinity (K_m for Na^+: 0.32 mM, vs. about 0.02 mM for K^+). In the presence of K^+, Na^+ transport by this mechanism is all but abolished. See Rains and Epstein (1965, 1967a) and Welch and Epstein (1968).

chloride or sulfate (Figure 6-12), despite the fact that chloride is itself absorbed at an appreciable rate whereas absorption of sulfate is extremely slow by comparison. Such experiments indicate that the operation of the potassium transport mechanism is to a degree autonomous, in keeping with the carrier hypothesis. Similar findings and conclusions apply to the absorption of other elements.

Nevertheless, absorption of the cation and anion of a salt at unequal rates demands metabolic adjustments on the part of the cell. These adjustments involve the synthesis and metabolism of organic acids. Specifically, when the cation of the salt is absorbed more rapidly than the anion, for example, when tissue is exposed to a solution of K_2SO_4, the anion deficit within the tissue resulting from the preferential absorption of potassium is made good by the synthesis of organic acid anions, most often of malate, but of citrate, oxalate and other organic acid anions also, their proportions depending mainly on the species of plant. This response was discovered by Ulrich (1941, 1942) in Hoagland's laboratory.

The synthesis of the organic acid is through dark fixation of carbon dioxide, as shown by several investigators. Hiatt and Hendricks (1967) established that the rate of fixation of carbon from ^{14}C-labeled CO_2 was adequate to account for the increase in organic acid content of barley roots absorbing potassium in excess of the anion of the salt. The initial trigger for the synthesis seems to be the rise in internal pH when cations are absorbed in excess. In other words, under these conditions hydrogen ions are extruded, causing a rise in internal pH (Hiatt, 1967a). Hiatt (1967b) has described the enzymic pathway of CO_2 fixation and organic acid synthesis. The initial fixation of CO_2 is presumably in the cytoplasm. When anions are absorbed in excess of cations, the reverse changes take place and the level of organic acid in the cell drops.

Mutual Effects of Ions.   In respect to the absorption of a given ion (the substrate ion), the presence in the solution of another ion of the same sign may slow down the absorption of the substrate ion, may accelerate it, or may have neither effect. The most common instance of the first kind is competition with the absorption of the substrate ion by a chemically related ion, as already discussed. Competition usually is established in experiments in which the rate of absorption of the substrate ion is measured over a range of concentrations of it in the solution, as in the experiments shown in Figures 6-12 and 6-14, both in the presence and absence of the potential competing ion. The data are then plotted by means of one of the straight-line transformations of the Michaelis–Menten equation (see Figure 6-13 and its legend). It is by means of such experiments that the strictly competitive behavior of the members of the pairs K^+–Rb^+, Ca^{2+}–Sr^{2+}, Cl^-–Br^-, and SO_4^{2-}–SeO_4^{2-} was established, although experiments with entire plants grown for extended periods had earlier given indications of it (see the review by Epstein, 1962).

Often, the rate of absorption of a given ion is entirely indifferent to the presence of another ion of the same sign, as discussed earlier and shown for absorption of potassium in the presence of excess sodium in Figure 6-14. This selectivity depends absolutely on the presence of calcium in the solution. In the absence of calcium, it breaks down and there ensues an indiscriminate, unselective mutual competition between the members of almost any pair of cations or pair of anions that one might investigate. The role of calcium in the functioning of the membrane is therefore a dual one: it maintains its impermeability to ions, minimizing diffusive permeation, and it maintains the selectivity of the ion transport mechanisms.

The interaction whereby calcium performs this second function cannot yet be explained in definitive terms. In work on the selectivity of potassium absorption by barley roots in the presence of sodium, Rains and Epstein (1967a) thought the following the most attractive hypothesis. In the presence of calcium the molecular configuration of the active site of the carrier is conducive to a preferential "fit" for potassium (and rubidium) ions. In the absence of calcium the geometry of the active site is loosened to the extent that other ions, sodium included, are also admitted.

The third effect an ion may have on the absorption of another is to accelerate it. The best known instance of this is the "Viets effect," so called after F. G. Viets, Jr., who discovered it while a student in Hoagland's laboratory (Viets, 1944). He immersed excised barley roots in aerated solutions of 5 mM KBr, according to the classical Hoagland–Broyer method, and found that at increasingly higher concentrations of a calcium salt, progressively more potassium (and bromide) ions were absorbed.

Similar results have since been frequently obtained, but questions remain. The main one has to do with the "control" treatment—monovalent salt in the absence of calcium. It is now known, as it was not in 1944, that the absence of calcium quickly causes derangements in membrane structure and function. The presence of calcium is the normal, physiological condition, and the "plus calcium" treatment in such experiments should therefore be considered the control. Looked at in this light, the rate of ion absorption is diminished in the absence of calcium, not increased in its presence (cf. Figure 6-10). However, when quite high concentrations of calcium accelerate the absorption of an ion beyond the rate observed at low and moderate concentrations of calcium a true stimulatory or "Viets effect" may be inferred (Epstein, 1962). Lithium, an element having considerable chemical affinity with calcium, also elicits a "Viets effect" (Epstein, 1962), and so do various divalent and trivalent cations (Viets, 1944). But these other ions cannot substitute for calcium in the maintenance of the structural and functional integrity of the membrane.

5. THE DUAL PATTERN OF TRANSPORT

Kinetics.   In Figure 6-12, the rate of potassium absorption levels off at the highest concentrations of potassium used in that experiment and at 0.20 mM, approaches the theoretical maximum value, V_{max}, indicated by the dashed line. The mechanism which effects the absorption of potassium over this range of concentrations has several well-defined properties as discussed above: it operates at very low concentrations of potassium, follows simple Michaelis–Menten kinetics, is highly specific for potassium (and rubidium), and is indifferent to the identity and rate of absorption of the anion.

If this mechanism were the only one to transport potassium into the cell, then even at much higher concentrations of potassium the rate of its absorption would not exceed the maximal rate asymptotically approached at the highest concentrations used. In terms of the carrier hypothesis, virtually all carrier sites available for transport of potassium are occupied at these concentrations, and raising the concentration still further cannot cause the rate to exceed the maximum corresponding to complete occupancy of all sites.

But what are the facts when the external concentration is varied over a very wide range, including concentrations much higher than the highest used in the experiments shown in Figures 6-12 and 6-14? Figure 6-15 shows the results of such an experiment, done with barley roots. Over the low range of concentrations, up to 0.20 mM, the results closely resemble those of the experiments shown in Figures 6-12 and 6-14. At 0.20 mM, the rate of potassium absorption is almost independent of the concentration, and closely approaches the maximum indicated by the dashed line. In order to accommodate the high range of concentrations, the horizontal (concentration) scale is broken between 0.20 and 0.50 mM, and the right side of the figure shows the concentration range, 0.50 to 50 mM potassium. The observed rate of potassium absorption over this range greatly exceeds the theoretical maximum predicted on the basis of the relation which applies to the range of low potassium concentrations (the dashed line).

It is apparent, therefore, that there is more than a single mechanism of potassium absorption; there are at least two. Our discussion of carrier-

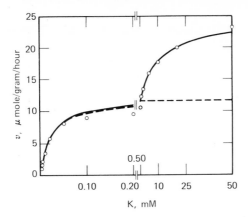

Figure 6-15. Rate, v, of absorption of potassium by barley roots as a function of the concentration of KCl in the solution. Concentration of CaCl$_2$, 0.5 mM. The horizontal (concentration) scale is broken between 0.20 and 0.50 mM. The solid line at the low concentrations, continued by the dashed line, is a plot of the Michaelis–Menten equation. K_m = 0.021 mM; V_{max} = 11.9 μmole/g fresh weight/hour. After Epstein *et al.* (1963).

mediated transport so far has been in terms of the mechanism which operates even at low concentrations. This is mechanism 1 of potassium absorption, and there are mechanisms of this type for all the ions listed in Table 6-1, and for others as well.

The mechanism which becomes operative at the high concentrations, over the range 1 to 50 mM potassium, mechanism 2, differs from mechanism 1 in respect to every one of the properties listed on page 132. Mechanism 1 operates at very low concentrations; potassium at a concentration as low as 0.002 mM (2×10^{-6} M, or 0.08 ppm) results in an appreciable rate of absorption (see the lowest concentration in Figure 6-12). Mechanism 2 only comes into play at concentrations of about 1 mM and above (Figure 6-15). That is, it has a much lower affinity for the ions than mechanism 1 does.

Mechanism 1 obeys simple Michaelis–Menten kinetics. The plot on the right side of Figure 6-15 might suggest that this is also true of mechanism 2, and indeed this was first thought to be the case (Epstein *et al.*, 1963). However, a closer examination of rates of potassium absorption over the

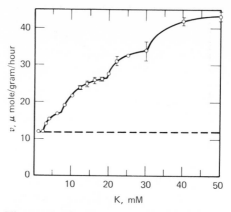

Figure 6-16. Rate, v, of absorption of potassium by barley roots as a function of the concentration of KCl in the solution. Concentration of $CaSO_4$, 0.5 mM. All treatments replicated; circles represent the means of two individual values indicated by the short horizontal lines. Horizontal lines are not drawn where the distance between them would have been equal to or less than the diameter of the circle. The dashed line represents the maximal rate of absorption, V_{max}, by mechanism 1 (cf. Figure 6-15). After Epstein and Rains (1965).

range of high concentrations revealed a more complex pattern (Epstein and Rains, 1965). Figure 6-16 shows that the plot of potassium absorption rates as a function of concentration has a number of inflections. This suggests that carrier mechanism 2 has several active sites which differ somewhat in their affinities for potassium: it is polyvalent.

We have seen that potassium absorption via mechanism 1 is specific or selective for potassium (and rubidium). Neither sodium nor other cations (lithium, calcium, magnesium) effectively compete with potassium for transport by this mechanism. In contrast to this, absorption of potassium via mechanism 2 is interfered with by other cations such as sodium and calcium (Epstein *et al.*, 1963; Rains and Epstein, 1967b). That is, unlike mechanism 1, mechanism 2 does not show a high degree of specificity; other cations can occupy the sites of this carrier mechanism.

The two mechanisms differ also in respect to the effect of the counterion. Mechanism 1 of potassium absorption is not markedly affected by substitution of sulfate for chloride as the anion (Figure 6-12). The operation of mechanism 2, on the other hand, is severely inhibited when sulfate is the anion, and the rate of potassium absorption then may not rise much above the maximal rate of mechanism 1 (Epstein *et al.,* 1963; Lüttge and Laties, 1966).

Type 2 mechanisms have been found not only for transport of potassium but for ions of many different elements including chlorine (Elzam *et al.,* 1964; Lüttge and Laties, 1966; Torii and Laties, 1966), sodium (Rains and Epstein, 1967a), ammonium (Fried *et al.,* 1965), boron (Bowen, 1968), and others. The dual pattern reflecting two different mechanisms of absorption for a given kind of ion has been shown in fibrous roots, in storage tissue, and in leaf tissue. It may well be universal in mature tissues of higher plants. A discussion of the most important findings and their implications has been given by Epstein (1966), and Table 6-2 summarizes many instances of this duality.

Location of the Two Mechanisms.   The recognition of two different mechanisms of absorption of a given element naturally raises the question, where are they located? Are both in the same membrane, and if so, what membrane is it? Or are the two in different membranes? There is little question about the location of mechanism 1, at least for absorption of potassium and chloride, and probably for the type 1 absorption mechanisms of other ions as well. Mechanism 1 is in the plasmalemma, the external cytoplasmic membrane. Several lines of evidence lead to this conclusion.

As discussed in Chapter 3, roots of plants normally confront a solution, the soil solution, in which the concentrations of many ions, potassium included, may be low in the extreme. The absorption mechanism in the plasmalemma, the first membrane to be negotiated by the ion, must therefore have a high enough affinity for the ion to bind and transport it appreciably even when its concentration is very low. Only the type 1 mechanisms possess such high affinities (cf. Table 6-1). These mechanisms must therefore be in the plasmalemma, in contact with the external solution. At the low concentrations often prevailing in this solution, the type 2 mechanisms make no measurable contribution to absorption.

A second piece of evidence favoring the plasmalemma as the locale of the type 1 mechanisms has to do with the effects of calcium. As discussed above, absorption of potassium, chloride, and other ions, if they are present at concentrations within the range of the type 1 mechanisms, is suboptimal when no calcium is present in the external solution. Furthermore, the rate of absorption is often not constant under such conditions

TABLE 6-2. Dual Pattern of Ion Absorption: Selected Experiments[1,2]

Substrate ion	Plant species	Plant organ	Remarks	Reference
K⁺	Barley, *Hordeum vulgare*	Roots		Epstein *et al.* (1963)
K⁺	Corn, *Zea mays*	Roots	Intact plants. ^{86}Rb-labeled K	Lüttge and Laties (1966)
K⁺	Wheatgrass, *Agropyron* spp.	Roots		Elzam and Epstein (1969)
K⁺	Beet, *Beta vulgaris*	Discs	Fresh and aged tissue	Osmond and Laties (1968)
K⁺	*Avicennia marina* (a tropical mangrove)	Leaf tissue	^{86}Rb-labeled K	Rains and Epstein (1967c)
Rb⁺	Barley	Roots		Epstein *et al.* (1963)
Rb⁺	Barley	Roots		Jackman (1965)
Rb⁺	Perennial ryegrass, *Lolium perenne*	Roots		Jackman (1965)
Rb⁺	Mung beans, *Phaseolus aureus*	Roots		Jackman (1965)
Rb⁺	Subterranean clover, *Trifolium subterraneum*	Roots		Jackman (1965)
Na⁺	Barley	Roots		Rains and Epstein (1967a)
Na⁺	Wheatgrass	Roots		Elzam and Epstein (1969)
Ca²⁺	Corn	Roots		Maas (1969)
Ca²⁺	Cotton, *Gossypium hirsutum*	Roots	Tissue culture	Johanson and Joham (1971)
Mg²⁺	Barley	Roots		Kaplan (1969)
Fe²⁺	Rice, *Oryza sativa*	Roots		Kannan (1971a)
Cl⁻	Barley	Roots		Elzam *et al.* (1964)
Cl⁻	Wheatgrass	Roots		Elzam and Epstein (1969)
Cl⁻	Beet	Discs	Fresh and aged tissue	Osmond and Laties (1968)
Cl⁻	Bean, *Phaseolus vulgaris*	Roots, leaf tissue		Jacoby and Plessner (1970)
Cl⁻	*Mnium cuspidatum* (a moss)	Branches		Lüttge and Bauer (1968)
H₂PO₄⁻	Barley	Roots		Leggett *et al.* (1965)
H₂PO₄⁻	Corn	Roots		Carter and Lathwell (1967)
H₂PO₄⁻	Wheat, *Triticum vulgare*	Roots	Entire plants	Edwards (1970)
H₂PO₄⁻	Waterweed, *Elodea densa*	Leaf		Jeschke and Simonis (1965)
SO₄²⁻	Waterweed	Leaf		Jeschke and Simonis (1965)
SO₄²⁻	*Mnium cuspidatum*	Branches		Lüttge and Bauer (1968)
H₂BO₃⁻	Sugar cane, *Saccharum officinarum*	Leaf tissue		Bowen (1968)

[1] No experiments are listed in which calcium was not included in the experimental solutions.
[2] Nissen (1971) and Thellier (1970) have analyzed kinetic data on ion absorption and offer interpretations which differ from the dual pattern (and from each other).

but rapidly declines, sometimes in the very first few minutes of the experimental absorption period. Upon addition of calcium, the rate quickly returns to the higher, constant rate characteristic of tissue kept in a solution containing calcium. And finally, the selectivity of absorption is impaired when no calcium is present in the solution.

In interpreting these findings we must keep in mind that the tissues used in these experiments were roots of seedlings grown, up to the time of the experiment, in solutions containing calcium, i.e., they were by no means calcium deficient. Nevertheless, the process of absorption of potassium and other ions responds at once to the absence of calcium from the external solution, and to its reintroduction into the solution. Since the cell interior is not deficient in calcium, the mechanism 1 absorption processes affected so intimately by the calcium status of the external solution must therefore be located in the membrane bathed by that solution, the plasmalemma (Epstein, 1965, 1966). Additional evidence leading to the same conclusion is discussed below.

As for the location of the type 2 mechanisms which come into play in the range of high concentrations (1 mM and above), Epstein *et al.* (1963) considered both types of mechanisms to operate in parallel. This implies that the type 2 mechanisms also reside in the plasmalemma. Laties and co-workers in an extensive series of investigations took a different view. Working mainly with corn roots, they confirmed the dual pattern of absorption for chloride and potassium. Considering, then, that plant cells have two principal ion transporting membranes, the plasmalemma and the tonoplast, and two mechanisms of absorption for a given ion, they thought it likely that one of these mechanisms might reside in one of these membranes, and the second one, in the other. For reasons already explained, they, like Epstein and co-workers, assigned the type 1 mechanisms to the plasmalemma. They proceeded with experiments from which they concluded that the type 2 mechanisms are located in the tonoplast.

The parallel operation of both mechanisms across the plasmalemma raises no conceptual difficulties. At low concentrations of the ions in the medium, only the type 1 mechanisms operate; at high concentrations, the type 2 mechanisms also make a contribution, and the rate of absorption at these concentrations is the sum of the rates of transport via both (top part of Figure 6-17). Placing the type 2 mechanisms in the tonoplast means that the two mechanisms operate in series (bottom part of Figure 6-17). If the rate of entry of the ions into the cytoplasm were limited by mechanism 1, the over-all rate of absorption could never exceed the maximal rate of mechanism 1 and no rates higher than that should be observed. But in fact they are.

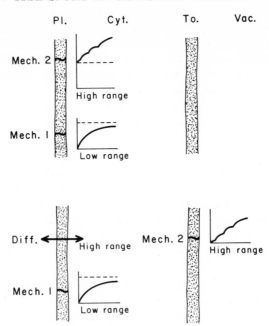

Figure 6-17. Diagrammatic representation of the parallel (top) and series (bottom) models of operation of the two types of ion transport mechanisms. The wavy lines in the membranes represent ion transport mechanisms. Pl., plasmalemma; Cyt., cytoplasm; To., tonoplast; Vac., vacuole; Diff., diffusion. According to the parallel model (top), rates of absorption from an external solution furnish no direct evidence concerning transport at the tonoplast.

To meet this difficulty with the series model, Laties therefore assumes that at concentrations in the high range, the range of mechanism 2, the ions diffuse through the plasmalemma at rates higher than the maximal rate of mechanism 1. Then mechanism 2, considered by him to lie in the tonoplast, becomes the rate-limiting step in absorption. This means that at the high concentrations, mechanism 1 is set aside as a rate-limiting step (bottom part of Figure 6-17), and the observed rate of absorption then is "solely the rate of ion movement into the vacuole" (Torii and Laties, 1966).

What is the evidence? In their first experiments, Torii and Laties (1966) measured the rate of absorption of chloride and of rubidium by tips of corn roots (the apical 2 mm) and by proximal, mature tissue farther back. The latter, where the cells have mature vacuoles, showed typical dual absorption isotherms as observed with barley roots by Epstein *et al.* (1963), Epstein and Rains (1965), and Elzam *et al.* (1964). The root tips showed typical type 1 kinetics over the range of low concentrations but at high concentrations absorption was related in a roughly linear manner to the external concentration, as might be expected for entry by diffusion. On the basis that tip (meristematic) cells are not vacuolated the authors concluded that they have no tonoplasts and hence, no type 2 mechanisms. The argument lacks force, for meristematic cells of root tips have numerous small vacuoles (Figure 5-1) and the total area of tonoplast in such cells may equal or exceed that of mature cells. In experiments with the moss, *Mnium cuspidatum,* Lüttge and Bauer (1968) found no correlation between linear absorption isotherms in the range of high concentrations and lack of vacuolation.

Other evidence from experiments on absorption of ions by corn roots and their translocation across the root into the conducting elements of the xylem (Lüttge and Laties, 1966, 1967a, b) is discussed in the next chapter. Laties (1969) and Lüttge (1969a) have expanded their arguments in detailed reviews.

Welch and Epstein (1968, 1969) have presented evidence that both types of absorption mechanisms operate in parallel across the plasmalemma. Evidence from experiments on absorption and translocation is presented in the next chapter. Their other experiments were of two kinds. When potassium is present at 10 mM the rate of its absorption should be the sum of the rates of transport via mechanisms 1 and 2, if the two operate in parallel (top of Figure 6-17). They did experiments with barley roots in which potassium was present at this concentration and sodium ranged from nil to 50 mM.

The total rate of absorption of potassium was indeed found to be the sum of the contributions from two mechanisms, one of them sensitive to sodium (Figure 6-18). This is the contribution from mechanism 2, in which sodium competes with potassium, as discussed under the previous heading. The other fraction of the total rate was not diminished by excess sodium—the hallmark of potassium absorption via mechanism 1, which is highly selective for potassium vis-à-vis sodium. The total rate of absorption at 10 mM potassium was thus "dissected" into two components, as expected on the basis of the parallel operation of the two mechanisms. According to the serial model, the rate of absorption should have been that due to operation of a single mechanism, viz. mechanism 2, since on

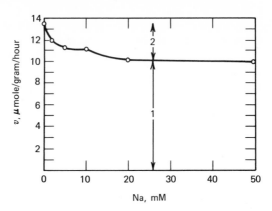

Figure 6-18. Rate, v, of absorption of potassium by barley roots as a function of the concentration of NaCl. Concentration of KCl, 10 mM, and of CaSO₄, 0.5 mM. After Welch and Epstein (1968).

that view, potassium ions would equilibrate across the plasmalemma, shunting out mechanism 1 (bottom of Figure 6-17).

Welch and Epstein's (1968, 1969) second type of evidence was to the effect that there is no diffusive equilibration of ions across the plasmalemma, such as is demanded for the range of high concentrations by the series model. This conclusion is in keeping with much other evidence as already discussed; in the present connection, the work by Johansen *et al.* (1970) and of Venrick and Smith (1967) is especially pertinent.

Independent evidence to the effect that both types of absorption mechanisms are in the plasmalemma has been furnished by Kannan (1971b). He observed a typical dual pattern in rubidium absorption by the non-vacuolate unicellular green alga, *Chlorella pyrenoidosa*. Since this alga lacks a tonoplast membrane the dual absorption mechanisms must reside in the plasmalemma. It is unlikely that the same kinetic pattern of ion absorption should, in this alga, reflect the operations of only the plasmalemma, and in cells of higher plants, those of both the plasmalemma and the tonoplast. Rather, the plasmalemma is implicated as the seat of the dual mechanisms in plant cells generally.

The conclusion that both types of absorption mechanism operate in parallel across the plasmalemma does not necessarily imply that both deliver the ions into the same cytoplasmic compartments or "inner" spaces (Epstein *et al.*, 1963). Welch and Epstein (1968) have speculated that one might transport ions into the aqueous phase of the cytoplasm and the other via the membranous network of the endoplasmic reticulum into the vacuole.

NATURE OF THE CARRIERS

No ion carriers of the kind postulated on the basis of kinetic evidence have been chemically identified in plants, let alone isolated *in vitro*. Epstein and Hagen (1952) thought it likely that they were proteins. Certainly, interference with protein synthesis by appropriate inhibitors interferes with ion absorption and with the generation of ion absorbing capacity in freshly cut discs of storage tissue, but this is no direct proof that the carriers themselves are proteins—virtually every performance a cell is capable of depends ultimately on the generation, maintenance, and turnover of proteins. Nevertheless, it remains likely that the carriers are directly connected with proteins or are themselves proteins (Pardee, 1968; also see Chapter 9, pp. 240–242).

According to the kinetic evidence, the carriers must have the property of selectively binding ions. Such binding has occasionally been observed. Epstein (1955) demonstrated selective sulfate binding (as contrasted with steady-state transport) in barley roots, and so did Pettersson (1966) and Persson (1969) in experiments with roots of sunflower, *Helianthus annuus,* and of wheat, *Triticum* sp., respectively. There is evidence that the concentration of such binding sites in plant tissue must be extremely low. When ion absorption is followed as a function of time (Figures 6-3, 6-7, and 6-9) the relationship is linear, and the data extrapolate to zero uptake at zero time. Theoretically, there should be a positive zero-time intercept denoting the (virtually instantaneous) binding of the ions to the carrier sites, but their concentrations appear to be too low to show up in such experiments. In other words, the rate of transport (carrier turnover) is very high, in relation to the number of binding sites, so that the number of ions occupying carrier sites at any one time is an exceedingly small fraction of the ions transported in even a very short period. This is quite analogous to the situation in enzymic catalysis.

What, then, might be the nature of these sites? Pressman (1968) and Pressman and Haynes (1969) have described certain polypeptide antibiotics which bind potassium ions with a high degree of specificity and accelerate their movement across biological membranes. That is, they act as "ionophores," or ion carriers. Valinomycin is one such compound. Elucidation of the structure of the valinomycin–potassium complex

(Ohnishi and Urry, 1970) shows that the polypeptide molecule forms a molecular "cage" in which the non-hydrated potassium ion is held. Other such ion binding polypeptides have been found (Pressman and Haynes, 1969). Thus there exist actual molecular configurations of precisely the kind envisioned on the basis of kinetic experiments on ion transport and concepts developed from them about trans-membrane ion carriers and their mode of operation.

REFERENCES

Arisz, W. H. 1964. Influx and efflux of electrolytes. Part II. Leakage out of cells and tissues. Acta Bot. Neerl. 13:1–58.

Barber, J. 1968. The efflux of potassium from *Chlorella pyrenoidosa*. Biochim. Biophys. Acta 163:531–538.

Benson, A. A. 1964. Plant membrane lipids. Ann. Rev. Plant Physiol. 15:1–16.

Benson, A. A. 1968. The cell membrane: a lipoprotein monolayer. In: Membrane Models and the Formation of Biological Membranes. L. Bolis and B. A. Pethica, eds. North-Holland Publishing Company, Amsterdam. Pp. 190–202.

Böszörményi, Z. 1966. The ion uptake of excised barley roots with special reference to the low concentration process. In: Advancing Frontiers of Plant Sciences. L. Chandra, ed. Institute for the Advancement of Science and Culture, New Delhi. Vol. 16, pp. 11–29.

Bowen, J. E. 1968. Borate absorption in excised sugarcane leaves. Plant and Cell Physiol. 9:467–478.

Bowen, J. E. 1969. Absorption of copper, zinc, and manganese by sugarcane leaf tissue. Plant Physiol. 44:255–261.

Branton, D. 1969. Membrane structure. Ann. Rev. Plant Physiol. 20:209–238.

Carter, O. G. and D. J. Lathwell. 1967. Effects of temperature on orthophosphate absorption by excised corn roots. Plant Physiol. 42:1407–1412.

Chaudhry, F. M. 1971. Zinc Absorption by Plants. Ph.D. Thesis, University of Western Australia, Nedlands.

Collander, R. 1959. Cell membranes: their resistance to penetration and their capacity for transport. In: Plant Physiology—A Treatise. F. C. Steward, ed. Academic Press, New York and London. Vol. 2, pp. 3–102.

Cram, W. J. 1968. Compartmentation and exchange of chloride in carrot root tissue. Biochim. Biophys. Acta 163:339–353.

Crossett, R. N. and B. C. Loughman. 1966. The absorption and translocation of phosphorus by seedlings of *Hordeum vulgare* (L). New Phytol. 65:459–468.

Dainty, J. 1969. The ionic relations of plants. In: The Physiology of Plant Growth and Development. M. B. Wilkins, ed. McGraw-Hill Publishing Company, Ltd., Maidenhead. Pp. 455–485.

Danielli, J. F. and H. Davson. 1935. A contribution to the theory of permeability of thin films. J. Cell. and Compar. Physiol. 5:495–508.

Edwards, D. G. 1970. Phosphate absorption and long-distance transport in wheat seedlings. Austral. J. Biol. Sci. 23:255–264.

Ellis, R. J. and I. R. MacDonald. 1967. Activation of protein synthesis by microsomes from aging beet disks. Plant Physiol. 42:1297–1302.

Elzam, O. E. 1966. Absorption of Sodium, Potassium, and Chloride by Two Species of *Agropyron* Differing in Salt Tolerance. Ph.D. Thesis, University of California, Davis.

Elzam, O. E. and E. Epstein. 1965. Absorption of chloride by barley roots: kinetics and selectivity. Plant Physiol. 40:620–624.

Elzam, O. E. and E. Epstein. 1969. Salt relations of two grass species differing in salt tolerance. II. Kinetics of the absorption of K, Na, and Cl by their excised roots. Agrochimica 13:196–206.

Elzam, O. E., D. W. Rains and E. Epstein. 1964. Ion transport kinetics in plant tissue: complexity of the chloride absorption isotherm. Biochem. Biophys. Res. Comm. 15:273–276.

Epstein, E. 1953. Ion absorption by plant roots. Proc. Fourth Ann. Oak Ridge Summer Symp. C. L. Comar and S. L. Hood, eds. U. S. Atomic Energy Commission TID-5115. Pp. 418–434.

Epstein, E. 1955. Passive permeation and active transport of ions in plant roots. Plant Physiol. 30:529–535.

Epstein, E. 1961. The essential role of calcium in selective cation transport by plant cells. Plant Physiol. 36:437–444.

Epstein, E. 1962. Mutual effects of ions in their absorption by plants. Agrochimica 6:293–322.

Epstein, E. 1965. Mineral metabolism. In: Plant Biochemistry. J. Bonner and J. E. Varner, eds. Academic Press, New York and London. Pp. 438–466.

Epstein, E. 1966. Dual pattern of ion absorption by plant cells and by plants. Nature 212:1324–1327.

Epstein, E. and C. E. Hagen. 1952. A kinetic study of the absorption of alkali cations by barley roots. Plant Physiol. 27:457–474.

Epstein, E. and D. W. Rains. 1965. Carrier-mediated cation transport in barley roots: kinetic evidence for a spectrum of active sites. Proc. Nat. Acad. Sci. 53:1320–1324.

Epstein, E., D. W. Rains and O. E. Elzam. 1963. Resolution of dual mechanisms of potassium absorption by barley roots. Proc. Nat. Acad. Sci. 49:684–692.

Epstein, E., D. W. Rains and W. E. Schmid. 1962. Course of cation absorption by plant tissue. Science 136:1051–1052.

Epstein, E., W. E. Schmid and D. W. Rains. 1963. Significance and technique of short-term experiments on solute absorption by plant tissue. Plant and Cell Physiol. 4:79–84.

Etherton, B. and N. Higinbotham. 1960. Transmembrane potential measurements of cells of higher plants as related to salt uptake. Science 131:409–410.

Foote, B. D. and J. B. Hanson. 1964. Ion uptake by soybean root tissue depleted of calcium by ethylenediaminetetraacetic acid. Plant Physiol. 39:450–460.

Fried, M., F. Zsoldos, P. B. Vose and I. L. Shatokhin. 1965. Characterizing the NO_3 and NH_4 uptake process of rice roots by use of ^{15}N-labelled NH_4NO_3. Physiol. Plantarum 18:313–320.

Greenway, H. 1967. Effects of exudation on ion relationships of excised roots. Physiol. Plantarum 20:903–910.

Hanson, J. B. 1960. Impairment of respiration, ion accumulation, and ion retention in root tissue treated with ribonuclease and ethylenediamine tetraacetic acid. Plant Physiol. 35:372–379.

Helder, R. J. 1952. Analysis of the process of anion uptake of intact maize plants. Acta Bot. Neerl. 1:361–434.

Helder, R. J. 1958. Studies on the absorption, distribution and release of labelled rubidium ions in young intact barley plants. Acta Bot. Neerl. 7:235–249.

Helder, R. J. 1964. The absorption of labelled chloride and bromide ions by young intact barley plants. Acta Bot. Neerl. 13:488–506.

Hendler, R. W. 1971. Biological membrane ultrastructure. Physiol. Rev. 51:66–97.

Hiatt, A. J. 1967a. Relationship of cell sap pH to organic acid change during ion uptake. Plant Physiol. 42:294–298.

Hiatt, A. J. 1967b. Reaction *in vitro* of enzymes involved in CO_2 fixation accompanying salt uptake by barley roots. Z. Pflanzenphysiol. 56:233–245.

Hiatt, A. J. and S. B. Hendricks. 1967. The role of CO_2 fixation in accumulation of ions by barley roots. Z. Pflanzenphysiol. 56:220–232.

Higinbotham, N., B. Etherton and R. J. Foster. 1967. Mineral ion contents and cell transmembrane electropotentials of pea and oat seedling tissue. Plant Physiol. 42:37–46.

Hoagland, D. R. 1944. Lectures on the Inorganic Nutrition of Plants. Chronica Botanica Company, Waltham.

Hoagland, D. R. and T. C. Broyer. 1936. General nature of the process of salt accumulation by roots with description of experimental methods. Plant Physiol. 11:471–507.

Hoagland, D. R. and A. R. Davis. 1923. Further experiments on the absorption of ions by plants, including observations on the effect of light. J. Gen. Physiol. 6:47–62.

Hopkins, H. T. 1956. Absorption of ionic species of orthophosphate by barley roots: effects of 2,4-dinitrophenol and oxygen tension. Plant Physiol. 31:155–161.

Jackman, R. H. 1965. The uptake of rubidium by the roots of some graminaceous and leguminous plants. New Zealand J. Agric. Res. 8:763–777.

Jacobson, L., D. P. Moore and R. J. Hannapel. 1960. Role of calcium in absorption of monovalent cations. Plant Physiol. 35:352–358.

Jacoby, B. and J. Dagan. 1967. A comparison of two methods of investigating sodium uptake by bean-leaf cells and the vitality of isolated leaf-cells. Protoplasma 64:325–329.

Jacoby, B. and O. E. Plessner. 1970. Some aspects of chloride absorption by bean leaf tissue. Ann. Bot. N. S. 34:177–182.

Jeschke, W. D. 1967. Die cyclische und die nichtcylcische Photophosphorylierung als Energiequelle der lichtabhängigen Chloridionenaufnahme bei *Elodea*. Planta 73:161–174.

Jeschke, W. D. 1970. Der Influx von Kaliumionen bei Blättern von *Elodea densa*, Abhängigkeit vom Licht, von der Kaliumkonzentration und von der Temperatur. Planta 91:111–128.

Jeschke, W. D. and W. Simonis. 1965. Über die Aufnahme von Phosphat-und Sulfationen durch Blätter von *Elodea densa* und ihre Beeinflussung durch Licht, Temperatur und Aussenkonzentration. Planta 67:6–32.

Johansen, C., D. G. Edwards and J. F. Loneragan. 1970. Potassium fluxes during potassium absorption by intact barley plants of increasing potassium content. Plant Physiol. 45:601–603.

Johanson, L. and H. E. Joham. 1971. The influence of calcium absorption and accumulation on the growth of excised cotton roots. Plant and Soil 34:331–339.

Jyung, W. H., S. H. Wittwer and M. J. Bukovac. 1965. Ion uptake by cells enzymically isolated from green tobacco leaves. Plant Physiol. 40:410–414.

Kahl, G., H. Lange and G. Rosenstock. 1969. Substratspiegel, Enzymaktivitäten und genetische Regulation nach Derepression in pflanzlichen Speichergeweben. Z. Naturforsch. 24b:911–918.

Kannan, S. 1971a. Kinetics of iron absorption by excised rice roots. Planta 96:262–270.

Kannan, S. 1971b. The plasmalemma: the seat of dual mechanisms of ion absorption in *Chlorella pyrenoidosa*. Science. In press.

Kaplan, O. B. 1969. Kinetics of Magnesium Absorption by Excised Barley Roots. Ph.D. Thesis, University of California, Davis.

Kearney, T. H. and F. K. Cameron. 1902. Some Mutual Relations between Alkali Soils and Vegetation. The effect upon seedling plants of certain components of alkali soils. U. S. Department of Agriculture, Washington. Report No. 71, pp. 7–60.

Korn, E. D. 1968. Structure and function of the plasma membrane—a biochemical perspective. J. Gen. Physiol. 52:No. 1, Part 2:257s–278s.

Laties, G. G. 1969. Dual mechanisms of salt uptake in relation to compartmentation and long-distance transport. Ann. Rev. Plant Physiol. 20:89–116.

Läuchli, A. and E. Epstein. 1970. Transport of potassium and rubidium in plant roots—the significance of calcium. Plant Physiol. 45:639–641.

Leggett, J. E., R. A. Galloway and H. G. Gauch. 1965. Calcium activation of orthophosphate absorption by barley roots. Plant Physiol. 40:897–902.

Lüttge, U. 1969. Aktiver Transport (Kurzstreckentransport bei Pflanzen.) In: Protoplasmatologia. M. Alfert, H. Bauer, C. V. Harding, W. Sandritter and P. Sitte, eds. Springer-Verlag, Wien, New York. Vol. VIII 7b, pp. 1–146.

Lüttge, U. and K. Bauer. 1968. Die Kinetik der Ionenaufnahme durch junge und alte Sprosse von *Mnium cuspidatum*. Planta 78:310–320.

Lüttge, U. and G. G. Laties. 1966. Dual mechanisms of ion absorption in relation to long distance transport in plants. Plant Physiol. 41:1531–1539.

Lüttge, U. and G. G. Laties. 1967a. Selective inhibition of absorption and long distance transport in relation to the dual mechanisms of ion absorption in maize seedlings. Plant Physiol. 42:181–185.

Lüttge, U. and G. G. Laties. 1967b. Absorption and long distance transport by isolated stele of maize roots in relation to the dual mechanisms of ion absorption. Planta 74:173–187.

Maas, E. V. 1969. Calcium uptake by excised maize roots and interactions with alkali cations. Plant Physiol. 44:985–989.

Maas, E. V. and J. E. Leggett. 1968. Uptake of ^{86}Rb and K by excised maize roots. Plant Physiol. 43:2054–2056.

MacDonald, I. R., P. C. DeKock and A. H. Knight. 1960. Variations in the mineral content of storage tissue disks maintained in tap water. Physiol. Plantarum 13:76–89.

Macklon, A. E. S. and N. Higinbotham. 1970. Active and passive transport of potassium in cells of excised pea epicotyls. Plant Physiol. 5:133–138.

MacRobbie, E. A. C. 1962. Ionic relations of *Nitella translucens*. J. Gen. Physiol. 45:861–878.

MacRobbie, E. A. C. 1966. Metabolic effects on ion fluxes in *Nitella translucens*. I. Active influxes. Austral. J. Biol. Sci. 19:363–370.

MacRobbie, E. A. C. 1968. Active transport in giant algal cells. In: Internationales Symposium Stofftransport und Stoffverteilung in Zellen höherer Pflanzen. K. Mothes, E. Müller, A. Nelles and D. Neumann, eds. Akademie–Verlag, Berlin. Pp. 179–186.

MacRobbie, E. A. C. and J. Dainty. 1958a. Ion transport in *Nitellopsis obtusa*. J. Gen. Physiol. 42:335–353.

MacRobbie, E. A. C. and J. Dainty. 1958b. Sodium and potassium distribution and transport in the seaweed *Rhodymenia palmata* (L.) Grev. Physiol. Plantarum 11:782–801.

Mengel, K. 1962. Die K- und Ca-Aufnahme der Pflanze in Abhängigkeit vom Kohlenhydratgehalt ihrer Wurzel. Z. Pflanzenern. Düng. Bodenk. 98:44–54.

Nissen, P. 1971. Uptake of sulfate by roots and leaf slices of barley: mediated by single, multiphasic mechanisms. Physiol. Plantarum 24:315–324.

Nobel, P. S. 1969. Light-dependent potassium uptake by *Pisum sativum*. Plant and Cell Physiol. 10:597–605.

Ohnishi, M. and D. T. Urry. 1970. Solution conformation of valinomycin-potassium ion complex. Science 168:1091–1092.

Ordin, L. and L. Jacobson. 1955. Inhibition of ion absorption and respiration in barley roots. Plant Physiol. 30:21–27.

Osmond, C. B. and G. G. Laties. 1968. Interpretation of the dual isotherm for ion absorption in beet tissue. Plant Physiol. 43:747–755.

Pallaghy, C. K., U. Lüttge and K. von Willert. 1970. Cytoplasmic compartmentation and parallel pathways of ion uptake in plant root cells. Z. Pflanzenphysiol. 62:51–57.

Pardee, A. B. 1968. Membrane transport proteins. Science 162:632–637.

Persson, L. 1969. Labile-bound sulfate in wheat-roots: localization, nature and possible connection to the active absorption mechanism. Physiol. Plantarum 22:959–976.

Pettersson, S. 1966. Active and passive components of sulfate uptake in sunflower plants. Physiol. Plantarum 19:459–492.

Pierce, W. S. and N. Higinbotham. 1970. Compartments and fluxes of K+, Na+, and Cl- in *Avena* coleoptile cells. Plant Physiol. 46:666–673.

Pitman, M. G. 1963. The determination of the salt relations of the cytoplasmic phase in cells of beetroot tissue. Austral. J. Biol. Sci. 16:647–668.

Pitman, M. G., S. M. Mertz, Jr., J. S. Graves, W. S. Pierce and J. N. Higinbotham. 1971. Electrical potential differences in cells of barley roots and their relation to ion uptake. Plant Physiol. 47:76–80.

Pitman, M. G. and H. D. W. Saddler. 1967. Active sodium and potassium transport in cells of barley roots. Proc. Nat. Acad. Sci. 57:44–49.

Polya, G. M. 1968. Inhibition of protein synthesis and cation uptake in beetroot tissue by cycloheximide and kryptopleurine. Austral. J. Biol. Sci. 21:1107–1118.

Poole, R. J. 1969. Carrier-mediated potassium efflux across the cell membrane of red beet. Plant Physiol. 44:485–490.

Pressman, B. C. 1968. Ionophorous antibiotics as models for biological transport. Fed. Proc. 27:1283–1288.

Pressman, B. and D. H. Haynes. 1969. Ionophorous agents as mobile ion carriers. In: The Molecular Basis of Membrane Function. D. C. Tosteson, ed. Prentice-Hall, Inc., Englewood Cliffs. Pp. 221–246.

Rains, D. W. 1967. Light-enhanced potassium absorption by corn leaf tissue. Science 156:1382–1383.

Rains, D. W. 1968. Kinetics and energetics of light-enhanced potassium absorption by corn leaf tissue. Plant Physiol. 43:394–400.

Rains, D. W. and E. Epstein. 1965. Transport of sodium in plant tissue. Science 148:1611.

Rains, D. W. and E. Epstein. 1967a. Sodium absorption by barley roots: role of the dual mechanisms of alkali cation transport. Plant Physiol. 42:314–318.

Rains, D. W. and E. Epstein. 1967b. Sodium absorption by barley roots: its mediation by mechanism 2 of alkali cation transport. Plant Physiol. 42:319–323.

Rains, D. W. and E. Epstein. 1967c. Preferential absorption of potassium by leaf tissue of the mangrove, *Avicennia marina*: an aspect of halophytic competence in coping with salt. Austral. J. Biol. Sci. 20:847–857.

Rains, D. W., W. E. Schmid and E. Epstein. 1964. Absorption of cations by roots. Effects of hydrogen ions and essential role of calcium. Plant Physiol. 39:274–278.

Robertson, R. N. and M. J. Wilkins. 1948. Studies in the metabolism of plant cells. VII. The quantitative relation between salt accumulation and salt respiration. Austral. J. Sci. Res. B. 1:17–37.

Schmid, W. E., H. P. Haag and E. Epstein. 1965. Absorption of zinc by excised barley roots. Physiol. Plantarum 18:860–869.

Scott, G. T. and H. R. Hayward. 1953. Metabolic factors influencing the sodium and potassium distribution in *Ulva lactuca*. J. Gen. Physiol. 36:659–671.

Smith, R. C. and E. Epstein. 1964a. Ion absorption by shoot tissue: technique and first findings with excised leaf tissue of corn. Plant Physiol. 39:338–341.

Smith, R. C. and E. Epstein. 1964b. Ion absorption by shoot tissue: kinetics of potassium and rubidium absorption by corn leaf tissue. Plant Physiol. 39:992–996.

Steward, F. C. and J. A. Harrison. 1939. The absorption and accumulation of salts by living plant cells. IX. The absorption of rubidium bromide by potato discs. Ann. Bot. N. S. 3:427–453.

Steward, F. C. and R. L. Mott. 1970. Cells, solutes, and growth: salt accumulation in plants reexamined. Internat. Rev. Cytol. 28:275–370.

Steward, F. C. and J. F. Sutcliffe. 1959. Plants in relation to inorganic salts. In: Plant Physiology—A Treatise. F. C. Steward, ed. Academic Press, New York and London. Vol. 2, pp. 253–478.

Thellier, M. 1970. An electrokinetic interpretation of the functioning of biological systems and its application to the study of mineral salts absorption. Ann. Bot. N. S. 34:983–1009.

Torii, K. and G. G. Laties. 1966. Dual mechanisms of ion uptake in relation to vacuolation in corn roots. Plant Physiol. 41:863–870.

Ulrich, A. 1941. Metabolism of non-volatile organic acids in excised barley roots as related to cation-anion balance during salt accumulation. Am. J. Bot. 28:526–537.

Ulrich, A. 1942. Metabolism of organic acids in excised barley roots as influenced by temperature, oxygen tension and salt concentration. Am. J. Bot. 29:220–227.

van den Honert, T. H. 1937. Over eigenschappen van plantenwortels, welke een rol spelen bij de opname van voedingszouten. Natuurk. Tijdschr. v. Nederl.–Ind. 97:150–162.

van den Honert, T. H. and J. J. M. Hooymans. 1955. On the absorption of nitrate by maize in water culture. Acta. Bot. Neerl. 4:376–384.

van Lookeren Campagne, R. N. 1957. Light-dependent chloride absorption in *Vallisneria* leaves. Acta Bot. Neerl. 6:543–582.

Venrick, D. M. and R. C. Smith. 1967. The influence of initial salt status on absorption of rubidium by corn root segments of two stages of development. Bull. Torrey Bot. Club 94:501–510.

Viets, F. G., Jr. 1944. Calcium and other polyvalent cations as accelerators of ion accumulation by excised barley roots. Plant Physiol. 19:466–480.

Welch, R. M. and E. Epstein. 1968. The dual mechanisms of alkali cation absorption by plant cells: their parallel operation across the plasmalemma. Proc. Nat. Acad. Sci. 61:447–453.

Welch, R. M. and E. Epstein. 1969. The plasmalemma: seat of the type 2 mechanisms of ion absorption. Plant Physiol. 44:301–304.

7

THE UPWARD MOVEMENT OF WATER AND NUTRIENTS

CELLULAR VS. LONG-DISTANCE TRANSPORT

Up to now, our discussion has dealt with the cellular physiology of transport. When the same cells perform the processes of both photosynthesis and the absorption of mineral nutrients from the external medium, distances over which chemical transport must occur do not exceed cellular dimensions. Such is the case in unicellular algae, and even in multicellular algae consisting of strands of cells or of flat sheets. In such organisms, no cell is farther than a few cells' width removed from the external medium. Such short distances are readily negotiated by solutes and gases through diffusion.

During the Silurian Period, about 425 million years ago, when some plants began to emerge from the oceans and invaded the land, the two functions previously performed by the same cells—the acquisition of carbon on the one hand, and of mineral nutrients and water on the other—became separated in space. Part of the plant had to be in the soil, deprived of light and hence incapable of photosynthesis, while another part had to be above ground, exposed to the light of the sun, and therefore without direct contact with the source of water and minerals. This spatial separation of the functions of organic and inorganic nutrition necessitated the evolutionary elaboration of structures and mechanisms for the long-distance transport of chemicals. Organic foodstuffs synthesized as the result of photosynthesis had to be transported to the portions of the plants confined in the soil, while water and inorganic nutrients acquired from the soil had to be conducted to the photosynthetic organs above. Furthermore, since the elaboration of a complex plant body requires the integration and regulation of many processes during growth, there arose the need for the movement of hormones and other growth regulators.

When liquids and dissolved substances are to move from compartment to compartment within a system, it is logical to look for pores in the partitions separating the compartments. If transport occurs over still larger distances, one would expect to find pipes or conduits. Such structures do in fact exist in plants. They have evolved as adaptations which make translocation possible. In this chapter, we shall briefly review the anatomical features specifically associated with translocation, and then describe the processes of translocation in the xylem and the hypotheses which have been formulated to account for them. Transport in the phloem will be discussed in the next chapter.

2. PORES, PIPES, AND PATHWAYS

Primary cell walls, those first formed in developing cells, have thin areas called primary pit fields. When secondary walls are laid down, they do not completely cover the primary wall. Rather, secondary walls have pits, usually but not always coinciding with a primary pit field of the primary wall. Pits represent areas where no secondary wall thickening occurs so that a front view reveals the primary wall at the bottom of the pit (the pit membrane).

Where there is a pit in the secondary wall of a cell there usually is a complementary pit in the adjacent cell, so that the two make up a twin structure, a pit pair. The pits of the two adjacent cells share a common pit membrane, consisting of the primary walls of the two neighboring cells (Figure 7-1). Pits and pit pairs occur in a variety of shapes. Their arrangement on the cell wall also may be quite variable in different plants and even in different tissues of the same plant (Esau, 1960, 1965a).

The significance of primary pit fields and of pits has to do with the subject of this chapter, translocation. At primary pit fields, the protoplasts of adjacent cells are separated by a partition of minimal thickness. Through this thin part of the wall, the pit membrane, extend fine strands of cytoplasm called plasmodesmata. Each plasmodesma represents a connection between the protoplasts of contiguous cells. Thus, the protoplasts of the cells of the plant are all interconnected and form a continuum, the

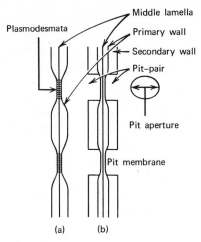

Figure 7-1. (a) Cell wall composed of middle lamella and the primary wall layers of two adjacent cells. The thin areas in the walls are pits traversed by plasmodesmata. (b) Cell wall composed of middle lamella and both primary and secondary wall layers of two adjacent cells. Secondary wall has not been laid down in pit regions. After Esau (1960).

symplast (Figure 7-2). Water and substances in solution may move from the protoplast of one cell into that of the next cell through the plasmodesmata, mainly by diffusion and osmosis, without having to cross the plasmalemma of either cell (Tyree, 1970).

Transport from cell to cell via plasmodesmata is adequate for small plants in which no cells are far removed from moisture and dissolved minerals. But for truly long-distance translocation, more efficient adaptations had to evolve. These have taken the form of long pipes or conduits, and therefore the advanced land plants all belong in the phylum *Tracheophyta,* or "plants with pipes." In contrast, the *Bryophyta* (the liverworts and mosses), lacking such efficient means for long-distance transport, are very small plants, restricted to moist habitats such as shady fens and bogs.

The system of conduits in tracheophytes is called the vascular system, consisting of the xylem and the phloem. These are both complex tissues. We shall confine ourselves to a brief review of those types of cells in each which are chiefly concerned with translocation.

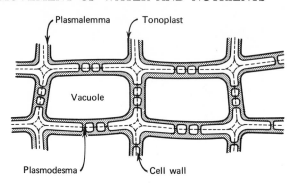

Figure 7-2. Schematic representation of a uniform tissue with plasmodesmata, showing the symplastic continuum. After Brouwer (1965).

In the xylem, two types of conduits are made up of tracheids and vessel members. The tracheids, which evolved first, are the more primitive of the two and are the only such pipe-like elements to be found in most of the lower vascular plants and most gymnosperms. Tracheids are narrow, elongated cells with numerous pits. In length they vary from a fraction of a millimeter to 10 or more. Near their ends they abut for varying distances on other tracheids, with pit pairs forming connections from one to the other. As conducting elements, tracheids suffer from the limitation that material moving through them over long distances has to negotiate numerous partitions as it moves from cell to cell across the pits.

In vessels, this limitation has been overcome. Vessels evolved from tracheids. Vessels are made up of vessel members representing cells which elongate little during their development but grow in girth. They therefore are shorter than tracheids but have a much greater diameter. Their more or less sloping end walls, where they abut on other vessel members, eventually become perforated and often disappear completely (Figure 7-3), so that numerous such vessel members come to form long tubes or pipes. These, and not the individual elements, are called vessels. Mature, functioning vessels do not retain the individuality of their constituent vessel members, and in the final stages of the development of vessels, the protoplasts responsible for their formation die. The mature vessel is thus a long pipe offering minimal resistance to the flow of water and solutes. In the angiosperms with highly developed vessel systems the upward flow of water and nutrients therefore can reach speeds far in excess of those in gymnosperms lacking vessels.

The phloem chiefly serves the function of transporting carbohydrates

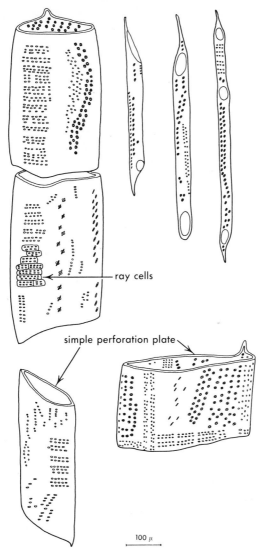

ray cells

simple perforation plate

100 μ

Figure 7-3. Vessel members, showing the perforations at the top and bottom where the end wall has disappeared, and pits in the walls. After Esau (1960).

and other organic metabolites from the sites of their synthesis in green tissues to regions of utilization. It consists of several types of cells of which the sieve elements are those most specifically adapted to translocation. The phloem and the xylem usually occur side by side, forming the "vascular tissue." It contains, in addition to the main conducting elements, parenchyma cells, fibers whose chief function is that of providing mechanical strength, and some other types of cells. Usually, the xylem and phloem are arranged radially in the root and the stem, the xylem lying closer to the center, the phloem toward the periphery. A strand of vascular tissue containing xylem, phloem, and associated parenchyma is called a "vascular bundle." In the root, the xylem of the vascular bundles coalesces into a wavy ring, the phloem strands lying in the invaginations outside this ring. The structure of the phloem, and specifically, of the sieve elements, is considered in the next chapter, which deals with translocation in the phloem.

3. TISSUES AND ORGANS

After this consideration of some cellular characteristics having to do with long-distance transport we now turn to the organization of tissues and organs across which substances move in their passage through the plant.

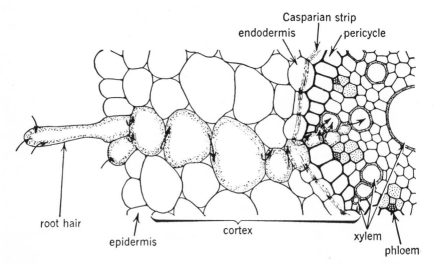

Figure 7-4. Part of transection of a wheat root, showing the symplastic pathway of ions from the external medium into the xylem vessels. Stipling indicates cytoplasm. Modified after Esau (1965a).

Figure 7-4 shows part of a transverse section of a wheat root. The epidermis is outermost and is underlain by the cortex which consists of relatively large and loosely arranged parenchyma cells, with many intercellular spaces. The innermost layer of the cortex, however, is different. This is the endodermis which forms a cylinder of closely arranged cells. The most characteristic feature of these cells is the Casparian strip. The strip represents an impregnation with suberin of those primary walls which would be cut crosswise by a tangential cut through the cell. It completely encircles the cells of the endodermis and is therefore encountered by any substance which moves within the walls across the root in a radial direction (Peirson and Dumbroff, 1969). Being impermeable, the strip is thought to block progress of these substances in the wall and to force solutes to negotiate the endodermis by passing through the protoplast of an endodermal cell (see section 5 of Chapter 5).

The stele represents the innermost cylinder of tissues of the root and contains the vascular tissue. Its outer layer is the pericycle which abuts on the endodermis. Lateral roots in angiosperms and gymnosperms are initiated by pericyclic cells and grow outward across the cortex and the epidermis.

The arrangement of the conducting vascular tissues, the xylem and phloem, varies much from plant to plant, and even within the same plant depending upon the stage of growth and the particular organ. A discussion of its disposition and the details of its complex routing throughout the plant body is beyond the scope of the present book. Excellent accounts are those by Esau (1960, 1965b) and Fahn (1967). Gunning et al. (1970) have described specialized xylem "transfer cells" often located at strategic traffic points, especially at departing foliar traces.

In the leaf, the vascular tissue is disposed in veins containing one or more vascular bundles. Single veins occur mainly in needles of conifers. In leaves of angiosperms there are two main patterns of venation, the reticulate (net-like) pattern common in dicotyledons and the parallel arrangement characteristic of monocotyledons. Even in these, however, the parallel veins are interconnected by cross veins, yielding a net-like pattern somewhat like rope ladders.

The internal structure of leaves varies greatly in different groups of plants, but it is a universal feature that the anastomosing network of veins and veinlets brings elements of the conducting system into close proximity of every photosynthesizing cell of the leaf. Therefore, the distance solutes have to travel by diffusion between any leaf cell and the nearest conducting tissue corresponds to no more than the diameters of a few cells (Esau, 1967; Wylie, 1939). We have already seen that in simple plants of aquatic or moist habitats, where no distances larger than these need to be traversed by solutes, elaborate conducting tissues are absent. Diffusion readily accounts for transport over such small distances. Small, terminal bundles are bordered by bundle sheaths consisting of specialized cells which are the first recipients of solutes conveyed into the leaf through the xylem and may facilitate their distribution to the mesophyll cells.

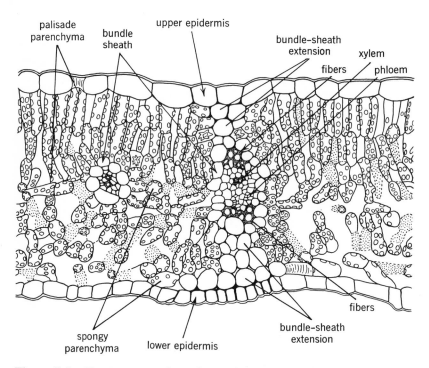

Figure 7-5.   Transverse section of a leaf of pear, *Pyrus communis*. The round structures in the mesophyll cells are chloroplasts. The cells of bundle sheaths have fewer chloroplasts (none are shown). After Esau (1965a).

Figure 7-5 shows a transverse section of a typical angiosperm leaf. The upper and lower epidermal layers enclose the mesophyll, the main photosynthetic tissue. It is often differentiated into palisade parenchyma near the upper epidermis and spongy parenchyma between the palisade layer and the lower epidermis. There are large intercellular spaces among the mesophyll cells, especially in the spongy parenchyma. As a result, the interface between the liquid water in cell surfaces and the gaseous atmosphere in the intercellular spaces of the leaf is very large. The total area of the internal cell surfaces of a leaf may be 10, 20, or even 30 times larger than the external surface of the leaf.

The epidermis is a compact layer (sometimes two or more layers) of cells. Two principal features related to transport characterize the epidermis: the cuticle and the stomates.

The cuticle is a layer of cutin, a lipid polymer embedded in wax, coating the external, often thickened walls of the epidermal cells. It is not very permeable to water and thus forms a barrier between the moist interior of the leaf and the air outside. In many species, the cuticle is covered with a bloom of wax which further cuts down on the rate of water loss. Figure 7-6 shows the disposition of the cuticle and Figure 7-7 shows a bloom of wax on the leaf of pea, *Pisum sativum.*

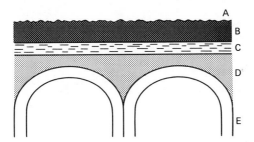

Figure 7-6.   Schematic representation of the structure of the leaf surface. A, surface wax; B, cutin embedded in wax; C, a mixed layer containing some cutin, wax and carbohydrate polymers possibly with traces of protein; D, pectin; E, cellulose wall of the epidermal cells. After Kolattukudy (1970).

While the cuticle minimizes the loss of water from leaf surfaces, nevertheless there must be a traffic of oxygen and carbon dioxide between the leaf and the atmosphere. This traffic is mainly via the stomates. Stomates occur on both leaf surfaces in some species, but typically are more abundant in the lower epidermis and often entirely confined to this surface. Details of stomatal structure and disposition in the epidermis vary, but typically the stomate consists of two kidney-shaped guard cells and sometimes associated cells. Through changes in their water content the two

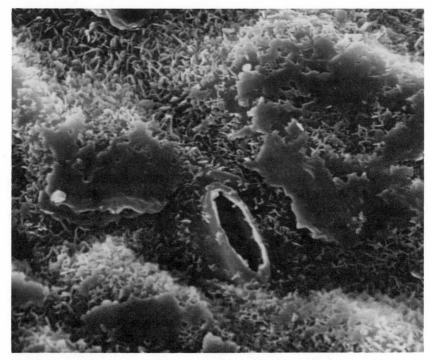

Figure 7-7. Leaf surface of pea, *Pisum sativum,* showing waxy plates and spicules and an annular waxy deposition overlying a stomate. Scanning electron micrograph, ×3,000, courtesy of Y. and J. Heslop–Harrison.

guard cells control the opening between them, and thereby, the degree of communication between the atmosphere and the leaf interior. There is usually a large vacant space directly below the stomate which is not occupied by mesophyll cells (the substomatal chamber). The mechanism whereby stomates open and close is discussed under the heading, Potassium, in section 4 of Chapter 11 (pp. 309–310).

4. THE MOVEMENT OF WATER THROUGH THE PLANT

Pathway. Of all the traffic of substances within the plant that of water is by far the most prominent in the quantities involved. This is a direct outcome of the evolutionary emergence of plants from the sea. On land, the photosynthetic organs, exposed to light and to air, became subject to desiccation. As a result, the cuticle evolved as a waterproofing cover over the leaves. However, carbon dioxide and oxygen had to move in and out, and the stomates represent the adaptations for this function. Water vapor can and does move through these openings. The water transpired is replenished by entry of water into the plant from the soil. As a result there is a stream of water from the soil into the root, through the stem and into the leaf. Eventually, at the surfaces of leaf mesophyll cells, the water evaporates and diffuses through the stomates into the atmosphere.

During the life of the plant, the entire mass of water moving through the plant and dissipated into the atmosphere in this way is very large. Briggs and Shantz (1914) measured the weight of water used by numerous species of plants in the production of unit dry weight of plant material. This "water requirement" varied from about 200 to over 1,000. Transpiration is a wasteful process, the result of an evolutionary compromise between two conflicting requirements—the need for exposure of moist, green cells to light, and that for open pathways for gas exchange between these cells and the atmosphere. It does, however, serve the function of conducting mineral nutrients from the roots to the tissues of the shoot, and that of cooling the leaves.

Initial entry of water into the root is mainly through that region which extends for a few centimeters behind the tip. This is the region of most active metabolism and also the region in which root hairs are most numerous. The older tissue back of this region becomes progressively suberized and hence, impermeable. This impermeability is by no means absolute, however, and a considerable flux of water and nutrients probably occurs through these less active regions of the root (Kramer, 1969).

Moving radially across the root toward the stele, water follows the path of least resistance, through the cell walls. Osmotic movement, across the plasmalemma and into protoplasts and vacuoles, represents, in the main, a diversion of a small fraction of the total water moving through the root. At the endodermis, flow through the cell walls is impeded by the Casparian strip and water has to move through endodermal cytoplasm into the stele. This is the reason why the flux of water through the root is severely affected by cold and poisons. Such sensitivity would not be expected for a flow through the "outer" space (the permeable cell walls). It is at the endodermis that the entering stream of water is forced to traverse "inner" (cytoplasmic) space and therefore becomes subject to factors which influence metabolism (Brouwer, 1954, 1965; Weatherley, 1963).

Beyond the endodermis water is free to move through the cell walls into the vascular tissue and then upward in the xylem. A more limited movement, however, occurs through the phloem and even through non-vascular tissue. Also, there is evidence for rapid lateral movement of water; for example, tritiated water (water labeled with the radioisotope ^3H) absorbed by three-year old willow trees, *Salix* sp., showed up prominently in the phloem (Wray and Richardson, 1964). Since the main pathway of water is in the xylem, rapid lateral movement is indicated by this finding.

The vascular tissue is the pathway of water in the root, through the stem, and into the petioles and the veins of leaves. In the leaf, the branching and anastomosing veins and veinlets bring elements of the vascular system within close proximity of all mesophyll cells. Water migrates throughout the mesophyll much as it does through the cortex of the root, i.e., mainly in the "outer" (cell wall) space of the tissue (Weatherley, 1963). It evaporates at the surface of mesophyll cells and water vapor then diffuses through the extensive intercellular spaces and eventually through stomates into the external atmosphere. Cuticular transpiration occurs to a limited extent. A third pathway of water out of leaves is via hydathodes. This matter is discussed below, in connection with root pressure.

Mechanisms. Transpiration is a physical process. Evaporation of water from leaves causes the water content of the leaves to drop. This produces a gradient of water potential from the root to the leaves. The resulting hydrostatic differential causes a flow of water through the interconnected system, soil–root–vascular tissue–leaf. This means that the water in the system is under tension (negative pressure). Very large cohesive forces between water molecules are implied.

This view of the process, called the cohesion theory of sap movement in plants, was first advanced independently by H. H. Dixon (Dixon and Joly, 1895) and E. Askenasy (1895), but much more elegant and botanically relevant evidence for it was furnished by O. Renner (1915). In ingenious experiments with fern sporangia Renner found that water could be under tension (negative pressure) as high as 300 atmospheres. In 1914, Dixon published a book, *Transpiration and the Ascent of Sap in Plants,* which summed up his own work and that of others on this important problem (without, however, any mention of Askenasy's and Renner's contributions). For more recent findings on the measurement of negative pressure in the xylem, see the enjoyable paper by Scholander and his associates at Scripps Institution of Oceanography of the University of California (Scholander *et al.,* 1965). Using cut twigs of higher plants put under gas pressure, they determined that pressure just sufficient to force sap from the cut end (Figure 7-8). This (with a negative sign) is taken as the pressure existing in the twig before it was severed. Tensions were not quite

Optics

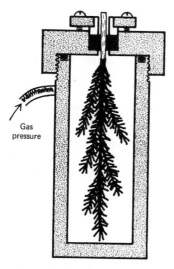

Gas
pressure

Figure 7-8. Pressure bomb for measurement of sap pressure in the xylem of a twig. After Scholander *et al.* (1965).

so spectacular as in Renner's fern sporangia but still exceedingly impressive—up to about 80 atmospheres. The most recent comprehensive accounts on the water relations of plants are those by Briggs (1967), Kozlowski (1964), Kramer (1969), and Slatyer (1967). Zelitch (1963) has edited contributions to a seminar on stomates and water relations in plants. A symposium held by the British Ecological Society on water relations of plants was edited by Rutter and Whitehead (1963). Water deficits and their bearing on the growth of plants have been discussed by experts in two volumes edited by Kozlowski (1968a, b), and in a book by Slavik (1969). There never is a deficit of books and articles on water deficits and water relations in general, although there are some who consider this a fairly dry subject.

Transpiration accounts for all but a small fraction of the total water movement through higher plants under most conditions. But there is another mechanism which is responsible for water movement into and through plants under conditions when transpiration is minimal and conditions for salt absorption by the roots are favorable. When the roots are in moist, warm soil containing ample concentrations of absorbable ions, and the leaves are exposed to humid air, water will enter the roots, move into the xylem and eventually be exuded through hydathodes, modified stomates which act as ports through which the "guttation" liquid is discharged. There have been many studies of the mechanisms of this water movement. The usual technique is to cut off the root and examine the rate of discharge of xylem exudate from the cut end and its composition under various conditions. This technique assures that transpiration is eliminated as a contributing factor.

It is commonly found that the exudate contains concentrations of salt in excess of that of the solution bathing the roots (Anderson and Reilly, 1968; House and Findlay, 1966; Läuchli and Epstein, 1971; Welch and Epstein, 1968). The mechanism of the movement is as follows. Salt is actively absorbed and transferred into the xylem elements, as described below, in section 5. As a result, an inward osmotic gradient is created: the water potential in the xylem is lower than that of the external solution, and water moves into the xylem osmotically, setting up a hydrostatic pressure there (root pressure). This condition which results in a positive pressure in the xylem is the exact opposite of that prevailing when the plant is actively transpiring, resulting in a negative pressure (tension) in the xylem. The movement of water is thus passive along an osmotic gradient set up through the active transport of ions. This has long been the favorite hypothesis to account for root pressure and exudation (Arisz et al., 1951; Crafts and Broyer, 1938; Sabinin, 1925), and more recent evidence confirms it (Anderson et al., 1970; House and Findlay, 1966; Klepper, 1967).

The point is often made that root pressure fulfills no important function in the water economy of plants, but there are some considerations which throw doubt on this opinion. One of these has to do with the restoration of breaks in the water columns in vessels and tracheids. When transpiration is at a high rate and the water columns in the xylem are under tension, breaks may occur in them through formation of gas bubbles. It is likely that once the rate of transpiration falls to a low value, as at night, these ruptures in the liquid columns in the xylem are eliminated and that root pressure furnishes the necessary force.

Another situation in which root pressure is important occurs in monocot trees. Davis (1961) measured root pressures in palms which indicated that root pressure could supply water to a height of more than 10 meters. This suggests a significant role for it in the water economy of these plants.

Experiments by White (1938) with tomato roots cultured *in vitro* showed that very high root pressures could develop, enough to raise a water column 200 feet tall (60 meters)—rather more than necessary to supply water to the topmost leaf of a tomato plant. If such root pressures are common, and this is likely, root pressure must be at the very least an important factor contributing to the ascent of sap in plants, in addition to transpiration (cf. Rufelt, 1956).

THE UPWARD MOVEMENT OF IONS

Pathway.   Like water, ions freely penetrate into the "outer" (cell wall) spaces of the root cortex as far as the endodermis where further progress in the walls is impeded by the Casparian strip of the endodermis (see Chapter 5). Perhaps limited movement through the endodermis cell walls may occur where branch roots arising from the pericycle pierce the otherwise intact endodermis, according to Dumbroff and Peirson (1971). Ions are removed from the "outer" space of the cortex by being accumulated in cortical cells, through the processes discussed in Chapter 6. Initial transport into these cells is across the plasmalemma into the cytoplasm. Once there, two main routes are followed, one across the tonoplast into the vacuole, the other within the cytoplasm through plasmodesmata into the contiguous cell (symplastic movement).

Clear evidence for such a "fork in the road" was first presented by T. C. Broyer (1950). Root systems of barley were put into a solution of 5 mM KBr for six hours and then transferred to a chemically identical solution in which the bromide was radioactively labeled. After a further period of six hours the concentration of radiobromide in the expressed sap of the roots and in the solution exuding from the cut end was determined. The expressed sap was taken to represent mainly the solution in the vacuoles of the root cells, and the exudate, the content of the xylem.

It turned out that the concentration of radiobromide in the xylem was almost twice that in the vacuolar sap. The interpretation was as follows: the vacuoles, having already absorbed much bromide during the initial period when the tissue was in the solution of nonradioactive bromide, absorb radiobromide sluggishly during the second period, since they are closed systems and their total capacity for accumulation is limited. The symplastic pathway, however, leading eventually to the stele and the conducting elements of the xylem, is an open system, since solution can continue to exude from the cut end. It is therefore visualized that ions, having initially been transported into the cytoplasm of an epidermal or cortical cell across the plasmalemma, migrate within the symplast, passing from cell to cell within the plasmodesmata without crossing any membrane. That is, they largely by-pass the vacuoles (Brouwer, 1959; Broyer, 1950; Helder, 1958, 1964; Läuchli and Epstein, 1971). There is, however, a slow release of ions from the vacuoles back into the cytoplasm (Hodges and Vaadia, 1964a, b; Läuchli and Epstein, 1971).

Once in the stele they eventually leave the symplast and reach the non-living conducting vessels of the xylem. The site of the exit from the symplast is the plasmalemma of stelar cells, but it is not known whether this release from the symplast occurs evenly throughout the stele or is restricted to the cells of certain regions within the stele. Having passed beyond the confines of the symplast, the ions move through stelar cell wall space into the xylem and upward to the shoot following mainly the routing of water in the transpiration stream. Stout and Hoagland (1939) demonstrated this in classical experiments with radioisotopes—among the first plant physiological experiments in which radioisotopes of physiologically important elements were used.

In one of their experiments, a small willow plant was grown in nutrient solution. The bark of a branch was slit longitudinally and a paraffined paper was inserted between the wood and the bark in such a way as to form an impervious sleeve separating the xylem and phloem. The branch was wrapped in paraffined paper to prevent loss of moisture. Radioactive potassium was then added to the nutrient solution, and after five hours the stem was sectioned as shown in Figure 7-9 and the sections of wood and bark were assayed for radiopotassium. The results are indicated in the figure. They show roughly equal concentrations in wood and bark below the strip and above it, but where the paraffined sleeve separated the bark (phloem) from the wood (xylem), almost all of the radiopotassium was in the xylem. The xylem was thus shown to be the primary pathway of potassium movement up the stem, but rapid lateral movement into the phloem normally occurs as shown by the results obtained with those sections of bark not separated from the wood by the impermeable sleeve. Similar findings were obtained with sodium and phosphorus.

Moving up in the xylem, ions eventually reach the terminal veinlets in leaves and from them are free to move into the "outer" (cell wall) spaces of the mesophyll cells. These cells are thus bathed by a solution of ions much as the cells of the root cortex are bathed by the solution in the "outer" space of the root. Like cortical cells of the root, leaf mesophyll cells accumulate ions from the solution bathing them, by mechanisms of active ion transport (Smith and Epstein, 1964a, b). Thus, most ions, before reaching the cytoplasm of a leaf cell, must be absorbed at least twice by cellular transport mechanisms: first by a root cell, and again by one of the cells of the leaf. Once within the cytoplasm of a leaf cell ions may

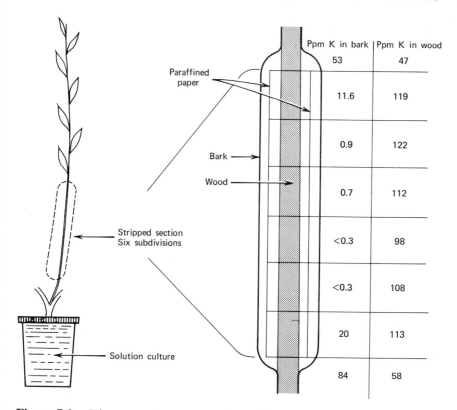

Figure 7-9. Diagrammatic representation of the Stout–Hoagland experiment on the upward movement of radioactively labeled potassium in the stem of willow. After Stout and Hoagland (1939).

move by the symplastic route from cell to cell, as discussed above for radial movement in the root.

The Dutch investigator, W. H. Arisz, and his collaborators have extensively investigated symplastic movement of ions in leaf tissue. They used strips of leaves of the submerged aquatic plant, *Vallisneria spiralis* (tape grass). In this plant, direct absorption of ions from the external solution is a normal function of the leaves. When part of a strip of leaf (the absorbing part) is in a solution and another part (the free part) in water, absorption of ions by the part kept in the solution and their translocation to the free part can be studied. When labeled ions absorbed by the absorbing part reached the free part there was no appreciable leakage from the free part into the water bathing it. This is evidence that the ions migrated within the cytoplasm; diffusion through the cell wall ("outer") space would lead to leakage into the water bathing the free part. Diffusion of ions within the "outer" (cell wall) space occurred, to be sure, but the diffusion path was so long, in these experiments, that virtually all movement observed was via the symplast. Helder (1967) has given a comprehensive review of this work.

Mechanisms: The Crafts–Broyer Hypothesis. "Despite the immense amount of research on absorption and movement of mineral solutes and water in plants, the mechanisms of the processes remain obscure" (Crafts and Broyer, 1938). "This process of secretion [into the xylem vessels], if it may be so called, is one of the least known of all processes in the realm of mineral translocation" (Biddulph, 1951). "It is regrettable to conclude that little more is known today than was known 2 or 3 decades ago about the mechanism of radial transport of salt into the xylem vessels" (Yu and Kramer, 1967).

In the paper from which the first of these melancholy sentiments is taken, Crafts and Broyer developed a hypothesis for lateral transfer of ions into the conducting elements of the xylem. Since xylem exudate commonly contains salt at higher concentration than that of the external solution, it is obvious that an active transport mechanism is operating somewhere between the external solution and the dead xylem vessels. Evidence for this comes also from experiments showing that interference with the metabolism of the root by means of poisons, anaerobic conditions, or cold inhibits transfer of ions to the vascular tissue and the shoot (Brouwer, 1965; Hoagland, 1944; Russell and Barber, 1960; Shone, 1969).

According to Crafts and Broyer (1938) absorption of ions into the cytoplasm of cortical cells is the active transport step responsible for the build-up of high xylem concentrations of salt. The authors draw attention to the favorable physiological conditions in the cortex: large intercellular spaces promoting aeration, and proximity to the external solution, favoring both aeration and the release of carbon dioxide resulting from respiration. In contrast the stele is encased in the cortex, farther removed from the source of oxygen, and exposed to higher concentrations of carbon dioxide because of its location and the tight packing of the cells in this region where intercellular spaces are much less prominent than in the cortex. Stelar cells, as a result of the less favorable O_2/CO_2 ratio, are less active physiologically than cortical cells, and in contrast to the membranes of actively accumulating cells of the cortex they have permeable plasmalemmas and therefore favor loss of solutes, according to this view.

Based on these considerations, the Crafts–Broyer hypothesis goes as follows. Ions are actively absorbed by the cells of the cortex. Once in the cytoplasm they diffuse within the symplast, moving from cell to cell through plasmodesmata. Having reached the stele they leak out of the cytoplasm across the permeable plasmalemmas of stelar cells into the cell walls and xylem vessels. They are kept from diffusing back into the external solution through cell wall space by the Casparian strip in the walls of the endodermis.

The cardinal feature of the hypothesis is the assumption that active transport into the cytoplasm of cortical cells represents the metabolic collection step in transfer of ions into the xylem. After that, movement within the symplast is passive, following a diffusion gradient. As a result, concentrations of salt in the xylem are lower than in the cytoplasm of the cortical cells but often higher than in the external solution. These conditions favor an osmotic flow of water into the xylem, resulting in the development of root pressure (see Mechanisms in section 4, pp. 163–165).

One assumption of this hypothesis is that the plasmalemma of the cells of the stele is leaky, permitting ions to diffuse out of the cytoplasm. It therefore lent support to the hypothesis when stele freshly dissected out of corn roots proved to be leaky (Laties and Budd, 1964). The cortical tissue deprived of stele was capable of active transport, like intact root. To be sure, low O_2/CO_2 ratios could not, in this case, be the explanation for the leakiness of the stele, since the stele in these experiments was no longer encased in the cortex. But the principal requirement of the Crafts–Broyer hypothesis was met: cortical tissue accumulated ions actively, and stele tissue leaked.

Despite the success of the Crafts–Broyer hypothesis in accounting for various features of the lateral transfer of ions across the root into the xylem, the statement by Yu and Kramer quoted above, made almost 30 years after the publication of the hypothesis, indicates a feeling that the problem is not yet solved. There are reasons for their dissatisfaction. When roots have absorbed radioactively labeled ions, the distribution of the radioactivity in a cross-section of the root can be examined. It has often been found that the concentration of label is higher in the central cylinder than in the cortex. Such results have been obtained for sulfate labeled with [35]S (Biddulph, 1967; Weigl and Lüttge, 1962) and with calcium labeled with [45]Ca by Biddulph (1967). In experiments of Yu and Kramer (1967, 1969) on the transport of ^{32}P, [36]Cl, and ^{86}Rb in corn roots, similar results were obtained with a different technique. They either separated the stele and cortex, as Laties and Budd (1964) had done, and let them absorb the labeled ions, or they immersed entire roots in the labeled solutions and separated the stele and cortex afterward and determined the amounts of radioions in each. By either method they found that stelar tissue absorbed ions as readily as cortical tissue did; often concentrations in the stele were higher than in the cortex. Thus these authors failed to confirm the findings of Laties and Budd (1964) which had appeared to favor the Crafts–Broyer hypothesis of a "leaky" stele.

An elegant method was brought to bear upon this problem by A. Läuchli (1967, 1968) who used the electron probe analyzer to examine the distribution of various elements along transects of corn roots. The instrument makes it possible to scan across a section of tissue with a narrow beam of electrons (0.5 to 1 micron wide) which upon hitting target atoms in the sample elicit the emission of X-rays characteristic of the target element (Läuchli, 1971). The sample is prepared by cryostat technique in such a way as to avoid artifacts from diffusion of soluble substances. This technique yielded results similar to those obtained by means of radioactive labeling: concentrations of potassium and of phosphorus were higher in the stele than in the cortex. These results were confirmed and extended by Läuchli et al. (1971).

Thus, three independent lines of evidence—autoradiography, separation of stele and cortex, and electron probe analysis—make the stele appear to be much more than merely a "leaky sink" for ions arriving via the symplastic pathway from the cortex. Cells of the stele seem to be actively involved in transferring ions into the extracellular continuum of the stelar cell wall spaces and the xylem. However, before pursuing this subject we must return to a problem discussed in the preceding chapter—the dual pattern of ion absorption, and its bearing on the Crafts–Broyer hypothesis.

Laties and his collaborators, having concluded that only the type 1 mechanisms of ion absorption are located in the plasmalemma, and having accepted the basic tenets of the Crafts–Broyer hypothesis, reasoned as follows. At low concentrations in the medium, ions will be absorbed via the type 1 mechanisms across the plasmalemmas of cortical cells and migrate within the symplast into the stele where they will leak out across the permeable plasmalemmas of stelar cells and thus reach the xylem. At these concentrations, transport into the xylem should faithfully mirror the kinetics of mechanism 1. This was found to be so (Lüttge and Laties, 1966), as indeed is expected on the basis that at these low concentrations, only the type 1 mechanisms deliver ions into the symplast.

The crucial question is, what happens at the high concentrations, in the range of the type 2 mechanisms? According to Laties, these mechanisms are in the tonoplast, and therefore not in the main pathway of ion movement across the root into the xylem, which is through the symplast (see Figures 6-17 and 7-4). Rather, at these concentrations the ions diffuse across the plasmalemmas of cortical cells into the symplast, according to the series model of the disposition of the two types of mechanisms (Figure 6-17). This, followed by symplastic movement into the stele and leakage out of the symplast through permeable stelar plasmalemmas, in keeping with the Crafts–Broyer hypothesis, leads to the following principal predictions in regard to transfer into the xylem of ions present in the medium at high concentrations.

(1) The relationship between external concentration and rate of transfer to the xylem (translocation) should be linear, according to diffusion kinetics (Lüttge and Laties, 1966). (2) There should be no competition between the members of ion pairs such as K^+–Rb^+ or Cl^-–Br^- (Lüttge and Laties, 1966). (3) Translocation of potassium in this range should not depend upon the identity of the anion, specifically on whether the anion is chloride or sulfate (Lüttge and Laties, 1966). (4) Antimetabolic conditions should have relatively little effect on translocation through the root, as compared with their effect on absorption by the root (Lüttge and Laties, 1967).

In respect to each of these predictions and some other related points, Laties and co-workers concluded that the experimental results matched the expectations and hence, supported the series model of the location of the two types of transport mechanisms (see review by Laties, 1969). However, the evidence and the conclusions do not bear critical examination, as pointed out by Welch (1969) on the basis of the original data, and by several investigators on the basis of independent evidence. A detailed analysis of these results is beyond the scope of this discussion but a few comments are in order on the four cardinal points mentioned.

(1) The isotherms for translocation in the range of high concentrations presented by Lüttge and Laties (1966) do not show a straight line relationship with concentration. (2) Their results show that even in the range of high concentrations, bromide interfered with chloride translocation no less, on a percentage basis, than with absorption (Lüttge and Laties, 1966). Evidence for competition between bromide and chloride in translocation has since been furnished by Läuchli and Epstein (1971) who worked with corn roots in the high as well as the low range of concentrations. (3) In the range of high concentrations, substitution of sulfate for chloride as the anion diminished translocation of potassium at least as much as it slowed absorption (Lüttge and Laties, 1966). (4) Antimetabolic conditions severely inhibited translocation as well as transport, in the experiments of Lüttge and Laties (1967) and those of Krichbaum et al. (1967) and Läuchli and Epstein (1971), at high as well as at low concentrations of the ions.

On the basis of the original evidence, therefore, and of much other work, the series model of the disposition of the two types of transport mechanism must be rejected, along with the concept of "leaky" plasmalemmas in both the cortex and the stele which are an integral feature of this hypothesis. Evidence of Yu and Kramer (1967, 1969) and Anderson and Reilly (1968) on the ion transport capacity of the stele, of Welch and Epstein (1968, 1969) on the kinetics of absorption in both concentration ranges and the impermeability of the plasmalemma, and of Läuchli and Epstein (1971) on inhibition of chloride translocation by bromide and by inhibitors of metabolism are especially damning to the series model of the disposition of the two types of ion transport mechanisms.

This matter has been discussed in some detail because of the far-reaching implications which the finding of appreciably permeable plasmalemmas of plant cells would have. A host of well-substantiated, important observations would be puzzling in the extreme if plasmalemmas, especially those of the cells of the root cortex, were found to be substantially "leaky." Many of these features have already been discussed; others are mentioned later in the book. Brouwer (1965) has rightly remarked, "Although this concept [of an impermeable outer membrane] emerged long ago, it seems necessary to defend it continuously."

Mechanisms: the Endodermis, Stele, and Stem.   Evidence reviewed under the previous heading raises serious questions about that part of the Crafts–Broyer hypothesis which postulates a "leaky" stele. This leads to the idea that the ions are initially absorbed as Crafts and Broyer (1938) postulated, by accumulation in the cytoplasm of cortical cells followed by symplastic movement, but that then the innermost layer of the cortex (the endodermis) or the stele tissue itself actively participates in the events that lead to their delivery into the conducting elements of the xylem. Various such proposals have been made, but the mechanism of secretion into the xylem has not been located or identified.

One idea is that the endodermis is the active agent which "pumps" ions into the stele and the xylem, but this is unlikely, except that perhaps the thin-walled "passage cells" often found in the endodermis opposite the xylem may be implicated. One would expect numerous and well-developed mitochondria in cells actively pumping ions, but such does not seem to be the case for the mitochondria of the endodermis (Bonnett, 1968). In older, woody roots the endodermis is presumed to have disappeared in the course of the development of secondary growth, and yet such roots have been shown to develop root pressure (O'Leary and Kramer, 1964). Since root pressure results from salt concentrations in the xylem in excess of their concentration in the medium, active pumping of salt into the xylem must have taken place in these roots lacking an endodermis.

Another solution to the problem of pumping ions into the xylem was proposed by Hylmö (1953). He drew attention to the events leading to the formation of vessels (see section 2). Vessels originate from cells arranged in long files. The transverse cell walls at the ends of individual vessel members eventually are perforated or virtually disappear. The cytoplasmic membranes would be ruptured there but would persist in the lower part of the vessel member. A test tube-shaped cytoplasmic system, open at the top, would therefore remain functional. It would absorb ions and deliver them through the upper end into the mature xylem elements above.

This means that the lumen of the xylem "pipe" would be the equivalent of the vacuole and the recipient of the ions absorbed by the functional vessel element. Instead of wall pressure there would develop a hydrostatic pressure due to osmotic influx of water resulting in "root pressure." A given vessel element would remain functional for a short time only. With the rupturing of the lower transverse wall it would become part of the dead vessel above, without cytoplasm, and the next lower vessel member would assume the active role of ion transport, in the same manner.

Although Hylmö produced little convincing evidence to support this ingenious scheme it was valuable in drawing attention to the possibility that cells in the stele might be actively involved in "pumping" ions into the xylem vessels. Anderson and House (1967) found that vessel members near the root tip of corn, *Zea mays,* contain cytoplasm bounded by membranes. With increasing distance from the tip, the percentage of such vessel members decreased, and at 10 cm from the tip the xylem was mature, i.e., devoid of cytoplasm. Absorption of ions and of water decreased progressively over the same distance. These are observations of a kind that would be expected on the basis of Hylmö's hypothesis, and therefore make the hypothesis more attractive than it was when first offered as a mere conjecture. Scott (1963) has also drawn attention to the presence of cytoplasm in the xylem elements of root tips and its probable role in the upward transport of solutes and water.

There are, however, other possibilities of ion accumulation in the vessels and root pressure which do not include active ion "pumping" by either the endodermis or the maturing vessel members themselves. Many observations lend themselves to the interpretation that stelar parenchyma cells excrete ions, by a process akin to absorption across the plasmalemma but in the opposite direction. The vessels would then be the recipients of these ions which would move upward, either by root pressure flow (Anderson *et al.,* 1970) or by transpirational flow. Such a secretion mechanism has been discussed repeatedly (Anderson and Reilly, 1968; Arisz, 1956; Läuchli *et al.,* 1971; Sutcliffe, 1962; Yu and Kramer, 1967, 1969). The hypothesis gains force from the findings of Läuchli *et al.* (1971), obtained by electron probe analysis, that stelar parenchyma cells of corn roots, and the parenchyma cells associated with the xylem in particular, contain potassium at much higher concentrations in their cytoplasm and vacuoles than do cortical cells. It is difficult to reconcile these results with the view that stelar cells play a merely passive role in ion transfer into the vessels.

Having reached the vessels, ions in solution in the lumen of the xylem elements are carried upwards in the transpiration stream or the root pressure flow. The walls of vessels and tracheids bear fixed negative charges on which cations exchange in accord with their concentrations and affinity for the charged sites. As a result, ions such as calcium and zinc, which are strongly adsorbed at negative sites because of their double positive valence, ascend not only by a simple mass flow mechanism but also in a series of exchanges. This is in effect a surface migration on the walls of xylem elements, the cations being alternately adsorbed and displaced at negative sites, as in a cation exchange column (Bell and Biddulph, 1963; Hewitt and Gardner, 1956). Heavy metal elements, especially iron, are kept from being immobilized on these exchange sites by moving in the form of organic complexes (Schmid and Gerloff, 1961; Tiffin, 1970).

The affinity of monovalent cations for exchange sites is very much less than that of divalent ones. Therefore, this chromatographic migration changes the composition of the solution ascending in the xylem (Epstein, 1962). However, a high degree of selectivity is unlikely to result from this, especially between ions of like charge, such as calcium and magnesium. The total concentration of the ascending solution is not changed by these cation exchange reactions because cations from the solution, in being adsorbed on wall exchange sites, displace equivalent amounts of cations into the solution.

All living cells of tissues not bathed by the solution representing the mineral substrate of the plant draw their mineral nutrients from the supply furnished by the conducting elements. This applies to cells of root tissue in the older, suberized portions of the root and to all the cells of the shoot, including those close to the xylem elements in stems, branches, petioles and fruit stalks. These cells withdraw ions from the solution in the xylem by the processes of active, selective transport discussed in Chapter 6. In doing so, they deplete the transpiration stream, and often profoundly change its ionic composition.

The cells of the upper (proximal) portion of the root and of the lower stem are those first encountered by the solution moving through the xylem, and their selective withdrawal of ions from the solution is therefore especially important in affecting its composition. Along with the selectivity of the transport processes responsible for the initial delivery of the ions into the xylem, this selective withdrawal from it determines in a major way the composition of the solution supplying the shoot.

The low concentration of sodium in the shoots of many species of plants is due in large measure to retention of the element by cells in the root and its withdrawal from the xylem by parenchyma cells flanking the xylem in the lower stem. For example, Wallace *et al.* (1965) kept bush bean plants, *Phaseolus vulgaris,* for 24 hours in a 0.1 mM aerated solution of NaCl, the Na being labeled with ^{22}Na. At the end of this time the roots contained 2.28 μmoles labeled sodium per gram dry weight, the stems, 0.04, and the leaves, 0.01. With higher concentrations of sodium in the solution a larger percentage reached the tops, presumably because the capacity of cells in the root and stem for its retention had been exceeded. LaHaye and Epstein (1971) extended these findings by use of much higher concentrations of salt (50 mM salt).

Jacoby (1965) forced a solution containing 1 mM NaCl through segments of bean stems 8 cm long for periods of up to 200 minutes. The exudate never contained sodium at a concentration higher than 1 per cent of that of the original solution. Metabolic inhibitors interfered with this retention, and as in Wallace's experiments, a smaller percentage of the sodium was retained at higher concentrations of sodium in the solution.

Rains (1969) separated pieces of stem of bean, *Phaseolus vulgaris,* and cotton, *Gossypium hirsutum,* into two fractions: the extracambial tissue (epidermis, cortex, and phloem), and the inner cylinder or xylary tissue (xylem and associated cells, and pith). The results of experiments on absorption of sodium and potassium by these fractions are shown in Figure 7-10. The outstanding feature is the marked accumulation of sodium in the stem of bean. The results show that the xylary tissue was responsible for this absorption. The findings suggest that it is through retention of sodium by cells associated with the xylem that so little sodium reaches the upper parts of the bean plant. This selective retention of sodium begins in the root. In excised roots of corn, Shone *et al.* (1969) found progressively less sodium to be exuded, the farther up the root the excision was made. As sodium moved up the vessels, more and more of it was withdrawn from them by the surrounding tissue.

The above discussion of discrimination involving sodium serves to stress an important point. When, as is clearly the case (Chapter 6, pp. 125–127), there is discrimination among ions at the cellular level, the possibility exists for still sharper discrimination when there is transport over long distances,

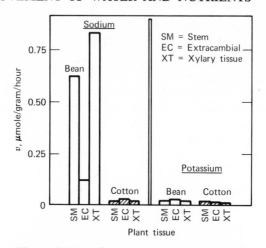

Figure 7-10. The rate, v, of absorption of sodium and potassium by stem sections and separated stem tissues of bean, *Phaseolus vulgaris*, and cotton, *Gossypium hirsutum*. Concentration of potassium and sodium in the medium, 0.1 mM, and of calcium, 0.5 mM. After Rains (1969).

because then many cells along the route participate in the process as the stream bearing ions in solution moves through the conduits. Evidently, in plants tending to exclude sodium from the shoots, the cells lining the conduits effectively deplete the xylem stream of sodium, while in plants which readily translocate sodium to the tops they do not. This was suggested long ago in the classical work of Collander (1941a, b). He grew plants of many species in solutions containing equal concentrations of potassium and sodium. Plants differed widely in the sodium contents of their shoots, which were exceedingly low in some plants and just as high as those of potassium in others. The differences among species in sodium contents were much more pronounced in shoots than in roots, emphasizing the importance of long-distance translocation in effecting this discrimination.

If, on the other hand, ions of two elements are so similar in their chemical affinities that the ion transport mechanisms at the cellular level fail to discriminate between them effectively, then the expectation would be that the elements would also tend to resemble each other in their long-distance distribution within the plant body. This expectation is borne out by the evidence, as first shown by Collander (1941a, b). He found that the distribution of potassium and rubidium in the plants he used was such as to suggest that in regard to their transport "the plants are in a certain sense unable to 'distinguish' between K and Rb and also between Ca and Sr in the same manner as they are unable to distinguish between two different isotopes of a given element." A wealth of more recent evidence attests to the soundness of this conclusion. Figure 7-11 shows the results of an experiment with sugar cane plants, *Saccharum officinarum*, which were grown in a complete nutrient solution to which radioactive rubidium, ^{86}Rb, had been added. In plant part after plant part, the distribution of rubidium closely mimics that of potassium.

However, rubidium (and strontium) are rare elements not normally of physiological significance. Plants transport the essential and even some other elements in a highly selective manner, as described above, and as a result the solution eventually reaching the ultimate terminals of the xylem in the leaves may be very different in composition from that which entered the xylem in the absorbing region of active rootlets. In the leaf, mesophyll cells withdraw minerals from the solution delivered by the xylem. The processes of absorption by these cells seem to be identical with those in other plant tissues, except that in the light the energy supply may come directly from photosynthesis instead of via mitochondrial respiration (see Chapter 9, pp. 242–247). As a result of this withdrawal, the solution exuded by guttating plants may be more dilute than that in the xylem of the petiole supplying the leaf; in passing through the leaf, salt is absorbed from the xylem sap (Klepper and Kaufmann, 1966).

A phenomenon still awaiting an adequate explanation is that of periodicity in translocation. The amounts of rubidium and phosphate translocated to the tops of bean plants, *Phaseolus vulgaris,* were maximal at about noon and minimal near midnight, in the experiments of Hanson and Biddulph (1953). There was some evidence of this periodicity even when the normal day-and-night cycle was abolished by keeping the plants dark; in other words, the periodicity was to a degree autonomous, as had been noted before (Grossenbacher, 1939; Skoog *et al.,* 1938).

With entire plants, autonomous periodic changes in water relations might be responsible for this phenomenon (Barrs and Klepper, 1968), although that would only shift the problem into the domain of water relations, without solving it. However, it is in any event likely that the primary events governing this response involve ion transport rather than water relations, as shown by the fact that such autonomic periodicity clearly shows up in experiments with detopped plants (Grossenbacher, 1939; Skoog *et al.,* 1938; Vaadia, 1960; van Andel, 1953). In Minshall's (1968) experiments with detopped tomato plants a rise in the osmotic pressure of the exudate preceded the increase in the rate of exudation, after potassium

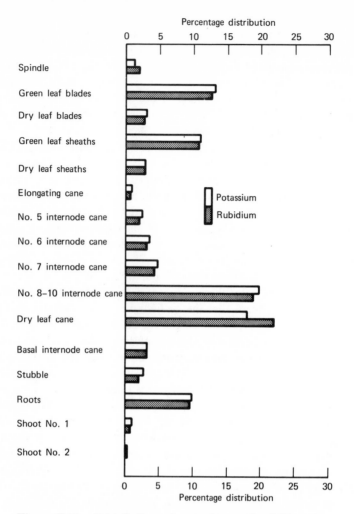

Figure 7-11. Distribution of potassium and rubidium in different tissues of sugar cane, *Saccharum officinarum*, which had been grown in a complete nutrient solution to which radioactively labeled rubidium (^{86}Rb) had been added. After Burr and Tanimoto (1955).

nitrate was added to the medium—evidence of the primacy of solute transport in the phenomenon. This, however, brings us no nearer to a solution of this refractory problem.

BEYOND THE LEAF

The mesophyll cells of leaves are the farthest receiving stations of mineral ions initially absorbed by roots. Ions within these cells, and those in the leaf xylem and cell wall spaces, can only take one of two possible routes. They may leave the plant, or they may be retranslocated within it. Retranslocation will be discussed in the next chapter, which deals with phloem transport. The mechanisms by which ions in the leaf are separated from the plant are the following: leaching, guttation, salt excretion, and leaf fall.

Leaching.   The mesophyll cells of the leaf are bathed by a solution of salts, much in the same way as the cortical cells of the root are bathed by a solution—the solution in the "outer" space which fills the cell wall spaces and forms a film lining the intercellular spaces. But while the "outer" space solution in the root cortex is in rapid diffusional communication with the soil solution of which it is an extension, the "outer" space solution in the leaf abuts on air. As long as this is the situation the ions within this solution are subject to withdrawal by living cells, but not to diffusional loss like those of the root cortex.

This situation changes, however, when it rains. The solution in the cell walls is now in communication with an external, constantly renewed "wash" solution. True, the communication is far from close because of the cuticle and waxy excretions on the leaf. Nevertheless, the cuticle is not entirely impermeable, and it has cracks and openings. As a result, ions may leach out of the leaf in rain or mist, or when sprinkler irrigation is used (Tukey, 1966, 1970; Tukey et al., 1965; Tukey and Tukey, 1962).

This explanation of leaching assumes that it is a passive, diffusional process and that the ions so removed come from the cell wall spaces of the leaf, not from the "inner" spaces of mesophyll cells. Evidence for that is the continuous functioning of leaves in prolonged rain which would not be possible if intracellular nutrients such as potassium were leached out. Tukey and Morgan (1963) found that freezing injury caused a large loss of calcium and potassium from leaves of *Chrysanthemum morifolium*. This shows that the injury had impaired the normal effectiveness of the leaf cell plasmalemmas as diffusion barriers. Other damaging agencies have similar effects, and so does senescence. The findings of Tukey *et al.* (1965) were also to the effect that the fraction leached is mainly the extracellular (cell wall) fraction of ions in the leaf, not the "inner" space fraction. Experiments by Rains (1968) with corn leaf tissue, similar to the one done with roots which is shown in Figure 6-7, support this conclusion.

Leaching of nutrients is of great importance in the mineral economy of plants, especially in the tropics with their repeated, prolonged, and violent downpours. It is furthermore a matter for concern wherever advanced agricultural methods include spraying, sprinkler irrigation, and use of artificial mists.

Guttation.   This subject has already been mentioned (see the heading, Mechanisms, in section 4, p. 164 of this chapter). The pores (hydathodes) through which the guttation liquid is exuded are the openings of modified stomates but unlike stomates they do not close. Guttation is observed only under conditions of high root pressure and low transpiration. The guttation fluid serves to remove some salt from the extracellular (cell wall or "outer" space) solution in the leaf, a loss which sometimes may have adverse effects on the mineral nutrition of the plant (Ivanoff, 1963). Stocking (1956) and Kramer (1969) have useful discussions.

Excretion of Salt.   Many plants normally exposed to high concentrations of salt, such as those of the sea coasts and saline deserts, possess structures and mechanisms for the excretion of salt onto the surface of the shoot, principally the leaf, from where the salt is finally removed through the action of wind and water. The specialized structures are salt glands. These are common in species of two families, the *Plumbaginaceae* and *Frankeniaceae,* and in some genera not belonging to these families: *Avicennia, Aegialitis, Spartina, Tamarix,* and some others. Leaves of such plants are often covered with a glistening bloom of salt crystals. The glands are numerous; for example, Atkinson *et al.* (1967) counted about 900 of them per square centimeter on the upper surface of leaves of the mangrove, *Aegialitis annulata.*

The structure of salt glands varies in different species. They consist of two or more cells at the surface of the leaf which together perform the various steps involved in secretion: withdrawal of salt from the leaf tissue in which the gland is embedded, its transfer to those cells of the gland from which the salt is eventually excreted, and the act of excretion itself. Figure 7-12 shows an electron micrograph of a salt gland of *Limonium vulgare.* Levering and Thomson (1971) have given a description, with electron micrographs, of the two-celled salt gland of the halophytic grass, *Spartina foliosa.*

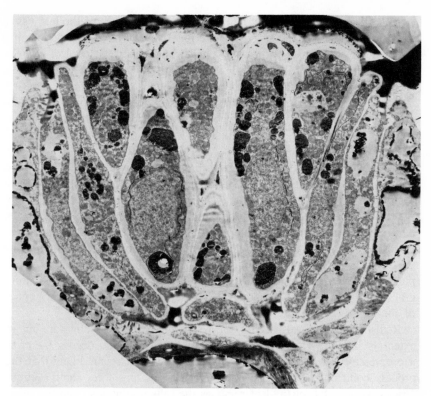

Figure 7-12. Electron micrograph of a section of a salt gland of *Limonium vulgare* cut at a right angle to the surface of the leaf (top). The two large cells to the left and right of the center of the picture are secreting cells, the other cells are accessory gland cells. Magnification, ×3,900. From Ziegler and Lüttge (1966).

Salt excretion by salt glands is in the nature of metabolically active transport. The concentration of salt in the excreted solution may be higher than that of the leaf sap, and antimetabolic agents inhibit excretion (Atkinson *et al.,* 1967; Helder, 1956). Like other active transport processes, excretion by salt glands shows selectivity. Thus, the solution excreted by salt glands of *Aegialitis annulata* had a ratio of sodium to potassium of about 13/1, higher than that in the bulk of the leaf tissue (3/1) and in the xylem sap (8/1) (Atkinson *et al.,* 1967). In effect, exudate was in the main a solution of NaCl.

The same authors found that when radioactively labeled chloride was supplied to the leaf through the petiole, the specific activity of the chloride excreted was higher than that of the chloride in the bulk of the leaf tissue. Evidently, much of the chloride entering the leaf from the petiole had reached the salt glands without passing through the main pool of chloride in the leaf (which probably is in the vacuoles of mesophyll cells). These extremely interesting glands are belatedly getting close scientific scrutiny. It is to be expected that as modern concepts of active transport and up-to-date techniques are brought to bear on their activities the findings obtained will add much to our understanding of ion transport by plant cells.

Leaf Fall. In animals, there is a through-put of liquid, water, and solutes. In plants, only guttation and salt excretion represent such a through-put of water without a change of phase, and these processes play no important role in the water and mineral economy of most plants. The major portion of the water which enters a terrestrial higher plant leaves it as water vapor. As discussed above, the solution from which the water evaporates has been depleted of salts in its ascent through the plant, and even within the leaf, but it is by no means completely demineralized. As the water evaporates salt is left behind in the "outer" space of the leaf.

Redistribution of salt occurs, as described in the next chapter, but its extent is limited. Throughout the growing season there occurs therefore a progressive mineralization of transpiring leaves. It seems likely that this process contributes to senescence of leaves and promotes their eventual abscission. The fall of the leaf removes the salts it contains from the plant. Leaf fall is, among other things, a mechanism of mineral waste disposal.

REFERENCES

Anderson, W. P., D. P. Aikman and A. Meiri. 1970. Excised root exudation—a standing-gradient osmotic flow. Proc. Roy. Soc. London B 174:445–458.

Anderson, W. P. and C. R. House. 1967. A correlation between structure and function in the root of *Zea mays*. J. Expt. Bot. 18:544–555.

Anderson, W. P. and E. J. Reilly. 1968. A study of the exudation of excised maize roots after removal of the epidermis and outer cortex. J. Expt. Bot. 19:19–30.

Arisz, W. H. 1956. Significance of the symplasm theory for transport across the root. Protoplasma 46:1–62.

Arisz, W. H., R. J. Helder and R. van Nie. 1951. Analysis of the exudation process in tomato plants. J. Expt. Bot. 2:257–297.

Askenasy, E. 1895. Ueber das Saftsteigen. Verhandl. d. Heidelb. Naturhist.-Med. Vereins, n.s. 5:325–345.

Atkinson, M. R., G. P. Findlay, A. B. Hope, M. G. Pitman, H. D. W. Saddler and K. R. West. 1967. Salt regulation in the mangroves *Rhizophora mucronata* Lam. and *Aegialitis annulata* R. Br. Austral. J. Biol. Sci. 20:589–599.

Barrs, H. D. and B. Klepper. 1968. Cyclic variations in plant properties under constant environmental conditions. Physiol. Plantarum 21:711–730.

Bell, C. W. and O. Biddulph. 1963. Translocation of calcium. Exchange versus mass flow. Plant Physiol. 38:610–614.

Biddulph, O. 1951. The translocation of minerals in plants. In: Mineral Nutrition of Plants. E. Truog, ed. The University of Wisconsin Press, Madison. Pp. 261–275.

Biddulph, S. F. 1967. A microautoradiographic study of Ca^{45} and S^{35} distribution in the intact bean root. Planta 74:350–367.

Bonnett, H. T., Jr. 1968. The root endodermis: fine structure and function. J. Cell Biol. 37:199–205.

Briggs, G. E. 1967. Movement of Water in Plants. Blackwell Scientific Publications, Ltd., Oxford.

Briggs, L. J. and H. L. Shantz. 1914. Relative water requirement of plants. J. Agric. Res. 3:1–63.

Brouwer, R. 1954. The regulating influence of transpiration and suction tension on the water and salt uptake by the roots of intact *Vicia faba* plants. Acta Bot. Neerl. 3:264–312.

Brouwer, R. 1959. Diffusible and exchangeable rubidium ions in pea roots. Acta Bot. Neerl. 8:68–76.

Brouwer, R. 1965. Ion absorption and transport in plants. Ann. Rev. Plant Physiol. 16:241–266.

Broyer, T. C. 1950. Further observations on the absorption and translocation of inorganic solutes using radioactive isotopes with plants. Plant Physiol. 25:367–376.

Burr, G. O. and T. Tanimoto. 1955. Absorption and distribution of nutrients in sugar cane. Part II: Potassium. Hawaiian Planters' Rec. 55:11–13.

Collander, R. 1941a. Selective absorption of cations by higher plants. Plant Physiol. 16:691–720.

Collander, R. 1941b. The distribution of different cations between root and shoot. Acta Bot. Fennica 29:3–12.

Crafts, A. S. and T. C. Broyer. 1938. Migration of salts and water into xylem of the roots of higher plants. Am. J. Bot. 25:529–535.

Davis, T. A. 1961. High root-pressures in palms. Nature 192:277–278.

Dixon, H. H. 1914. Transpiration and the Ascent of Sap in Plants. Macmillan and Company, Ltd., London.

Dixon, H. H. and J. Joly. 1895. On the ascent of sap. Proc. Roy. Soc. London 57:3–5.

Dumbroff, E. B. and D. R. Peirson. 1971. Probable sites for passive movement of ions across the endodermis. Can. J. Bot. 49:35–38.

Epstein, E. 1962. Mutual effects of ions in their absorption by plants. Agrochimica 6:293–322.

Esau, K. 1960. Anatomy of Seed Plants. John Wiley and Sons, Inc., New York and London.

Esau, K. 1965a. Plant Anatomy. 2nd ed. John Wiley and Sons, Inc., New York.

Esau, K. 1965b. Vascular Differentiation in Plants. Holt, Rinehart and Winston, New York.

Esau, K. 1967. Minor veins in *Beta* leaves: structure related to function. Proc. Am. Phil. Soc. 11:219–233.

Fahn, A. 1967. Plant Anatomy. Pergamon Press, Oxford.

Grossenbacher, K. A. 1939. Autonomic cycle of rate of exudation of plants. Am. J. Bot. 26:107–109.

Gunning, B. E. S., J. S. Pate and L. W. Green. 1970. Transfer cells in the vascular system of stems: taxonomy, association with nodes, and structure. Protoplasma 71:147–171.

Hanson, J. B. and O. Biddulph. 1953. The diurnal variation in the translocation of minerals across bean roots. Plant Physiol. 28:356–370.

Helder, R. J. 1956. The loss of substances by cells and tissues (salt glands). In: Encyclopedia of Plant Physiology. W. Ruhland, ed. Springer-Verlag, Berlin. Vol. 2, pp. 468–488.

Helder, R. J. 1958. Studies on the absorption, distribution and release of labelled rubidium ions in young intact barley plants. Acta Bot. Neerl. 7:235–249.

Helder, R. J. 1964. The absorption of labelled chloride and bromide ions by young intact barley plants. Acta Bot. Neerl. 13:488–506.

Helder, R. J. 1967. Translocation in *Vallisneria spiralis*. In: Encyclopedia of Plant Physiology. W. Ruhland, ed. Springer-Verlag, Berlin. Vol. 13, pp. 20–43.

Hewitt, W. B. and M. E. Gardner. 1956. Some studies of the adsorption of zinc sulfate in Thompson Seedless grape canes. Plant Physiol. 31:393–399.

Hoagland, D. R. 1944. Lectures on the Inorganic Nutrition of Plants. Chronica Botanica Company, Waltham.

Hodges, T. K. and Y. Vaadia. 1964a. Uptake and transport of radio-chloride and tritiated water by various zones of onion roots of different chloride status. Plant Physiol. 39:104–108.

Hodges, T. K. and Y. Vaadia. 1964b. Chloride uptake and transport in roots of different salt status. Plant Physiol. 39:109–114.

House, C. R. and N. Findlay. 1966. Water transport in isolated maize roots. J. Expt. Bot. 17:344–354.

Hylmö, B. 1953. Transpiration and ion absorption. Physiol. Plantarum 6:333–405.

Ivanoff, S. S. 1963. Guttation injuries of plants. Bot. Rev. 29:202–229.

Jacoby, B. 1965. Sodium retention in excised bean stems. Physiol. Plantarum 18:730–739.

Klepper, B. 1967. Effects of osmotic pressure on exudation from corn roots. Austral. J. Biol. Sci. 20:723–735.

Klepper, B. and M. R. Kaufmann. 1966. Removal of salt from xylem sap by leaves and stems of guttating plants. Plant Physiol. 41:1743–1747.

Kolattukudy, P. E. 1970. Biosynthesis of cuticular lipids. Ann. Rev. Plant Physiol. 21:163–192.

Kozlowski, T. T. 1964. Water Metabolism in Plants. Harper and Row, New York.

Kozlowski, T. T., ed. 1968a. Water Deficits and Plant Growth. I. Development, Control, and Measurement. Academic Press, New York and London.

Kozlowski, T. T., ed. 1968b. Water Deficits and Plant Growth. II. Plant Water Consumption and Response. Academic Press, New York and London.

Kramer, P. J. 1969. Plant and Soil Water Relationships: A Modern Synthesis. McGraw-Hill Book Company, Inc., New York.

Krichbaum, R., U. Lüttge and J. Weigl. 1967. Mikroautoradiographische Untersuchung der Auswaschung des "anscheinend freien Raumes" von Maiswurzeln. Ber. Deutsch. Bot. Ges. 80:167–176.

LaHaye, P. A. and E. Epstein. 1971. Calcium and salt toleration by bean plants. Physiol. Plantarum. In press.

Laties, G. G. 1969. Dual mechanisms of salt uptake in relation to compartmentation and long-distance transport. Ann. Rev. Plant Physiol. 20:89–116.

Laties, G. G. and K. Budd. 1964. The development of differential permeability in isolated steles of corn roots. Proc. Nat. Acad. Sci. 52:462–469.

Läuchli, A. 1967. Untersuchungen über Verteilung und Transport von Ionen in Pflanzengeweben mit der Röntgen-Mikrosonde. I. Versuche an vegetativen Organen von *Zea mays*. Planta 75:185–206.

Läuchli, A. 1968. Untersuchung des Stofftransports in der Pflanze mit der Röntgen-Mikrosonde. Vorträge aus dem Gesamtgebiet der Botanik N.F. 2:58–65.

Läuchli, A. 1972. Electron probe analysis. In: Methods of Microautoradiography and Electron Probe Analysis. U. Lüttge, ed. Springer, Berlin–Heidelberg–New York. In press.

Läuchli, A. and E. Epstein. 1971. Lateral transport of ions into the xylem of corn roots. I. Kinetics and energetics. Plant Physiol. 48:111–117.

Läuchli, A., A. R. Spurr and E. Epstein. 1971. Lateral transport of ions into the xylem of corn roots. II. Evaluation of a stelar pump. Plant Physiol. 48:118–124.

Levering, C. A. and W. W. Thomson. 1971. The ultrastructure of the salt gland of *Spartina foliosa*. Planta 97:183–196.

Lüttge, U. and G. G. Laties. 1966. Dual mechanisms of ion absorption in relation to long distance transport in plants. Plant Physiol. 41:1531–1539.

Lüttge, U. and G. G. Laties. 1967. Selective inhibition of absorption and long distance transport in relation to the dual mechanisms of ion absorption in maize seedlings. Plant Physiol. 42:181–185.

Minshall, W. H. 1968. Effects of nitrogenous materials on translocation and stump exudation in root systems of tomato. Can. J. Bot. 46:363–376.

O'Leary, J. W. and P. J. Kramer. 1964. Root pressure in conifers. Science 145:284–285.

Peirson, D. R. and E. B. Dumbroff. 1969. Demonstration of a complete Casparian strip in *Avena* and *Ipomoea* by a fluorescent staining technique. Can. J. Bot. 47:1869–1871.

Rains, D. W. 1968. Kinetics and energetics of light-enhanced potassium absorption by corn leaf tissue. Plant Physiol. 43:394–400.

Rains, D. W. 1969. Cation absorption by slices of stem tissue of bean and cotton. Experientia 25:215–216.

Renner, O. 1915. Theoretisches und Experimentelles zur Kohäsionstheorie der Wasserbewegung. Jahrb. wiss. Bot. 56:617–667.

Rufelt, H. 1956. Influence of the root pressure on the transpiration of wheat plants. Physiol. Plantarum 9:154–164.

Russell, R. S. and D. A. Barber. 1960. The relationship between salt uptake and the absorption of water by intact plants. Ann. Rev. Plant Physiol. 11:127–140.

Rutter, A. J. and F. H. Whitehead, eds. 1963. The Water Relations of Plants. A Symposium of the British Ecological Society. John Wiley and Sons, Inc., New York.

Sabinin, D. A. 1925. On the root system as an osmotic apparatus. Bull. Inst. Rech. Biol. Univ. Perm. Vol. 4. Suppl. 2. Pp. 1–136. (English Summary, pp. 129–136).

Schmid, W. E. and G. C. Gerloff. 1961. A naturally occurring chelate of iron in xylem exudate. Plant Physiol. 36:226–231.

Scholander, P. F., H. T. Hammel, E. D. Bradstreet and E. A. Hemmingsen. 1965. Sap pressure in vascular plants. Science 148:339–346.

Scott, F. M. 1963. Root hair zone of soil-grown roots. Nature 199:1009–1010.

Shone, M. G. T. 1969. Origins of the electrical potential difference between the xylem sap of maize roots and the external solution. J. Expt. Bot. 20:698–716.

Shone, M. G. T., D. T. Clarkson and J. Sanderson. 1969. The absorption and translocation of sodium by maize seedlings. Planta 86:301–314.

Skoog, F., T. C. Broyer and K. A. Grossenbacher. 1938. Effect of auxin on rates, periodicity, and osmotic relations in exudation. Am. J. Bot. 25:749–759.

Slatyer, R. O. 1967. Plant–Water Relationships. Academic Press, London and New York.

Slavik, B. 1969. Water Deficit in Plants as a Physiological Factor. W. Junk N. V., Publishers, The Hague.

Smith, R. C. and E. Epstein. 1964a. Ion absorption by shoot tissue: technique and first findings with excised leaf tissue of corn. Plant Physiol. 39:338–341.

Smith, R. C. and E. Epstein. 1964b. Ion absorption by shoot tissue: kinetics of potassium and rubidium absorption by corn leaf tissue. Plant Physiol. 39:992–996.

Stocking, C. R. 1956. Guttation and bleeding. In: Encyclopedia of Plant Physiology. W. Ruhland, ed. Springer-Verlag, Berlin. Vol. 3, pp. 489–502.

Stout, P. R. and D. R. Hoagland. 1939. Upward and lateral movement of salt in certain plants as indicated by radioactive isotopes of potassium, sodium, and phosphorus absorbed by roots. Am. J. Bot. 26:320–324.

Sutcliffe, J. F. 1962. Mineral Salts Absorption in Plants. Pergamon Press, New York.

Tiffin, L. O. 1970. Translocation of iron citrate and phosphorus in xylem exudate of soybean. Plant Physiol. 45:280–283.

Tukey, H. B., Jr. 1966. Leaching of metabolites from above-ground plant parts and its implications. Bull. Torrey Bot. Club 93:385–401.

Tukey, H. B., Jr. 1970. The leaching of substances from plants. Ann. Rev. Plant Physiol. 21:305–324.

Tukey, H. B., Jr., R. A. Mecklenburg and J. V. Morgan. 1965. A mechanism for the leaching of metabolites from foliage. In: Isotopes and Radiation in Soil–Plant Nutrition Studies. International Atomic Energy Agency,Vienna. Pp. 371–385.

Tukey, H. B., Jr. and J. V. Morgan. 1963. Injury to foliage and its effect upon the leaching of nutrients from above-ground plant parts. Physiol. Plantarum 16:557–564.

Tukey, H. B., Jr. and H. B. Tukey, Sr. 1962. The loss of organic and inorganic materials by leaching from leaves and other above-ground plant parts. In: Radioisotopes in Soil–Plant Nutrition Studies. International Atomic Energy Agency, Vienna. Pp. 289–302.

Tyree, M. T. 1970. The symplast concept—a general theory of symplastic transport according to the thermodynamics of irreversible processes. J. Theor. Biol. 26:181–214.

Vaadia, Y. 1960. Autonomic diurnal fluctuations in rate of exudation and root pressure of decapitated sunflower plants. Physiol. Plantarum 13:701–717.

van Andel, O. M. 1953. The influence of salts on the exudation of tomato plants. Acta Bot. Neerl. 2:445–521.

Wallace, A., N. Hemaidan and S. M. Sufi. 1965. Sodium translocation in bush beans. Soil Sci. 100:331–334.

Weatherley, P. E. 1963. The pathway of water movement across the root cortex and leaf mesophyll of transpiring plants. In: The Water Relations of Plants. A. J. Rutter and F. H. Whitehead, eds. John Wiley and Sons, Inc., New York. Pp. 85–100.

Weigl, J. and U. Lüttge. 1962. Mikroautoradiographische Untersuchungen über die Aufnahme von $^{35}SO_4^{--}$ durch Wurzeln von Zea mays L. Die Funktion der primären Endodermis. Planta 59:15–28.

Welch, R. M. 1969. The Plasmalemma as the Seat of the Dual Ion Carrier Mechanisms of Plant Cells. M.S. Thesis, University of California, Davis.

Welch, R. M. and E. Epstein. 1968. The dual mechanisms of alkali cation absorption by plant cells: their parallel operation across the plasmalemma. Proc. Nat. Acad. Sci. 61:447–453.

Welch, R. M. and E. Epstein. 1969. The plasmalemma: seat of the type 2 mechanisms of ion absorption. Plant Physiol. 44:301–304.

White, P. R. 1938. "Root pressure"—an unappreciated force in sap movement. Am. J. Bot. 25:223–227.

Wray, F. J. and J. A. Richardson. 1964. Paths of water transport in higher plants. Nature 202:415–416.

Wylie, R. B. 1939. Relations between tissue organization and vein distribution in dicotyledon leaves. Am. J. Bot. 26:219–255.

Yu, G. H. and P. J. Kramer. 1967. Radial salt transport in corn roots. Plant Physiol. 42:985–990.

Yu, G. H. and P. J. Kramer. 1969. Radial transport of ions in roots. Plant Physiol. 44:1095–1100.

Zelitch, I., ed. 1963. Stomata and Water Relations in Plants. The Connecticut Agricultural Experiment Station, New Haven. Bulletin 664.

Ziegler, H. and U. Lüttge. 1966. Die Salzdrüsen von *Limonium vulgare*. I. Mitteilung. Die Feinstruktur. Planta 70:193–206.

8

THE DOWNWARD
MOVEMENT OF FOODS
AND NUTRIENTS

THE BASIC DILEMMA

Just as the upward movement of water and solutes was found to be effected through specialized tube-like structures, the tracheids and vessels, so the downward movement of substances from the leaves is via files of elongated pipe-like cells, called sieve tube elements or sieve elements in angiosperms and sieve cells in gymnosperms. These cells were discovered in the bark of woody species well over a hundred years ago. It had been known even much earlier that when a stem or twig is "ringed," that is, when a ring of bark is removed, the part above the ring would enlarge and that below it would stop growing.

The conclusion was that the bark serves as an avenue of transport of food from the upper to the lower parts of the plant. Interrupting the flow of food by ringing causes a buildup of growth promoting substances above the ring and a deficiency of them below it, resulting in the unequal growth that is observed. When pipe-like cells (the sieve tube elements) were discovered in the bark, it was logical to conclude that they were involved in this downward transport.

Later evidence has supported this view. The experiments of Mason and Maskell (1928a) in Trinidad with cotton plants, *Gossypium barbadense,* were outstanding. They showed by chemical analysis that ringing caused a buildup of sugars, mainly sucrose, in the bark and the wood above the ring, and a depletion below. Since these effects were produced promptly after removal of a ring of bark it was evident that the bark rather than the wood served in the downward movement of sugar. The parallel response in the sugar content of the wood showed that there is lateral movement between bark (phloem) and wood (xylem). Figure 8-1 shows the results of one of their experiments.

Recently, Trip and Gorham (1967) in Canada and Schmitz and Willenbrink (1968) and Schmitz (1970) in Germany produced direct

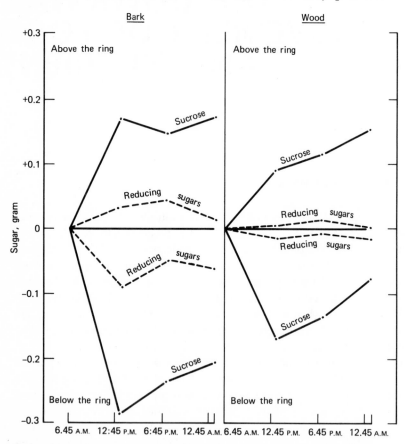

Figure 8-1. The effect of ringing stems of cotton, *Gossypium barbadense,* on the sugar content of wood and bark above and below the ring. The results are expressed as the difference in the weight of sugar between ringed and normal plants in the stem sections, 6.5 inches long (16.5 centimeters), above and below the ring. After Mason and Maskell (1928a).

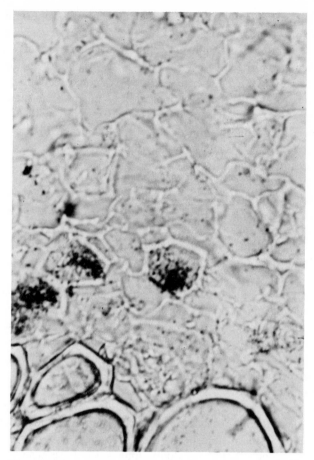

Figure 8-2. Cross section of leaf vein of *Cucurbita pepo* after application of ^{3}H-labeled glucose above it. The focus is on the photographic emulsion overlying the section of tissue, showing the exposure (blackening) due to the presence of the radioactive label. It coincides with the location of sieve tubes, shown slightly out of focus beneath the emulsion. From Schmitz and Willenbrink (1968).

evidence that specifically the sieve elements are the routes of phloem transport. Both teams applied to leaves of squash, *Cucurbita* sp., or cucumber, *Cucumis* sp., an aqueous solution of glucose or sucrose labeled with radioactive hydrogen, ^{3}H (tritium). After a period no longer than four hours in any of their experiments they examined sections of the conducting tissues below the point of application by means of autoradiography. The blackening of the film caused by the radiation from the radioactive label coincided with the location of sieve tube elements. Figure 8-2 shows the cross-section of a vascular bundle. Label is shown to be confined to the sieve tubes. Longitudinal sections are shown in Figure 8-3.

As we have seen, mature xylem vessels consist of cells which have died and lost their cell contents: they make up long, unobstructed tubes evidently well suited to the long-distance transport of a solution. Not so the sieve elements that make up the long files, called sieve tubes, in the phloem. The sieve elements, and the sieve cells of gymnosperms, are living cells containing protoplasts. Furthermore, instead of having large perforations in their end walls, or end walls that have nearly disappeared, the sieve elements have much smaller pores through which cytoplasmic connecting strands extend.

Thus, while vessels are obviously well adapted to serve as pipes for the flow of a solution with a minimum of obstruction, sieve tubes are not equally plausible as conduits for a flowing solution. Their cell contents, and the narrow pores connecting one with the other, would seem to be obstacles to any free flow. As a result, these features have been obstacles to the universal acceptance of the principal hypothesis of long-distance transport in the phloem, because this hypothesis envisions just such a flow.

Were it not for these objections, the "pressure flow" or "mass flow" hypothesis of Münch would no doubt long since have won general acceptance. In this chapter we shall examine first the structure of the sieve tubes and some parameters of the movement of solutes through them, next discuss the Münch hypothesis, and then ask ourselves: can the hypothesis be reconciled with the observed structure, or is it necessary, in view of all the evidence, to look for other explanations? "A mass flow obviously does occur . . ." (Zimmermann, 1957); ". . . it seems likely that there is no mass flow of solutions" (Milthorpe and Moorby, 1969). Let us examine the matter.

2. AGAIN: PORES, PIPES, AND PATHWAYS

The Cell Walls. The individual sieve tube elements or sieve elements are elongated tube-like cells. Where these cells lying in file abut

Figure 8-3. Longitudinal sections through phloem of *Cucumis sativus* after application of ³H-labeled glucose above it. A, focus on the sieve tube, showing its structure; B, focus on the photographic emulsion overlying the section of tissue. From Schmitz (1970).

end to end, the connecting walls are characterized by "sieve areas." These are portions of the walls which have pores through which connecting strands of cytoplasm extend. Sieve areas also occur along the length of the tubular walls of sieve tube elements, thus connecting cells lying side by side and presumably permitting lateral movement. The most highly developed sieve areas are those of the end walls of sieve tube elements of angiosperms. The sieve areas are often located in groups on parts of the walls called sieve plates. The pores and the cytoplasmic connecting

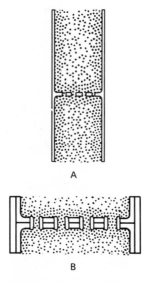

A

B

Figure 8-4. Structure of the sieve plate between two contiguous sieve tube elements, shown as in longitudinal section. A, general view; B, detail, showing sleeves of callose around pores. After Esau (1969).

strands passing through them are most prominent here, and appear most highly adapted to the function of translocation. Figure 8-4 shows portions of two sieve tube elements and the sieve plate between them.

A characteristic feature in pictures of the pores between sieve elements is the sleeve of callose with which they are lined (Figure 8-4). Callose, a polysaccharide, is seen to be deposited around the edge of the pore opening, narrowing the size of the pore. Sometimes it completely closes the pore, and may even form a massive covering over the entire sieve plate.

In this extreme situation, it blocks all movement across the sieve plate, but even the annular deposition around the pore would seem, on the face of it, to constitute a considerable impediment to movement of substances from one sieve element into the next.

However, a consideration of research and findings on callose drives home a lesson which is important not only in connection with phloem transport but in plant physiological research generally. This has to do with the risk of experimental artifacts and misinterpretations that may be drawn from them. Callose is widely distributed in plants and its association with sieve areas is not in doubt. However, recent research, notably in Currier's laboratory at Davis, has revealed that various kinds of agents, including foreign chemicals, borate, ultrasound, and heat, cause abnormally great depositions of callose, resulting in partial or complete blocking of the pores (McNairn and Currier, 1968).

Now the initial procedures of preparing phloem tissue for light or electron microscopy are themselves injurious to the tissue and quickly induce extra callose formation. The final picture, therefore, reveals more callose than existed in the tissue before the metabolic shock initiated by the experimenter (Cronshaw and Anderson, 1969). Much less callose is observed after rapid freezing of the tissue because this method of killing does not allow enough time for extra formation of callose induced by injury. On this evidence, then, the pores do not constitute as serious an obstacle to the flow of a solution as was thought at the time when callose was considered to block much of the cross-sectional area of the openings.

The Cell Contents.   The protoplast of a sieve element or sieve cell undergoes a characteristic development as the cell matures. The outstanding change in most species is the disorganization and eventual loss of the nucleus. Along with this, profound alterations of the cytoplasm occur: it becomes less dense and the tonoplast disappears as a discrete entity. This does not mean that the distinction between the cytoplasmic and the vacuolar regions disappears altogether. Rather, in the mature sieve tube a "parietal layer" of cytoplasm next to the cell wall grades imperceptibly into a less dense, more fluid inner region, the "lumen." The mitochondria occur mainly in the parietal layer; however, they are relatively few in number and simpler in their internal structure than are mitochondria of parenchyma cells (Esau and Cheadle, 1962) and especially, of companion cells.

In angiosperms, sieve elements occur in close association with companion cells. Each companion cell is derived from the same meristematic cell which gave rise to the sieve element which it flanks. Unlike the sieve elements, the companion cells retain their nuclei and possess large numbers of well developed mitochondria. The existence of this extensive cytoplasmic apparatus, together with the presence of numerous, branched plasmodesmata between the companion cell and the sieve element, lends force to the idea that the companion cell plays an important role in maintaining or promoting the conducting function of the sieve element (Esau and Cheadle, 1962; Shih and Currier, 1969). Plasmodesmata are much less common between sieve elements and ordinary phloem parenchyma cells.

The fine structure of the sieve tube protoplast has been much investigated, with both the light and the electron microscope. These investigations have been dogged by difficulties of artifact and uncertainties of interpretation. On the one hand, various structures such as strands and "slime" (or "P-protein") have often been observed in the sieve tube elements and specifically, in or on the connecting pores, and it has therefore been held likely that "there is no mass flow of solutions" (Milthorpe and Moorby, 1969). On the other hand, strands or filaments extending through the sieve plate pores have been thought to serve as tubes through which the solution bearing assimilate solute might flow longitudinally through files of sieve elements. And finally the existence and disposition of various such cytoplasmic features as observed by some investigators have been doubted by others, usually on the grounds that the observations were due to artifacts created in the course of killing, dehydrating, and embedding the tissue for microscopy. Differences among plants, and among sieve elements of the same plant at different stages during their ontogeny, add further complications. We can do no more here than briefly review the evidence and the interpretations that have been put forward. For more detail, reference is made to the excellent recent reviews by Biddulph (1969), Weatherley and Johnson (1968), and Zimmermann (1969b). Concerning the structure of the phloem, far and away the most comprehensive and authoritative current account of it is the book by Esau (1969). She also includes a discussion of the structure of phloem in relation to its function.

Because of the unusual nature of sieve element protoplasts it was thought by early investigators that they lacked a normal plasmalemma. However, Currier *et al.* (1955) showed that sieve elements can be plasmolyzed, strong evidence that they possess functional plasmalemmas. During plasmolysis, the cytoplasm does not withdraw from the sieve plates because the plasmalemma, in the form of a tube, extends through the pores from one sieve element to the next, much as is the case with plasmodesmata.

The remainder of the sieve element protoplast, however, differs profoundly from that of other cells. As the sieve element matures the dedifferentiation of the cytoplasm including loss of nucleus and tonoplast and thinning of the cytoplasmic matrix in the lumen, as well as the relatively wide cytoplasmic strands traversing the sieve plate pores, seem to render the sieve tubes more suitable to the function of longitudinal conduction. The relatively dense membranous network of the endoplasmic reticulum lies mainly in the parietal layer and may be intimately connected with the trans-pore cytoplasmic connections (Esau and Cronshaw, 1968).

Discussions of the sieve element as a conducting cell often stress the presence in both light and electron micrographs of elongated, thread-like structures which are arranged longitudinally, either singly or in bundles. Not only their function but even their very existence in undisturbed, conducting sieve tube elements have been much debated. Thaine (1962) described transcellular, tubular strands in sieve elements of several species. His observations were made with the light microscope. The diameter of the strands varied from 1 to 7 microns, and they extended through the lumina of sieve elements, and from cell to cell through the sieve plate pores. He thought that these strands, and similar ones in parenchyma cells, served as pathways of transcellular movement of solutes (Thaine, 1964); that is, the contents of these strands, rather than the bulk fluid in the sieve element lumen, are the mobile phase in phloem transport, according to that view. The two phases are separated by the membrane forming the outer layer of each strand. However, there has been considerable debate concerning the structure of these transcellular strands. The evidence for a tubular membrane delimiting them is not considered conclusive and some have thought that they consist of proteinaceous filaments (Weatherley and Johnson, 1968), or that they are experimental artifacts (Esau *et al.*, 1963).

A characteristic feature of sieve tubes is a proteinaceous material usually called "slime," but more recently "P-protein" (Cronshaw and Esau, 1968a, b). Although the sequence of events is not invariable, most commonly the P-protein bodies which originate in the cytoplasm enlarge during the maturation of the sieve element and eventually disperse and become more or less evenly distributed throughout the cell. In electron micrographs the material appears in the form of filaments or "plasmatic filaments" (Weatherley and Johnson, 1968). They are often seen to extend in a longitudinal direction leading away from sieve plate pores. When aggregated in bundles or strands they resemble the transcellular strands described in the preceding paragraph, and may be identical with them. It is not certain whether the strand-like disposition of P-protein extending through the pores, plugging them, is an artifact of preparation or reflects the condition in undisturbed sieve elements. Anderson and Cronshaw (1969) and Cronshaw and Anderson (1969) believe that in undisturbed, functional sieve tubes the pores are not plugged by bundles or strands of P-protein. But it is an indication of the unsettled state of this problem that in the same recent issue of *Planta,* Anderson and Cronshaw (1970) furnish evidence to the effect that the sieve plate pores are normally unplugged while Siddiqui and Spanner (1970) come to the opposite conclusion.

It should be apparent from the above discussion that we do not have, at this time, an unequivocal view of the structure of sieve tubes. The field is rife with contradictions, widely different interpretations of what is seen or believed to be seen are being offered, and there is much evidence which fails to command general acceptance. The observation of the structure of these cells does not furnish any patent clues as to their mode of action. We must therefore turn to physiological investigations on the function of the sieve tubes.

3. THE SPEED OF PHLOEM TRANSPORT

One of the earliest and most telling experiments on the speed of phloem transport was made by Dixon and Ball (1922). A potato tuber developed at the end of a branch in about 100 days. It contained approximately 50 grams of carbohydrate, all of which, by our present reckoning, must have reached the tuber by passing through the phloem of the branch. The phloem had a cross-sectional area of 0.422 mm². Thus the amount of carbohydrate, in grams, passing unit cross-sectional area (square centimeter) of phloem per unit time (hour) was:

$$\frac{50}{0.422 \times 10^{-2} \times 24 \times 100} = 4.9 \text{ g/cm}^2/\text{hr}$$

Canny (1960) has called this expression "specific mass transfer," having the dimensions mass per unit area per unit time. It has the advantage that it contains only one measurement which is equivocal: that of the cross-sectional area involved in the transfer. Dixon and Ball (1922) based their calculation on the total cross-sectional area of the phloem. If we consider that transfer probably takes place only in the sieve tubes the value of Dixon and Ball should be increased by a factor of about 5. Values for specific mass transfer in stems of a number of species were shown by Canny (1960) to be comparable to the figure calculated from the data of Dixon and Ball.

"Specific mass transfer" is not a velocity, as it involves no dimension of length or distance. But it can be made to yield a measure of velocity if it is assumed that the transport takes place through flow of a solution, and if the concentration of that solution is known or a value for it is assumed. We can then write

$$\text{specific mass transfer} = \text{velocity} \times \text{concentration}$$
$$(\text{g/cm}^2/\text{hr}) \qquad (\text{cm/hr}) \qquad (\text{g/cm}^3)$$

This is indeed what Dixon and Ball did. They assumed the concentration of the sugar solution moving into the tuber to be 10 per cent and on that basis calculated a rate of flow of about 40 centimeters per hour. Ironically, they thought that this was much too fast for movement in the phloem, and concluded that organic substances move in the xylem. It was not the first time that wrong conclusions were drawn from an excellent experiment. Considering that the sieve tubes occupy only about one-fifth of the cross-sectional area of the phloem, the velocity of flow through the sieve tubes in the Dixon and Ball experiment must have been 40 multiplied by 5, or about 200 centimeters per hour.

Their assumption of a 10 per cent concentration for the solution in the sieve tubes was reasonable; subsequent direct measurements have borne it out. For such measurements, the aphid stylet technique developed by Kennedy and Mittler (1953) has been invaluable. Certain aphids have mouth parts (stylets) which they insert into plants in such a way that the tip of the stylet taps the interior of a sieve element as a source of food. To this day we do not know the mechanism of this beautiful evolutionary adaptation, that is, the mechanism whereby the insect "finds" the interior of a sieve tube. But plant physiologists have used it to good advantage as follows.

The insect is snipped off, the stylet remaining attached as before (Figure 8-5). Solution from the sieve element now exudes from the open end of the stylet for hours or days on end and can be collected, its volume measured and its contents analyzed. Using this technique, Weatherley *et al.* (1959) examined the sap exuding from sieve tubes of several species of willow, *Salix*. The sap was principally a solution of sucrose at a concentration of between 5 and 15 per cent—close to the 1922 guess of Dixon and Ball. Other investigators have found similar values. If movement of sugar in the sieve tubes is via a flow of solution, the original estimate of Dixon and Ball (1922), corrected by basing the calculation on the cross section of the sieve tubes only, is therefore valid, and sap would seem to move through sieve tubes at rates as high as about 2 meters per hour. Another way of measuring the rate of translocation is by following the movement of a radioactive substance after it has been supplied at a given point, usu-

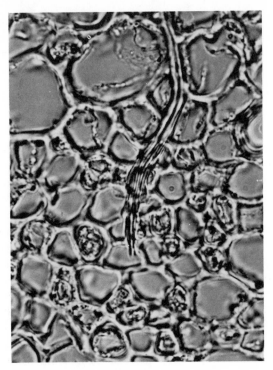

Figure 8-5. Stylet of the aphid, *Aphis fabae,* inserted into a sieve tube of St. Bernard's lily, *Anthericum liliago.* From Ziegler (1963).

ally a leaf. But interpretation of such experiments is beset by difficulties, as Canny (1960, 1971) has shown.

Zimmermann (1969a) recently devised a clever method for measuring the speed of transport in the phloem. In the white ash, *Fraxinus americana,* several sugars in addition to sucrose are transported in the phloem. Their ratios change somewhat during the day, and therefore a "ratio wave" travels down the stem and the rate of its advance can be measured. Its speed was found to be 30–70 centimeters per hour. This value is quite consistent with other, independent measurements, and we may take it that the rate of translocation in the phloem is of this order of magnitude. Undoubtedly, the time of year and various other conditions greatly influence the rate in a given species and even in individuals of the same species.

THE DIRECTION OF PHLOEM TRANSPORT

This chapter is called The Downward Movement of Foods and Nutrients but use of the word "downward" is loose and inexact. It is valid only in the sense that in higher green plants, the roots, being in darkness, receive assimilates originating in the leaves above. A major movement of food-stuffs is therefore downward, but organic solutes may move upward and sideways, depending upon the location of the exporting organ, or "source," and the receiving organ, or "sink." Mature leaves are the primary sources, but a sprouting potato tuber also constitutes a source from which organic metabolites are exported into the new growth, which constitutes the sink. Regardless of the relative location, then, a source is any organ or tissue from which organic material moves via the phloem, and a sink is any organ or tissue into which organic material moves through the phloem.

Whether a given organ or tissue acts as a source or a sink at a given time will depend on its own content of translocatable organic material, mainly soluble sugars, in relation to that of other regions. This was shown in the classical experiments of Mason and Maskell (1928b) with cotton, *Gossypium barbadense.* In one of their experiments they removed the branches and leaves from the middle region of the stems of a group of plants. This region therefore had no source of sugars of its own. When the leaves below this middle region were removed, those above remaining, carbohydrate was shown to move into the middle region from above. But when the leaves above the middle region were removed, those below being left on the plant, carbohydrate moved into the middle region from below. That is, the direction of the movement of carbohydrate depended on the position of the source region (the leaves left on the plant) in relation to the sink (the middle region stripped of leaves). More recently, Nakata and Leopold (1967) have experimentally shifted the relative locations of source and sink regions in a single leaf and shown that the direction of phloem transport shifted accordingly. Neales and Incoll (1968) have a useful discussion of methods for regulating source-sink relationships.

In the experiments just discussed, source and sink regions were experimentally manipulated, but changes in direction of phloem movement occur naturally, during the growth of the plant. Hale and Weaver (1962) exposed a young, rapidly growing leaf of a grapevine, *Vitis vinifera,* to CO_2 labeled with ^{14}C. After six hours the shoot on which this leaf was growing was cut off and a radioautograph was taken. The ^{14}C which had been assimilated by the leaf exposed to $^{14}CO_2$ was confined to the treated leaf.

This young expanding leaf was using the photosynthate it produced in its own metabolism and did not export any.

When the experiment was repeated with an older leaf, the radioactivity from ^{14}C showed up not only in the treated leaf but also in the younger, still rapidly growing leaves above it. Together these findings show that young leaves near the growing tips of branches use photosynthate from

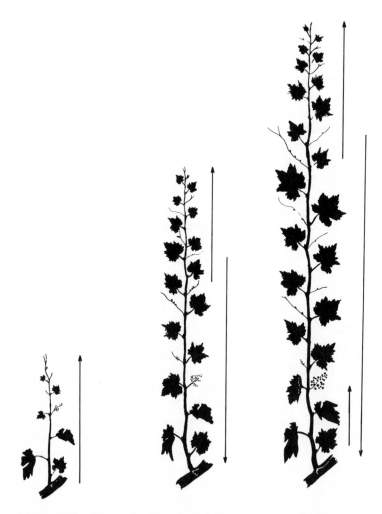

Figure 8-6. The main direction of the movement of photosynthate at three stages in the development of a rapidly growing grape shoot, *Vitis vinifera*. From Hale and Weaver (1962).

older leaves below, in addition to that which they themselves produce; that is, they act as sinks. As they mature, they cease importing and become sources instead, supplying photosynthate to younger, still expanding leaves, as well as to other sink regions, notably the clusters of grapes. Figure 8-6 shows the routing of sugar at three stages in the development of a rapidly growing grape shoot.

Often, a source, say a mature or nearly mature leaf, may be located between two sinks, for example, a young, rapidly growing leaf on the branch above it, and a developing fruit below. Can photosynthate from the source leaf move in the direction of both sinks? In other words, can there be bidirectional movement of carbohydrate in the phloem of the stem?

Mason *et al.* (1936) and Phillis and Mason (1936) recognized that the answer to this question is: yes. Later work has confirmed this for a variety of solutes and plant species (Baker, 1969; Biddulph *et al.*, 1958; Crafts, 1967; Eschrich, 1968; Trip and Gorham, 1968). Figure 8-7 shows

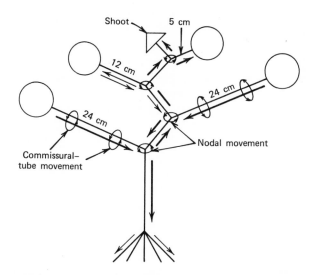

Figure 8-7.   A diagrammatic pattern of the movement of carbon-14 throughout a 15-day-old squash plant, *Cucurbita melopepo torticollis,* with mature primary and second leaves. The lengths of the petioles are indicated and the quantities of carbon-14 moving in any one direction are suggested by appropriately thickened arrows. After Webb and Gorham (1964).

the movements of sugar in a squash plant, *Cucurbita* sp. The authors concluded from their experiments that translocation of assimilated carbon in the squash plant is "a distinctly channeled multidirectional movement associated with the phloem. This movement appears to be determined by the demands of growth and is switched readily from leaf to leaf as each matures and ceases to require carbon assimilates synthesized elsewhere." Wardlaw (1968) has discussed various factors which control the pattern of movement of carbohydrates in plants. We shall return to this subject in the next section.

5. THE MECHANISM OF PHLOEM TRANSPORT

The Pressure Flow Hypothesis.   A model or mechanism of phloem transport must account for at least the following features of the process. (1) Movement in the phloem is rapid, on the order of tens of centimeters or even two meters per hour. (2) The direction of movement is from source to sink regions and may change in a given region of phloem tissue during the development of organs above and below it, or in response to experimental manipulation. Movement may be bidirectional in a given region. (3) Although sucrose is the principal carbohydrate transported by the phloem in most species that have been examined, various other carbohydrates, nitrogenous compounds, and other organic metabolites, as well as inorganic solutes, are transported in the phloem (Crafts and Crisp, 1971; Kursanov, 1966).

In 1926, Münch proposed a "pressure flow hypothesis" of phloem transport (Münch, 1930). Despite a great deal of controversy ever since, it remains the principal hypothesis and whatever its ultimate fate, it behooves us at this point to give it a careful hearing. To introduce the subject, let us regress for a moment to our discussion of the upward movement of water and solutes in the xylem. The conclusion emerged that transpiration is the principal mechanism of long-distance transport of both the solvent (water) and the solutes (ions). Transpiration, a purely physical process, sets up an ascending stream of solution in the xylem—the transpiration stream. But we found that under conditions when transpiration is negligibly small or absent, another mechanism driven by active metabolic processes of the plant can cause an upward movement of water and ions in the xylem. This is the mechanism of root pressure which depends upon active ion transport by root cells. This causes a buildup of ions in the xylem of the roots at concentrations higher than those in the external solution. As a result, the osmotic potential in the xylem becomes lower than that of the external medium and there is an osmotic movement of water into the xylem, setting up the root pressure which drives the solution in the xylem upward.

In turning now to the movement of sugars and other metabolites synthesized in leaves, it should be evident at once that to account for this long-distance transport we cannot invoke a purely physical process such as transpiration which depends ultimately on the evaporation of water from moist leaf cells. We know of no corresponding physical process that would drive a solution in the opposite direction, from the leaves toward the roots or into a tuberous stem such as a potato.

Discarding therefore anything akin to transpiration as a possible mechanism for the downward movement of solutes, we are naturally tempted to recall the other mechanism of xylem transport—the mechanism of root pressure. In the leaf, organic solutes, the products of photosynthesis, might be accumulated in sieve tube elements, and as a result, water from surrounding tissue would enter the sieve elements osmotically, setting up a "leaf pressure," in analogy to the "root pressure" in xylem vessels of the root. In 1926, ideas about root pressure were still vague. Yet in that year, Münch published his hypothesis of phloem transport which in many features parallels the mechanism of root pressure as we have described it.

Carbohydrate synthesized in leaf cells is transferred into the living sieve tube elements which as a result come to contain a fairly concentrated solution of sugar. This causes an osmotic movement of water into the sieve tubes and their turgor pressure rises. At the sink regions, sugar is removed from the phloem and utilized by the tissues of the sink. The resulting difference in hydraulic pressure between the source and sink regions of the sieve tubes causes a flow of solution from source to sink. The flow is maintained so long as sugar keeps being accumulated in the sieve tubes at the source end of the line and getting removed at the sink end. Water taken into the phloem transport system at the source end is withdrawn from the surrounding tissue and is resupplied through the xylem (Münch, 1930).

A model demonstrating such a mechanism is readily set up in the laboratory (Figure 8-8). The compartment at the left end of the tube represents a sieve element in the source region. It contains a solution of sucrose to which the wall of the compartment is impermeable. Water from the vessel enters the compartment by osmosis and the hydraulic pressure developed within it drives the solution through the tube which represents a sieve tube. The compartment at the right represents the sink region. If there were a continued supply of sugar into the source compartment and a mechanism of extracting the sugar from the solution in the sink the flow would continue.

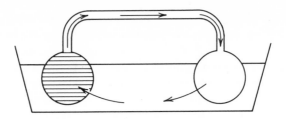

Figure 8-8. Model of the Münch pressure flow mechanism of phloem transport. After Münch (1930).

In the plant, sugar keeps being supplied to the sieve elements in the leaf as long as photosynthesis goes on, and active cells in sink regions withdraw sugar from the sieve tubes. Thus, in a growing, photosynthesizing plant conditions are such as to maintain the pressure flow. When the source is a non-photosynthetic storage organ, for example, a sprouting tuber, the source of the sugar delivered into the sieve elements is stored carbohydrate; but otherwise, the same pressure flow mechanism of phloem transport can be visualized for this situation.

In our discussion, one modification of the original hypothesis of Münch will be implicit. He assumed that the gradient in sugar concentration necessary for pressure flow to occur extended not only through the sieve tubes themselves but also from leaf mesophyll cells into the sieve tubes. However, it was recognized early that concentrations of sugar in leaf mesophyll cells may be lower than in the sieve tubes (Mason *et al.,* 1936). These authors insisted that sugar must be actively transported into the sieve tubes but admitted that "we are as yet quite ignorant of the mechanism by which the concentration is effected." This is still true, but it is in any event clear that an active "loading" of the sieve element occurs in source regions (Barrier and Loomis, 1957; Nakata and Leopold, 1967). In leaves of many species, specialized "transfer cells" evidently play a role in this process (Gunning *et al.,* 1968; Gunning and Pate, 1969; Pate and Gunning, 1969; Trip, 1969). Almost certainly, "unloading" in the sink regions is also by some process of active transport or secretion. True pressure flow, then, can only take place within the sieve tubes.

Let us now return to the criteria listed at the beginning of this section and see whether the Münch hypothesis of a pressure flow in the phloem can meet them. First, as to the velocity of flow, the answer depends on the gradient of osmotic potential between source and sink and on the resistance encountered within the sieve tubes along the way. Osmotic potentials as low as -15 atmospheres have been measured in sieve tube exudate obtained by the aphid stylet technique (Weatherley et al., 1959). If near the sink, the concentration of solutes in the sieve tubes is very low, a maximal difference in osmotic potential between the two regions of about 15 atmospheres is therefore possible, or 1 atmosphere per meter in a tree where the distance between a source (leafy branch) and sink (root tips in the soil) is 15 meters.

Assuming a viscosity, η, of 1.5×10^{-2} poise for the solution in the sieve tubes, a tube radius, r, of 12 μ (1.2×10^{-3} cm), and a pressure drop, Δp, of 0.01 atmosphere/cm, we can use the Poiseuille equation for laminar flow to calculate the velocity, v, in cm/second. The equation may be written

$$\Delta p = \frac{8\eta v \times 10^{-6}}{r^2}$$

The result is $v = 0.12$ centimeters per second, or 4.32 meters per hour. True, this calculation neglected the sieve plates, but reasonable values for their spacing and the number and dimensions of the pores in them do not increase the resistance by more than a factor of 2 to 3 (Weatherley and Johnson, 1968). On this basis, then, we may conclude that velocities on the order of 2 meters per hour may be achieved by the pressure flow mechanism. For longer distances, the pressure drop per unit distance would be less, and so, proportionally, would be the calculated rate of flow, other things being constant.

The second criterion by which to test the pressure flow mechanism has to do with the direction of transport. It must be from source to sink regions. When either by experimental manipulation or the natural development of the plant the source-to-sink relation of two regions is reversed, the direction of flow in the phloem between them should be reversed. We have already seen that this happens (section 4, p. 203).

But what about simultaneous flow in opposite directions in the phloem of a single stem or petiole? We have seen in section 4 that this occurs; the question is whether it is compatible with the pressure flow hypothesis. As long as the opposing streams move in different sieve tubes there is no contradiction, but simultaneous movement in opposite directions in the same sieve tube could not be reconciled with the hypothesis. Attempts have been made to demonstrate such counterflow in single sieve tubes (Eschrich, 1967; Ho and Peel, 1969; Trip and Gorham, 1968). The technical difficulties are formidable, however, and for the time being we must conclude that the evidence is not sufficiently convincing to force us to abandon the pressure flow hypothesis (Peterson and Currier, 1969).

The third test of the pressure flow hypothesis also has to do with the direction of movement in the phloem. If the hypothesis is right it follows that the concentration gradient of the major solute in the sieve tubes should determine the direction of movement, and that other solutes should ride along in the flowing solution, following the same source-to-sink path as the major solute. There is much evidence that this is so. Even virus particules introduced by aphids into the sieve tubes move within them in the same direction that sugars are presumed to move, as recognized early by plant pathologists and virologists (Bennett, 1940; Esau, 1961, 1968). Other esoteric substances such as synthetic herbicides follow the same rule: they move along in the stream of assimilate in the phloem (Crafts, 1967; Crafts and Crisp, 1971).

One would expect that once in a given sieve tube all materials would travel at the same rate, but actual demonstration of this is difficult. Rates of entry of different materials into the sieve tube may vary, and so may rates of withdrawal from it along the way. These and other factors may easily account for apparent differences in rates of translocation of different materials, even if it can be shown that they move in the same sieve tube, and that in itself is a problem.

Summing up, we have considered evidence concerning the rate of translocation, the direction of movement of a given solute as influenced by source-and-sink relations, and the direction of movement of minor sieve tube constituents in relation to that of the major component. All of it seems to be compatible with the hypothesis first proposed by Münch in 1926 (Münch, 1930). Why, then, do many plant physiologists still have reservations about it?

The principal reason has to do with the structure of the sieve tubes and the nature of their contents (see p. 194). Time and again objections have been raised to the effect that the narrowness of the sieve plate pores and of the cytoplasmic strands passing through them militate against the hypothesis that a mass flow of a solution passes through the sieve tubes. But we must remember that the microscopic and electron microscopic evidence comes mainly from static pictures of cells that have been killed and manipulated before the pictures are taken. The danger of artifact is ever present and no pictures can give a measurement of the degree of freedom of movement a solution in the sieve tubes would have. In the end, physiological evidence cannot be denied on the basis of appearances which are generally acknowledged to depend much on the operations performed in securing them. To be sure, the physiological evidence, too, depends on the activities of the experimenter, but his interference with the normal operation of the plant is often more subtle than the knock-out blow delivered by the microscopist.

Other Hypotheses. Let us briefly consider other hypotheses of phloem transport. Protoplasmic streaming has long been invoked and been advocated by Thaine (1962, 1964). Streaming of the bulk of the sieve tube contents would cause mixing but not directional transport. However, Thaine focused attention on transcellular strands which he believes to pass longitudinally through the sieve elements and from one to the other through the sieve plate pores. Streaming through these strands, Thaine (1969) has recently speculated, may be by a rhythmic contraction and relaxation of protein filaments in the outer layer of the transcellular strands.

The hypothesis leaves many questions unanswered. It is particularly difficult to account for the source-to-sink direction of flow: how would the direction of the pulsing be reversed upon a change in the source-sink relation of two regions? One feature of this model is appealing, however; it would permit bidirectional flow within a single sieve tube, since there are many strands in each, and the direction of flow might be one way in some of the strands and in the opposite direction in others. If unequivocal evidence for bidirectional flow in a single sieve tube is ever obtained, the Münch hypothesis of a mass flow through the bulk volume of the lumen of sieve elements would have to be abandoned, and some scheme of flow through transcellular strands or filaments would then become a very enticing idea.

According to another line of thought the sieve plates, far from offering resistance to the movement of solutes through the sieve tubes, are pumping stations whereby the solutes are transferred from one sieve element into the next. This idea is central in the "electroosmotic" hypothesis of Fensom (1957) and Spanner (1958). According to these authors, an electric potential difference is set up across the sieve plate, either through its differential permeability to H^+ and HCO_3^- ions from respiration (Fensom, 1957) or as a result of active transport of K^+ ions across it (Spanner, 1958; Spanner and Jones, 1970). The difference in electropotential across the sieve plate would cause an electroosmotic movement of water. However, the scheme raises more questions than it answers; for example, the source-to-sink relation would be as difficult to account for by it as by the hypothesis of pumping through strands. Convincing experiments in support of the electroosmotic hypothesis are lacking.

Whatever the mechanism of transport in the phloem, it depends on active metabolism (Kursanov, 1966; Weatherley and Johnson, 1968; Zimmermann, 1969b). This is true of the pressure flow mechanism as much as for pumping mechanisms. Loading and unloading at the source and sink regions demand energy for the active transport processes that are involved there. So does maintenance of the sieve element cytoplasm, including particularly its limiting membrane. It seems probable that the companion cells largely are the seat of the essential metabolic processes, and that there is an active traffic of metabolites between them and the adjacent sieve elements.

It is likely that in the evolution of the phloem as a conducting system, the sieve tubes became specialized for conduction proper by losing much of the normal, highly structured cytoplasmic apparatus (vacuoles, mitochondria, etc.) which would interfere with longitudinal movement. The companion cells, in this view, represent adaptations whereby the essential cytoplasmic machinery was shifted out of the path of flow, performing their essential functions "from the sidelines."

Three seemingly unrelated observations suggest a mechanism whereby the unique state of the protoplasts of sieve tubes may be brought about. (1) During maturation, sieve tube protoplasts lose much of the usual organization and internal structuring of cytoplasm. (2) As we shall see in section 6 of this chapter, calcium ions do not readily move in the phloem because of their failure to enter sieve tubes. (3) Calcium deficiency results in a breakdown of normal intracellular organization and compartmentation (Chapters 6 and 11, pp. 117–120 and 305–308).

Perhaps, therefore, exclusion of calcium from sieve tubes is part and parcel of the process whereby they attain the dedifferentiated, relatively structureless state of their protoplasts. The characteristic immobility of calcium in the phloem would be an incidental consequence of this control mechanism, and one not without some disadvantageous side effects because this immobility often leads to local calcium deficiencies and inefficient utilization of the element (section 6 of this chapter and section 4 of Chapter 11).

6. PHLOEM TRANSPORT OF INORGANIC SOLUTES

According to the view adopted here, assimilates from the leaves take pride of place in phloem transport; hence the emphasis on them in the preceding discussion of its mechanism. Minor constituents seem to get swept along in the same flow that carries the sugars. Any substance would be subject to phloem transport once it had entered a sieve element; the controlling step would be the entry itself. Inorganic ions would therefore be expected to travel in the phloem in basically the same way as other quantitatively minor sieve tube components, such as nitrogen compounds, viruses, and esoteric compounds like synthetic herbicides.

Mason and Maskell found early in their investigations that phosphorus and potassium accumulate above a ring and are depleted below it, while the distribution of calcium was unaffected by ringing. Phosphorus and potassium, they concluded, were transported in the phloem and calcium was not. In further work it was found that sulfur, magnesium, and chlorine are mobile in the phloem (Mason and Phillis, 1937). Since that time, the availability of radioisotopes of almost all physiologically important mineral elements has made it possible to follow their movement through the plant, including transport in the phloem. These investigations have revealed veritable "circulation patterns" (Biddulph et al., 1958). Initially acquired by the roots, mineral elements move upward in the xylem. Many of them are then subject to redistribution via the phloem; a few are not. Immobility in the phloem presumably is caused by failure of these elements to enter the sieve tubes.

Bukovac and Wittwer (1957) at East Lansing systematically studied the mobility of many radioactively labeled mineral elements applied to leaves of bean plants. Their results are included in Table 8-1. To the extent

TABLE 8-1.   Mobility of Mineral Elements in the Phloem

Mobile	Intermediate	Immobile
Potassium	Iron	Lithium
Rubidium	Manganese	Calcium
Sodium	Zinc	Strontium
Magnesium	Copper	Barium
Phosphorus	Molybdenum	Boron
Sulfur		
Chlorine		

that elements studied by them had also been included in the early, classical experiments of Mason and collaborators (Mason and Phillis, 1937), the newer evidence fully corroborates the earlier work. Lithium (Kent, 1941) and boron (McIlrath, 1965; Oertli, 1963) are among the phloem-immobile elements. The behavior of lithium is interesting in that this element in many respects chemically resembles calcium more than it does the alkali metals of which it is one (Epstein, 1960). It is therefore not too surprising that in regard to its mobility in the phloem lithium mimics the alkaline earth cation, calcium.

The heavy metal micronutrients are assigned an intermediate position by Bukovac and Wittwer, but on the whole experience indicates that they are mobile in the phloem, as suggested by experiments on redistribution and export from leaves, although the degree of mobility varies from species to species (Brown *et al.,* 1965, and Eddings and Brown, 1967, on iron; Uriu and Koch, 1964, on manganese; Sudia and Linck, 1963, and Wallihan and Heymann–Herschberg, 1956, on zinc).

The findings on mobility of inorganic nutrients in the phloem are consistent with evidence on the responses of plants to deficiencies. A striking example is the calcium nutrition of the peanut, *Arachis hypogaea*. The plant produces flowers above ground; the flowers after being fertilized form "gynophores" or "pegs"—stalk-like structures which grow downward, push into the soil and there develop into fruits. They are therefore not in the path of the transpiration stream and would seem to depend on import of mineral nutrients from the plant via the phloem, or on absorption from the soil. Harris (1949) found that when calcium was supplied to the roots but not to the gynophores, few peanuts developed. This is as expected if calcium is not mobile in the phloem, so that the developing fruit depends on calcium in the medium surrounding the gynophores. These findings were confirmed in experiments with radiocalcium (Bledsoe *et al.,* 1949). When the rooting medium contained labeled calcium but the medium in which the gynophores grew did not, very little labeled calcium reached the gynophores and fruits. When the labeling of the two media was reversed (gynophores in labeled medium and roots in unlabeled) the gynophores and fruits absorbed the labeled calcium from their medium. Figure 8-9 shows the experimental arrangement.

Not only subterranean fruits like peanuts, but also fruits borne on shoots in the more common fashion may be subject to stresses connected with the fact that little water reaches them via the transpiration stream, for fruits in general transpire water at a very low rate. It is likely that part at least of their import of water comes via the phloem, since developing fruits are of course major sinks. This flow would supply sugars and all other essential metabolites provided they are mobile in the phloem. Calcium is not, and this explains why fruits typically contain very little calcium (Wiersum, 1966). Like fruits, swelling buds are not in the mainstream of the transpirational movement of water, and import nutrients via the phloem, as shown for phosphorus by Burström (1948).

The consequences of the immobility of a mineral nutrient in the phloem were again demonstrated in an experiment on the boron nutrition of tobacco (Michael *et al.,* 1969). When the plants grew in a very humid atmosphere, newly developing leaves failed to grow normally and turned out

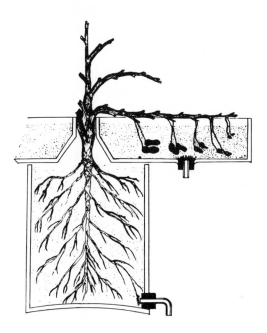

Figure 8-9. Diagrammatic view of the arrangement whereby either the rooting medium or the fruiting medium of a peanut plant, *Arachis hypogaea,* was supplied with radioactively labeled calcium, ^{45}Ca. After Bledsoe *et al.* (1949).

to suffer from boron deficiency. The situation resembles that discussed above in respect to the peanut gynophores. In the present case, transpiration was minimized because the humidity of the atmosphere was kept high, and boron, an element depending for mobility on the xylem, was therefore not reaching the young leaves at an adequate rate (cf. Oertli, 1963).

Although some mineral elements have been spoken of as "immobile" in the phloem, some movement in the phloem of even these elements can be shown with sufficiently sensitive methods. For example, some calcium labeled with ^{45}Ca was shown to move in the phloem of oats, *Avena sativa* (Ringoet *et al.,* 1968), and Millikan and his co-workers showed labeled calcium to be redistributed in subterranean clover, *Trifolium subterraneum* (Millikan and Hanger, 1967). Läuchli (1968) examined cross sections of the fruit stalk of the pea. *Pisum sativum,* with the electron probe analyzer which permits the precise localization of mineral elements at the cellular level. He found calcium in the sieve tubes and concluded that its transport into the pod is via the phloem. These findings, however, do not alter the general conclusion that the mobility of calcium in the phloem is very low.

The pathway of nutrients which are mobile in the phloem would be expected to follow the source-sink relationships determined by the movement of sugar. In a classical paper, Biddulph *et al.* (1958) showed that the movement of phosphorus in kidney bean plants, *Phaseolus vulgaris,* shows a pattern much like that which we have discussed in connection with sugar movement, except for the obvious difference that the initial uptake of phosphorus takes place in the roots and its upward movement from them is via the xylem. Once in a leaf, however, it may move back toward the roots and upward into the developing young growth, depending on its proximity to either. It was also shown by autoradiography that labeled sulfur and phosphorus applied to a leaf moved out via the phloem (Biddulph, 1956). Biddulph and Biddulph (1959) and Weatherley (1969) have given general accounts of the routing of nutrients in the plant, and the comprehensive treatment of transport in the phloem by Crafts and Crisp (1971) includes a discussion of the transport of inorganic ions.

REFERENCES

Anderson, R. and J. Cronshaw. 1969. The effects of pressure release on the sieve plate pores of *Nicotiana.* J. Ultrastructure Res. 29:50–59.

Anderson, R. and J. Cronshaw. 1970. Sieve-plate pores in tobacco and bean. Planta 91:173–180.

Baker, D. A. 1969. Transport pathways in sprouting tubers of the potato. J. Expt. Bot. 20:336–340.

Barrier, G. E. and W. E. Loomis. 1957. Absorption and translocation of 2,4-dichlorophenoxyacetic acid and P^{32} by leaves. Plant Physiol. 32:225–231.

Bennett, C. W. 1940. The relation of viruses to plant tissues. Bot. Rev. 6:427–473.

Biddulph, O. 1969. Mechanisms of translocation of plant metabolites. In: Physiological Aspects of Crop Yield. J. D. Eastin, F. A. Haskins, C. Y. Sullivan and C. H. M. van Bavel, eds. American Society of Agronomy, Inc., and Crop Science Society of America, Inc., Madison. Pp. 143–164.

Biddulph, O., S. Biddulph, R. Cory and H. Koontz. 1958. Circulation patterns for phosphorus, sulfur and calcium in the bean plant. Plant Physiol. 33:293–300.

Biddulph, S. F. 1956. Visual indications of S^{35} and P^{32} translocation in the phloem. Am. J. Bot. 43:143–148.

Biddulph, S. and O. Biddulph. 1959. The circulatory system of plants. Scientific American 200(2):44–49.

Bledsoe, R. W., C. L. Comar and H. C. Harris. 1949. Absorption of radioactive calcium by the peanut fruit. Science 109:329–330.

Brown, A. L., S. Yamaguchi and J. Leal–Diaz. 1965. Evidence for translocation of iron in plants. Plant Physiol. 40:35–38.

Bukovac, M. J. and S. H. Wittwer. 1957. Absorption and mobility of foliar applied nutrients. Plant Physiol. 32:428–435.

Burström, H. 1948. The rate of the nutrient transport to swelling buds of trees. Physiol. Plantarum 1:124–135.

Canny, M. J. 1960. The rate of translocation. Biol. Rev. 35:507–532.

Canny, M. J. 1971. Translocation: mechanisms and kinetics. Ann. Rev. Plant Physiol. 22:237–260.

Crafts, A. S. 1967. Bidirectional movement of labeled tracers in soybean seedlings. Hilgardia 37:625–638.

Crafts, A. S. and C. E. Crisp. 1971. Phloem Transport in Plants. W. H. Freeman and Company, San Francisco.

Crownshaw, J. and R. Anderson. 1969. Sieve plate pores of *Nicotiana*. J. Ultrastructure Res. 27:134–148.

Cronshaw, J. and K. Esau. 1968a. P-protein in the phloem of *Cucurbita*. I. The development of P-protein bodies. J. Cell Biol. 38:25–39.

Cronshaw, J. and K. Esau. 1968b. P-protein in the phloem of *Cucurbita*. II. The P-protein of mature sieve elements. J. Cell Biol. 38:292–303.

Currier, H. B., K. Esau and V. I. Cheadle. 1955. Plasmolytic studies of phloem. Am. J. Bot. 42:68–81.

Dixon, H. H. and N. G. Ball. 1922. Transport of organic substances in plants. Nature 109:236–237.

Eddings, J. L. and A. L. Brown. 1967. Absorption and translocation of foliar-applied iron. Plant Physiol. 42:15–19.

Epstein, E. 1960. Calcium–lithium competition in absorption by plant roots. Nature 185:705–706.

Esau, K. 1960. Anatomy of Seed Plants. John Wiley and Sons, Inc., New York.

Esau, K. 1961. Plants, Viruses, and Insects. Harvard University Press, Cambridge.

Esau, K. 1968. Viruses in Plant Hosts. The University of Wisconsin Press, Madison.

Esau, K. 1969. The Phloem. Gebrüder Borntraeger, Berlin and Stuttgart.

Esau, K. and V. I. Cheadle. 1962. Mitochondria in the phloem of *Cucurbita*. Bot. Gaz. 124:79–85.

Esau, K. and J. Cronshaw. 1968. Endoplasmic reticulum in the sieve element of *Cucurbita*. J. Ultrastructure Res. 23:1–14.

Esau, K., E. M. Engleman and T. Bisalputra. 1963. What are transcellular strands? Planta 59:617–623.

Eschrich, W. 1967. Bidirektionelle Translokation in Siebröhren. Planta 73:37–49.

Eschrich, W. 1968. Translokation radioaktiv markierter Indolyl-3-essig-säure in Siebröhren von *Vicia faba*. Planta 78:144–157.

Fensom, D. S. 1957. The bio-electric potentials of plants and their functional significance. I. An electrokinetic theory of transport. Can. J. Bot. 35:573–582.

Gunning, B. E. S. and J. S. Pate. 1969. "Transfer cells." Plant cells with wall ingrowths, specialized in relation to short distance transport of solutes—their occurrence, structure, and development. Protoplasma 68:107–133.

Gunning, B. E. S., J. S. Pate and L. G. Briarty. 1968. Specialized "transfer cells" in minor veins of leaves and their possible significance in phloem translocation. J. Cell Biol. 37:C-7–C-12.

Hale, C. R. and R. J. Weaver. 1962. The effect of developmental stage on direction of translocation of photosynthate in *Vitis vinifera*. Hilgardia 33:89–131.

Harris, H. C. 1949. The effect on the growth of peanuts of nutrient deficiencies in the root and the pegging zone. Plant Physiol. 24:150–161.

Ho, L. C. and A. J. Peel. 1969. Investigation of bidirectional movement of tracers in sieve tubes of *Salix viminalis* L. Ann. Bot. N.S. 33:833–844.

Kennedy, J. S. and T. E. Mittler. 1953. A method of obtaining phloem sap via the mouth-parts of aphids. Nature 171:528.

Kent, M. A. 1941. Absorption, translocation, and ultimate fate of lithium in the wheat plant. New Phytol. 40:291–298.

Kursanov, A. I. 1966. Le transport des substances organique-métaboliques chez les plants. Agrochimica 11:1–52.

Läuchli, A. 1968. Untersuchungen mit der Röntgen-Mikrosonde über Verteilung und Transport von Ionen in Pflanzengeweben. II. Ionentransport nach Früchten von *Pisum sativum*. Planta 83:137–149.

Mason, T. G. and E. J. Maskell. 1928a. Studies on the transport of carbohydrates in the cotton plant. I. A study of diurnal variation in the carbohydrates of leaf, bark, and wood, and of the effects of ringing. Ann. Bot. 42:189–253.

Mason, T. G. and E. J. Maskell. 1928b. Studies on the transport of carbohydrates in the cotton plant. II. The factors determining the rate and the direction of movement of sugars. Ann. Bot. 42:571–636.

Mason, T. G., E. J. Maskell and E. Phillis. 1936. Further studies on transport in the cotton plant. III. Concerning the independence of solute movement in the phloem. Ann. Bot. 50:23–58.

Mason, T. G. and E. Phillis. 1937. The migration of solutes. Bot. Rev. 3:47–71.

McIlrath, W. J. 1965. Mobility of boron in several dicotyledonous species. Bot. Gaz. 126:27–30.

McNairn, R. B. and H. B. Currier. 1968. Translocation blockage by sieve plate callose. Planta 82:369–380.

Michael, G., E. Wilberg and K. Kouhsiahi–Tork. 1969. Durch hohe Luftfeuchtigkeit induzierter Bormangel. Z. Pflanzenern. Bodenk. 122:1–3.

Millikan, C. R. and B. C. Hanger. 1967. Redistribution of ^{45}Ca in Trifolium subterraneum L. and Antirrhinum majus L. Austral. J. Biol. Sci. 20:1119–1130.

Milthorpe, F. L. and J. Moorby. 1969. Vascular transport and its significance in plant growth. Ann. Rev. Plant Physiol. 20:117–138.

Münch, E. 1930. Die Stoffbewegungen in der Pflanze. Verlag von Gustav Fischer, Jena.

Nakata, S. and A. C. Leopold. 1967. Radioautographic study of translocation in bean leaves. Plant Physiol. 54:769–772.

Neales, T. F. and L. D. Incoll. 1968. The control of leaf photosynthesis rate by the level of assimilate concentration in the leaf: a review of the hypothesis. Bot. Rev. 34:107–125.

Oertli, J. J. 1963. The influence of certain environmental conditions on water and nutrient uptake and nutrient distribution in barley seedlings with special reference to boron. In: Advancing Frontiers of Plant Sciences. L. Chandra, ed. Institute for the Advancement of Science and Culture, New Delhi. Vol. 6, pp. 55–85.

Pate, J. S. and B. E. S. Gunning. 1969. Vascular transfer cells in angiosperm leaves. A taxonomic and morphological survey. Protoplasma 68:135–156.

Peterson, C. A. and H. B. Currier. 1969. An investigation of bidirectional translocation in the phloem. Physiol. Plantarum 22:1238–1250.

Phillis, E. and T. G. Mason. 1936. Further studies on transport in the cotton plant. IV. On the simultaneous movement of solutes in opposite directions through the phloem. Ann. Bot. 50:161–174.

Ringoet, A., G. Sauer and A. J. Gielink. 1968. Phloem transport of calcium in oat leaves. Planta 80:15–20.

Schmitz, K. 1970. Der autoradiographische Nachweis tritiierter wasser-löslicher Substanzen in den Siebröhren von *Cucurbita* und *Cucumis.* Planta 91:96–110.

Schmitz, K. and J. Willenbrink. 1968. Zum Nachweiss tritiierter Assimilate in den Siebröhren von *Cucurbita.* Planta 83:111–114.

Shih, C. Y. and H. B. Currier. 1969. Fine structure of phloem cells in relation to translocation in the cotton seedling. Am. J. Bot. 56:464–472.

Siddiqui, A. W. and D. C. Spanner. 1970. The state of the pores in functioning sieve plates. Planta 91:181–189.

Spanner, D. C. 1958. The translocation of sugar in sieve tubes. J. Expt. Bot. 9:332–342.

Spanner, D. C. and R. L. Jones. 1970. The sieve tube wall and its relation to translocation. Planta 92:64–72.

Sudia, T. W. and A. J. Linck. 1963. The absorption and translocation of zinc-65 in *Pisum sativum* in relation to fruit development. Plant and Soil 19:249–254.

Thaine, R. 1962. A translocation hypothesis based on the structure of plant cytoplasm. J. Expt. Bot. 13:152–160.

Thaine, R. 1964. The protoplasmic-streaming theory of phloem transport. J. Expt. Bot. 15:470–484.

Thaine, R. 1969. Movement of sugars through plants by cytoplasmic pumping. Nature 222:873–875.

Trip, P. 1969. Sugar transport in conducting elements of sugar beet leaves. Plant Physiol. 44:717–725.

Trip, P. and P. R. Gorham. 1967. Autoradiographic study of the pathway of translocation. Can. J. Bot. 45:1567–1573.

Trip, P. and P. R. Gorham. 1968. Bidirectional translocation of sugars in sieve tubes of squash plants. Plant Physiol. 43:877–882.

Uriu, K. and E. C. Koch. 1964. Response of Yellow Newton apple leaves to foliar applications of manganese and zinc. Proc. Am. Soc. Hort. Sci. 84:25–31.

Wallihan, E. F. and L. Heymann–Herschberg. 1956. Some factors affecting absorption and translocation of zinc in *Citrus* plants. Plant Physiol. 31:294–299.

Wardlaw, I. F. 1968. The control and pattern of movement of carbohydrates in plants. Bot. Rev. 34:79–105.

Weatherley, P. E. 1969. Ion movement within the plant and its integration with other physiological processes. In: Ecological Aspects of the Mineral Nutrition of Plants. I. H. Rorison, ed. Blackwell Scientific Publications, Oxford. Pp. 323–340.

Weatherley, P. E. and R. P. C. Johnson. 1968. The form and function of the sieve tube: a problem in reconciliation. Internat. Rev. Cytol. 24:149–192.

Weatherley, P. E., A. J. Peel and G. P. Hill. 1959. The physiology of the sieve tube. Preliminary experiments using aphid mouth parts. J. Expt. Bot. 10:1–16.

Webb, J. A. and P. R. Gorham. 1964. Translocation of photosynthetically assimilated ^{14}C in straight-necked squash. Plant Physiol. 39:663–672.

Wiersum, L. K. 1966. Calcium content of fruits and storage tissues in relation to the mode of water supply. Acta Bot. Neerl. 15:406–418.

Ziegler, H. 1963. Der Ferntransport organischer Stoffe in den Pflanzen. Naturwiss. 50:177–186.

Zimmermann, M. H. 1957. Translocation of organic substances in trees. II. On the translocation mechanism in the phloem of white ash (*Fraxinus americana* L.). Plant Physiol. 32:399–404.

Zimmermann, M. H. 1969a. Translocation velocity and specific mass transfer in the sieve tubes of *Fraxinus americana* L. Planta 84:272–278.

Zimmermann, M. H. 1969b. Translocation of nutrients. In: The Physiology of Plant Growth and Development. M. B. Wilkins, ed. McGraw-Hill Publishing Company, Ltd., Maidenhead. Pp. 381–417.

PART III

ASPECTS OF ENERGETICS AND THE METABOLISM OF INDIVIDUAL ELEMENTS

9

ENERGETICS AND METABOLIC COUPLING OF ACTIVE TRANSPORT

INTRODUCTION

Energy. The evidence considered in Chapter 6 strongly pointed to the conclusion, now generally accepted, that in ion absorption by plant cells, carriers effect the transfer of ions across membranes which are barriers to ionic diffusion. Under suitable conditions, internal concentrations of salt may exceed the external concentration by a factor of 10,000/1, or even more. Obviously, energy must be expended to pump salt in a direction which is thermodynamically "uphill," and a minimal value for the required energy can be calculated by the equation

$$\Delta G^\circ = RT \ln \frac{C_2}{C_1}$$

where ΔG° is the free energy change required to bring about the ratio of concentrations (strictly, activities) C_2/C_1, R is the gas constant, and T the absolute temperature. Taking a ratio of concentrations of 10,000/1, and a temperature of 20°C, we get

$$\Delta G^\circ = 1.98 \times 293 \times 2.3 \log \frac{1 \times 10^4}{1}$$

$$= 1,330 \times 4$$
$$= 5,340 \text{ calories/mole}$$

If there is appreciable back leakage across the membrane more energy will be needed to effect this ratio. Under normal physiological conditions, the plasmalemma of plant cells seems to be highly impermeable to passive leakage of ions, and in that sense ion transport across it is a thermodynamically efficient operation. On the other hand, this high degree of impermeability to diffusive ion movement also means that ion absorption has to be via the active transport mechanisms even when the gradient for passive ion movement is inward.

Granting then that metabolic energy must be expended to move ions across the cellular membranes, we are faced with two questions. First, what is the source of the energy and its chemical form, and second, how is the energy applied to move ions across membranes? Inasmuch as we already have adopted the view that transport is mediated by carriers, the second question can be put a little more precisely: how is metabolic energy applied to energize the carrier-mediated ion transport? We shall, in this section, review in utmost brevity pertinent aspects of the general energetics of green plants, that is, photosynthesis and respiration, and in the remainder of the chapter, the energetics of ion transport in non-green tissues, the energetics of the process in photosynthetic tissues, and the ion transport processes in mitochondria and chloroplasts. Lehninger (1965) has written a general introduction to bioenergetics, with a glossary of frequently used terms and suggestions for further reading.

Photosynthesis.   As already discussed in Chapter 2, the energy input into the biosphere is almost entirely from the sun. (Certain bacteria derive energy from the oxidation of inorganic compounds and use the energy of oxidation to synthesize organic compounds from CO_2, which they reduce in a process called chemosynthesis.) The light energy trapped in the initial reactions of photosynthesis is eventually converted into the chemical energy contained in stable, reduced carbon compounds, principally carbohydrates.

In both the generation and the degradation of these compounds a key role is played by high-energy phosphate compounds which act as chemical intermediates, receiving energy from higher-energy phosphate compounds and yielding it to lower-energy phosphate compounds. The principal compound to serve this function is adenosine triphosphate (ATP), the structure of which is shown in Figure 9-1. The symbol $\sim$ indicates the so-called high-energy phosphate bonds. This does not mean that the energy resides in those bonds. All that is meant is that when ATP hydrolyzes to inorganic phosphate and adenosine diphosphate, ADP, the difference in energy content between ATP and the products of the reaction is high—about 7,000 calories/mole standard free energy at pH 7.0 and 25°C. We may note,

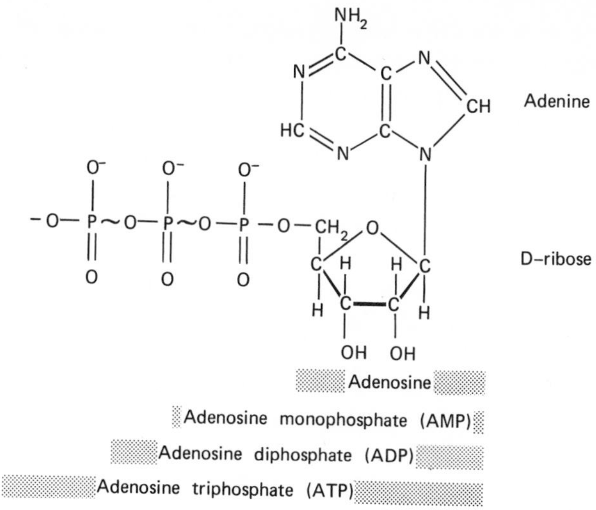

Figure 9-1. The structure of adenosine triphosphate, adenosine diphosphate, and adenosine monophosphate.

in passing, that this is on the order of the energy requirement for maintenance of a concentration gradient of 10,000/1 for one mole of solute, as discussed above.

Until recently, photosynthesis was considered to have as its central theme the fixation of CO_2 and the synthesis of reduced carbon compounds. This is still true in the sense that these are the ultimate results of the total sequence of events summed up under the term photosynthesis. Nevertheless, the focus has shifted as a result of discoveries which show that only a few, initial reactions are involved in, or closely related to, the capture of light energy. Thereafter, the sequences of reactions leading to the synthesis of carbohydrates and other compounds do not require light; they are "dark reactions." These discoveries were greatly aided by the development of techniques whereby the photosynthetic apparatus of green cells, the chloroplasts, could be separated from the cells and studied in isolation, or *in vitro* (Arnon, 1955).

The essential picture of photosynthesis that has emerged from these studies is as follows. When light strikes a chloroplast, part of its energy is absorbed by chlorophyll pigment and results in the excitation and ejection of electrons from it. The electrons subsequently travel along chains of electron carriers including ferredoxin and cytochromes, each of which is reduced by the electrons and reoxidized upon yielding them to the next member of the chain. Passage of the electrons down these photosynthetic electron chains results in the liberation of energy, some of which is conserved through phosphorylation of ADP to yield ATP, a process called "photosynthetic phosphorylation" or "photophosphorylation," discovered by Arnon and his collaborators (Arnon *et al.*, 1954).

What is the eventual fate of the electrons? There are two possibilities, and as it turns out, both are realized. The electrons might return to chlorophyll and restore it to its original condition. They would thus travel in a circuit or cycle, from chlorophyll via the electron chain back to chlorophyll, the net result being only the generation of ATP. This occurs; it is the route electrons take in "photosystem I," and this process of phosphorylation is called "cyclic photophosphorylation" which yields two molecules of ATP per pair of electrons making the cycle (Figure 9-2). Photosystem I operates best at long wavelengths of light (greater than 685 nanometers).

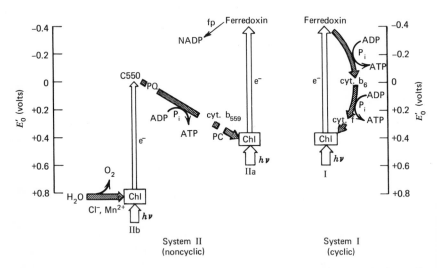

Figure 9-2. Photosystems I and II of photosynthesis in higher green plants. In cyclic photophosphorylation (system I) electrons ejected from chlorophyll (Chl) by light ($h\nu$) are accepted by ferredoxin and return via a "dark" electron transport chain including cytochromes, two molecules of ATP being generated in the process. In noncyclic phosphorylation (system II) the electron deficit created by the action of light on chlorophyll (system IIb) is made good by electrons from water, both chloride and manganese being required; the electrons ejected from system IIb travel along another "dark" electron transport chain to photosystem IIa and thence to ferredoxin and $NADP_{ox}$. After Knaff and Arnon (1969).

The other possible fate of the electrons is to be transferred to a compound in its oxidized state and reduce it. Photosynthesis results in the reduction of CO_2, but so far we have not encountered a reductant to serve this function. The needed reductant is formed in "photosystem II." According to the current concept of Arnon, this system entails two light reactions in series, connected by an electron transport chain as shown in Figure 9-2 (Knaff and Arnon, 1969). In this system, too, electron flow is attended by phosphorylation of ADP to ATP ("noncyclic photophosphorylation"), and results in the reduction of ferredoxin. But ferredoxin does not channel the electrons back to chlorophyll, as happens in photosystem I. Instead, electrons from ferredoxin are transferred to the oxidized form of nicotinamide adenine dinucleotide phosphate, $NADP_{ox}$, forming reduced NADP or $NADP_{red}$. This compound serves as an electron donor in many biosynthetic reductions—it is not uniquely associated with the photosynthetic apparatus. In photosynthesis it is the reductant which, in cooperation with the other product of photosynthesis, ATP, powers the sequence of dark reactions in which CO_2 is reduced and carbohydrate formed. Together, ATP and $NADP_{red}$ have been called "assimilatory power" by Arnon (for a review, see Arnon, 1967). Photosystem II operates best at short wavelengths (less than 685 nanometers).

In cyclic phosphorylation, electrons ejected by light energy from chlorophyll return to it (see system I in Figure 9-2). But when electrons in system II serve ultimately to reduce $NADP_{ox}$, how is the chlorophyll restored to its initial state? What, in this case, is the source of the needed electrons? The answer to this question is that in photosystem II of higher green plants, the electron deficit is made good by electrons derived from water. Robert Hill, in classical work summarized by Hill and Whittingham (1955), has shown that in the presence of chloroplasts, light, and a suitable electron acceptor, this electron acceptor becomes reduced, with the evolution of molecular oxygen. The process represents an oxidation of water brought about through the photochemical activity of the chloroplasts:

$$H_2O \rightarrow H^+ + OH^-$$

$$\underline{OH^- \xrightarrow[\text{chloroplasts}]{\text{light}} \tfrac{1}{2}O_2 + H^+ + 2e^-}$$

$$\text{Sum: } H_2O \xrightarrow[\text{chloroplasts}]{\text{light}} \tfrac{1}{2}O_2 + 2H^+ + 2e^-$$

In the "Hill reaction," the electrons yielded in this process reduce a non-physiological electron acceptor used for experimental purposes; in the complete photosynthetic system, they serve to resupply the electrons ejected from photosystem II and subsequently instrumental in noncyclic photophosphorylation and reduction of $NADP_{ox}$ (Figure 9-2).

The three stable end products of this process, products which can readily be recovered in substantial quantities, are ATP, $NADP_{red}$, and O_2. Of these, the last is a waste product and lost from the system. The other two, the "assimilatory power" of Arnon, lead to the formation of carbohydrate in sequences of dark reactions worked out in detail by Calvin and his collaborators (Calvin and Bassham, 1962) and Hatch and Slack (1970). A recent, very readable introduction to photosynthesis is that by Rabinowitch and Govindjee (1969). The book edited by San Pietro *et al.* (1967) presents contributions by specialists on various aspects of photosynthesis.

Respiration.   With the exception of chemosynthetic bacteria, cells incapable of photosynthesis depend on organic substrates furnished by photosynthetic cells. This goes not only for the whole of the animal world but for the non-green parts of plants as well. Animals are parasitic on plants; roots and other non-green parts of plants are associated with the photosynthetic portions in what would be called a symbiotic relationship if the two were different organisms. Roots furnish water and nutrients to the photosynthetic organs above ground, and receive from them products of photosynthesis, principally sugars.

The initial reactions in the breakdown of sugar are catalyzed by soluble enzymes in the cytoplasm. But subsequent events in which most of the energy yielded in respiration is conserved occur in specialized organelles, the mitochondria. Just as the clarification of the process of photosynthesis in chloroplasts depended upon their isolation from the cell and study *in vitro,* so our knowledge of the mechanism of aerobic respiration derives from experiments in which mitochondria were removed from the cells and their function studied separately (Hackett, 1955; Hall and Palmer, 1969).

In aerobic cells, three major sequences of reactions lead to the complete oxidation of sugar and the conservation of a fraction of the energy liberated thereby. The first of these is glycolysis, a sequence of reactions not requiring oxygen and leading in a number of enzyme-catalyzed steps from glucose to the three-carbon pyruvic acid, $CH_3 \cdot CO \cdot COOH$. In the second major sequence, pyruvic acid is transformed into organic acids which undergo a cyclic interconversion, the "Krebs" or "tricarboxylic acid" cycle.

The last of the three sequences is the only one we shall briefly discuss here, for it is the one where the bulk of the energy liberated in the process of respiration is conserved. Organic acids of the Krebs cycle are oxidized by dehydrogenase enzymes and all but one serve to reduce nicotinamide adenine dinucleotide, NAD, similar to NADP which we have encountered in our discussion of photosynthesis. The electrons from the oxidized Krebs cycle intermediates are then funneled into a chain of electron carriers most of which are cytochromes, though not identical with those of the photosynthetic electron chains discussed above. The mitochondrial electron chain is called the respiratory chain, or the respiratory electron transport chain. Passage of the electrons along this chain results in the alternate oxidation and reduction of each member of the chain. As a result of this passage, which energetically is "downhill" all the way, part of the energy becomes conserved in the form of the universal "energy currency" of living cells, ATP. This process of phosphorylation in the respiratory chain is called "oxidative phosphorylation," the reason for this term being an aspect of the process we have so far ignored—the need for molecular oxygen, O_2, in this process of respiration carried on by mitochondria.

When tissue which normally respires aerobically is deprived of oxygen the first sequence of reactions in the breakdown of sugar, glycolysis, can still proceed, but products of the reactions soon poison the cells. (Glycolysis does not involve the mitochondria.) For sustained operation of the entire respiratory apparatus mitochondrial electron transport through the respiratory chain of electron carriers is essential. The question therefore arises what the ultimate electron acceptor is. It is oxygen, and that is why we speak of "aerobic" respiration and of the phosphorylations in the respiratory chain as "oxidative" phosphorylation. The terminal member of the mitochondrial cytochrome chain activates molecular oxygen and effects its reduction to water:

$$\tfrac{1}{2}O_2 + 2H^+ + 2e^- \rightarrow H_2O$$

In discussing photosynthesis by green plants we saw that electrons are abstracted from water in a light energized step (the "Hill reaction" part of photosynthesis in photosystem II). We now see that water becomes the end product in the reverse process of respiration, the sequence of events whereby the reduced products of photosynthesis (carbohydrates and other carbon compounds) are oxidized, with conservation of part of the energy in a chemical form (ATP) which is usable in cellular metabolism.

The formation of ATP in the respiratory chain is not an incidental consequence of electron flow but is normally essential to it; that is, respiration (electron flow) and phosphorylation are "coupled," and when for any reason phosphorylation is slowed down, the rate of respiration as measured by disappearance of substrate or utilization of oxygen will also slow down. Reasons for a diminished rate of phosphorylation might be a deficiency of inorganic phosphate or of phosphate acceptor, ADP.

The normal, obligatory coupling between respiratory electron flow and phosphorylation can be abolished by "uncouplers" of which 2,4-dinitro-phenol (DNP) is the best known. In the presence of DNP, respiration is released from the phosphorylative constraint which normally puts a brake on the rate of respiration. Uncoupled from phosphorylation, respiration proceeds apace, often at an elevated rate, substrate is utilized, electrons traverse the respiratory chain and oxygen is consumed, but phosphoryl-ation is abolished so that there is no metabolically useful conservation of energy. Therefore, uncoupling inhibits energy requiring processes which depend on continual availability of ATP. Hall and Palmer (1969) have reviewed mitochondrial research and oxidative phosphorylation, and Lehninger (1965) has given a general account of cellular energy metabolism.

Our brief discussion of photosynthesis and respiration has pointed up the obvious conclusion that they are contrasting processes. Photosynthesis converts light energy into chemical energy; respiration expends chemical energy. Photosynthesis abstracts electrons from water and forms powerful reductants; respiration is an oxidative sequence, with water as one of its products. Photosynthesis in green plants results in the evolution of oxygen; in respiration, oxygen is used up. Carbon dioxide is a substrate of photo-synthesis but a product of respiration.

Just the same, this view is pat and misleading. It is only in the initial, photochemical events of photosynthesis having to do with the capture of light energy that photosynthesis is unique. Once these processes are over, with the formation of powerful reductants (the initial electron acceptors of the two photosystems), all the subsequent chemical events in photosyn-thesis are energetically "downhill," like the reactions of respiration. From then on, these sequences of reactions resemble each other in essential fea-tures. In both, reduced intermediates become oxidized, the electrons given up serving to reduce the next intermediate in the sequence. Specifically, electron flow in both includes passage through sequentially arranged elec-tron carriers including cytochromes, the seat of these chains being special-ized organelles (the chloroplast and the mitochondrion, respectively). And in both, electron flow through the electron transport chains is attended by phosphorylation of ADP to ATP (photosynthetic and oxidative phos-phorylation, respectively).

2.   ION TRANSPORT IN NON-GREEN TISSUES

Difficulties, Inherent and Experimental.   The energy required for ion transport in non-green tissues comes from the oxidation of substrates in metabolism. We have already seen (Chapter 6) that anaerobic conditions, poisons, and other antimetabolic agencies inhibit ion transport. The question we now must tackle has to do with the linkage between the energy-yielding processes of respiration and the operation of the mechanisms that transport ions across cellular membranes, that is, the carriers. At the outset we shall have to grant that the researchers who study this linkage between metabolism and ion transport are not agreed on any one scheme. Progress has been and continues slow, and it is worth considering briefly the reasons for this.

We have noted how the spectacular advances in photosynthesis and respiration have depended on the isolation of the cellular particulates—the chloroplasts and mitochondria, respectively—which carry on these specialized functions. This has not been possible with the entities, the carriers, which effect the transport of ions, because of the technical difficulties of isolating functional cellular membranes of plants and the additional factor that the carriers for a given element, potassium for example, are likely to represent only a subunit of the membrane, so that looking for them makes the search for a needle in a haystack seem child's play in comparison.

Another difficulty is the fact that we are attempting to study, not a single function of a given subcellular entity, but the connection between two, each itself incompletely understood: the membrane carriers, on the one hand, and the energy yielding processes in mitochondria, on the other. Disrupting the cell—the traditional approach of the biochemist—destroys the normal pathways of intracellular metabolic connections. And what is even more important, it wreaks havoc with the topology of membrane transport, which implies the existence of two compartments separated by a membrane across which the transport mechanism operates. Such difficulties are certainly not insurmountable, but they have so far not been overcome. James (1962) has discussed some of the problems of mitochondria and their relations with other subcellular entities.

Other difficulties in understanding the energy metabolism of ion transport have to do with features of the experiments themselves. In many investigations on this problem, plant tissues were exposed to solutions of salts of monovalent ions, KCl for example, in the absence of calcium, often for hours on end. But it has become apparent during the last decade or so that exposure of plant tissue to solutions containing only monovalent salts impairs cellular membranes and ion transport mechanisms, the damage being demonstrable within minutes after immersion of the tissue in such solutions, as already discussed in Chapter 6.

On the basis of this and other evidence to the same effect Epstein

(1965) insisted that a solution containing calcium at a low concentration represents a minimal "physiological saline" for plant tissue, the implication being that experiments done in solutions containing no calcium subject the tissue to unphysiological conditions which specifically and immediately damage the very system to be examined: the membranes and ion transport system. "It should be generally realized that ion accumulation studies in the absence of Ca^{++} may reflect the functioning of a partially disorganized transport system" (Hanson, 1965).

Another aspect of experiments on the energy relations of ion transport must be mentioned. In many if not most of the investigations dealing with this subject, tissue was put in solutions containing salt at what are now regarded extremely high concentrations, often 40 or even 100 mM and higher. Even when calcium is present (but it usually was not) monovalent cations at such high concentrations may displace calcium from functional sites of cells (Hanson, 1960; LaHaye and Epstein, 1971, and references cited there).

High salt concentrations present another difficulty. As we saw in Chapter 6, at concentrations of salt above 1 mM absorption of a given ion, say potassium, is effected not by one but by two mechanisms, each contributing an increment of the total rate of absorption. In other words, at high concentrations of salt, the rate of ion absorption measured represents the sum of the contributions of two very different absorption mechanisms. But in many experiments, the rate of absorption measured under these conditions has been considered a unitary value. Under different experimental conditions, the increments to the total rate furnished by the type 1 and type 2 mechanisms can differ a great deal. Undoubtedly, in some instances, the rate measured was largely that of one, and in others, that of the second type, and in most cases, it probably was an unresolved value to which each of the two types of transport mechanisms contributed an appreciable but unknown fraction (Epstein, 1965).

The experimental considerations just discussed, having to do with the role of calcium and the use of high concentrations of salt, apply to many if not the majority of experiments done before the early sixties. We shall later discuss other aspects of certain experiments and their interpretation.

Electrochemical Hypotheses. The earliest scheme to provide a rationale for the linkage between respiration and ion transport was the hypothesis of "anion respiration" which Lundegårdh in Sweden developed initially in the 1930's, with successive modifications since then (Lundegårdh, 1966; see Steward and Sutcliffe, 1959). It was initially based on the observation that addition of salt to tissue respiring in water increases the rate of respiration of the tissue, the increase being quantitatively related to the amount of anion taken up. Furthermore, the added increment of respiration, the "anion respiration," was inhibited by cyanide, in contrast to the "ground respiration." Cyanide also inhibited salt accumulation.

Together, these experiments suggested an intimate association between salt absorption and the cytochrome system of respiration of which cyanide is a potent inhibitor. Since the prime process in respiratory chain electron transport is the flow of electrons from the substrate end via cytochrome electron carriers toward ultimate acceptance by oxygen, the hypothesis envisaged a counter flow of anions in the opposite direction. The cation of the salt was thought to move along passively, following the electropotential gradient set up by the movement of the anion.

Although the Lundegårdh hypothesis in this form is no longer tenable, the scheme illustrates some basic features which it has retained in all its subsequent transformations. It is an electrochemical hypothesis based on the separation of charge when a substrate molecule becomes oxidized through removal of a proton and electron from it. Only one ion of a salt, namely the anion, is moved in the primary transport process. Finally, it predicts a stoichiometry between the rate of respiration and of ion transport, as pointed out by Robertson and Wilkins (1948a, b). In aerobic respiration, four electrons pass through the chain for each molecule of O_2 used up:

$$O_2 + 4H^+ + 4e^- \rightarrow 2H_2O$$

The Australian workers concluded from this that if the hypothesis were valid, no more than four moles of monovalent anions could be transported per mole O_2 used in the anion or salt respiration. They observed no ratios higher than that.

We shall briefly mention some of the difficulties inherent in this hypothesis, leaving the evaluation of specific experimental evidence for later consideration. During the early gestation of the idea, the association of the respiratory chain with the mitochondria was not understood. Now, the scheme might be considered directly relevant to ion transport by mitochondria, but transport at cellular membranes and especially at the plasmalemma is not readily accounted for on this basis.

Another even more basic difficulty has to do with the postulated mechanism itself rather than its site. Biological oxidations of substrates leading to electron transport through the respiratory chain are dehydrogenations. For each electron passing through the chain, a proton is liberated into the solution. Electrical neutrality is maintained and no counterflow of anions would result unless additional postulates are introduced (Laties, 1957).

As a rule, experiments on anion or salt respiration have been done at fairly high concentrations of salt. It is unlikely that at a salt concentration as low as 0.02 mM there would be any measurable salt respiration; yet at this concentration, the type 1 mechanisms of potassium and of chloride absorption operate roughly at half their maximal rate. Does this mean, as Robertson (1968) believes, that the ion transport mechanism related to the salt respiration is of the second (high concentration) type? And if so, what does the work on salt respiration teach us about the energy relations of the type 1 (low concentration) absorption mechanisms? We must remember that the concentrations of ions normally encountered in soil solutions are in the range where only the type 1 mechanisms operate (i.e., below 1 mM), and that it is only at these low external concentrations that really spectacular accumulation ratios are attained. Energetically, accumulation of salt from a solution of, say 10 or 40 mM, and its concentration within the cells to values at most a few times higher, are relatively undemanding.

The most recent and incisive elaboration of the concept of salt respiration in relation to ion transport is due to Robertson (1968). Indeed, his present picture represents such a departure from the earlier concepts that it should be regarded as a new hypothesis. Directly, the events associated with electron flow are confined to the cellular organelles known to contain

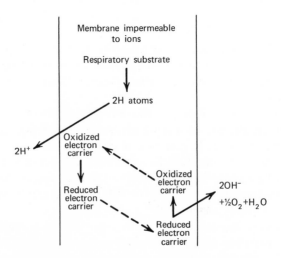

Figure 9-3.   The principle of charge separation, i.e., of the separation of hydrogen ions and electrons, in a membrane. After Robertson (1968).

the electron transport chains, i.e., the mitochondria and, in green cells, the chloroplasts. The basic event is a charge separation which occurs when hydrogen from respiratory substrate gives rise to hydrogen ions and electrons separated from each other in space, presumably by being directed to opposite sides of a membrane. The charge separation yields hydrogen ions on one side of this membrane (considered impermeable to ions). On the other side, the electrons ejected result in the formation of hydroxyl ions as follows: $2e^- + \frac{1}{2}O_2 + H_2O \rightarrow 2OH^-$. The scheme is shown in Figure 9-3.

Robertson visualizes that this separation of charges, effected initially in the mitochondrion, is maintained beyond the confines of the organelle. The cytoplasm is pervaded by the membranous network of the endoplasmic reticulum, and the electrically charged products of the mitochondrial charge separation, i.e., hydrogen and hydroxyl ions, may move to other cellular sites via this network, perhaps within vesicles formed by and being associated with the endoplasmic reticulum. The separated protons and hydroxyls might then drive ion movements, for example, by protons exchanging across a membrane with potassium ions, or hydroxyls with chloride. The energy which was expended initially to bring about the charge separation is conserved until the hydrogen and hydroxyl ions come together again in the same compartment, with the formation of water.

The definitive feature of this hypothesis and similar "redox pump" (Conway, 1953) models of ion transport is that it links transport more or less directly with the process of charge separation by which hydrogen ions and electrons derived from substrate take separate routes. In accord with Mitchell's ideas, Robertson believes that ion transport resulting from the initial charge separation is an alternative to the formation of ATP by the same process (Robertson, 1968). His approach has directed interest to the mitochondria and in green cells, the chloroplasts, and to the connection between ion transport and the generation of ATP by these organelles. This focus on the energy transducing organelles has been one of the most valuable results of the approach, regardless of the validity of the rather tenuous ideas about the connection between the initial charge separation in the organelles and ion transport taking place elsewhere in the cell.

Hypotheses Involving ATP.   The general recognition that ATP is a high-energy intermediate metabolite in many energy requiring processes has led to the obvious idea that it may also furnish the energy necessary for ion transport (Sutcliffe and Hackett, 1957). If, as the evidence indicates, ion transport is brought about by the activity of transmembrane carriers, it is tempting to assume that these agents, in the process of transport, are phosphorylated by ATP through a phosphate transfer enzyme and subsequently dephosphorylated. These processes would be accompanied by binding of the ion, reorientation of the carrier-ion complex within the membrane, and the eventual dissociation of this complex and release of the ion to the far side of the membrane, as shown in Figure 9-4 (and cf. Figure 6-11). The principal source of the ATP is the mitochondrion (and in green cells, the chloroplast). ATP is known to be exported from the organelles, so that transport occurring at cellular membranes, including the plasmalemma and the tonoplast, can readily be visualized as being driven by an ATP-dependent mechanism.

Since the bulk of cellular ATP in non-green aerobic cells comes from phosphorylation of ADP associated with the electron transport chain, inhibition of respiration resulting in inhibition of oxidative phosphorylation would be expected to interfere with ion transport. Respiration (utilization of substrate, electron flow, and concomitant uptake of O_2) not resulting in phosphorylation should not support ion transport.

Consideration of Evidence.   One of the earliest and most telling findings concerning the linkage between respiration and ion transport was the demonstration by Robertson and co-workers (1951) of the effect of 2,4-dinitrophenol, DNP. As discussed above, DNP is an "uncoupler" which permits respiratory electron flow to proceed without concomitant respiratory chain phosphorylation of ADP to ATP. Robertson et al. (1951) showed in experiments with discs of carrot, *Daucus carota,* that DNP increased the rate of O_2 consumption while virtually abolishing the absorption of KCl. The finding is most readily interpreted as evidence that phosphorylation rather than electron flow per se is involved in ion transport although Robertson (1960, 1968) himself did not favor this interpretation.

There are other reports to the effect that ion transport may be inhibited by agents used at concentrations at which they do not interfere, or interfere much less, with respiratory chain electron transport but block phosphorylation—evidence supporting a coupling between respiration and transport via ATP formation (Hackett et al., 1960; Higinbotham, 1959; Ordin and Jacobson, 1955; Shealtiel and Ducet, 1959; Weigl, 1964).

The experiments by Bledsoe et al. (1969) were revealing. They studied the absorption of labeled inorganic phosphate by corn roots, *Zea mays,* and the effect of the antibiotic, oligomycin, which inhibits the synthesis of ATP in mitochondria. They compared ATP synthesis and absorption of inorganic phosphate during a 4-minute period in the absence and presence of oligomycin. Oligomycin inhibited the synthesis of ATP by a third

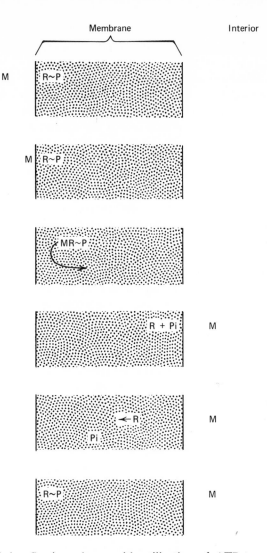

Figure 9-4. Carrier scheme with utilization of ATP to phosphorylate the carrier and activate transport. M, the ion; R, unphosphorylated carrier; R ~ P, phosphorylated carrier; MR ~ P, the phosphorylated carrier-ion complex; P_i, inorganic phosphate. The last two steps entail the phosphorylation of ADP to ATP and the phosphorylation of the carrier, R, by ATP:

$$
\begin{array}{rl}
& ADP + P_i \rightarrow ATP \\
& ATP + R \rightarrow ADP + R \sim P \\
\hline
\text{Sum:} & P_i + R \rightarrow R \sim P
\end{array}
$$

See Figure 6-11 for comparison.

without effect on the absorption of phosphate. During longer periods, oligomycin inhibited the synthesis of ATP even more, and also interfered with the absorption of inorganic phosphate. The results indicate that for a short period, the ATP available in the tissue suffices to drive phosphate absorption, but without generation of more ATP (in the presence of the inhibitor), the supply is quickly depleted and absorption of phosphate is then slowed down. Jacoby and Plessner (1970) found oligomycin to reduce the ATP content of barley roots and to diminish the uptake of potassium, sodium, and chloride. A different kind of evidence implicating ATP in ion absorption is that obtained by Soemarwoto and Jacobson (1965) and by Weigl (1963) who found the turnover or utilization of ATP to be stimulated during salt uptake.

In contrast to these authors, Cram (1969) and Polya and Atkinson (1969) have favored a redox pump coupling between respiration and ion transport. Cram used carrot discs; his results were not incompatible with the interpretation that the ion fluxes are driven by ATP.

Polya and Atkinson (1969) studied the uptake of potassium, sodium and chloride by discs of beet tissue, *Beta vulgaris*. The ions were present at 0.5 mM so that the experiments specifically dealt with the type 1 (high-affinity) absorption mechanisms (Osmond and Laties, 1968). Ion uptake and the ATP level of the tissue were not depressed by various poisons in parallel fashion. Polya and Atkinson interpreted these findings as evidence for a direct connection between electron transport and the type 1 mechanisms of ion absorption. Actually, no results on electron transport were reported, and a less sanguine interpretation than the authors' would be that the evidence failed to establish a linkage between ATP and ion absorption. If, as seems likely, only a small fraction of cellular ATP is involved in ion transport at any one time, correlations between total ATP levels in the tissue and the rates of ion transport need not be close. A compartmentation of ATP in metabolic pools is suggested by the authors' findings, though not by the authors.

In 1957, Skou in Denmark isolated an enzyme or enzyme system from crab nerves which hydrolyzes ATP to ADP and inorganic phosphate. This ATPase required magnesium, sodium, and potassium for full activity, and it fitted the specifications expected in a system mediating the reciprocal movements of potassium and sodium which take place across the nerve cell membrane. Since then, such "Na–K transport ATPases" have been found in many tissues of animals; Skou (1965) has reviewed the evidence of which by now there is a great deal, and research on the subject continues.

Briefly, these animal ATPases related to ion transport have the following characteristics: (1) The enzyme system is virtually ubiquitous in animal cells. (2) It is bound to membranes. (3) It depends for full activity on the presence of both potassium and sodium. (4) It is inhibited by the glycoside, ouabain, which is a potent inhibitor of ion transport in animal tissues.

The scheme of Figure 9-4 shows how such an ATPase might fit into a mechanism of carrier-mediated, ATP-driven ion transport since it envisions the hydrolysis of ATP to ADP linked to the formation of the phosphorylated carrier-ion complex, $M \cdot R \sim P$. Attempts have therefore been made to look for such transport-related ATPases in plant tissues. ATP hydrolyzing enzymes which are stimulated by monovalent cations have indeed been found (Atkinson and Polya, 1967; Brown and Altschul, 1964; Dodds and Ellis, 1966; Hansson and Kylin, 1969); however, these investigations did not produce convincing evidence that these cation-stimulated ATPases are related to ion transport. Neither have consistent inhibitions of ion transport and ATPases by ouabain been shown for plant tissues (Cram, 1968; Gruener and Neumann, 1966; Hodges, 1966; McClurkin and McClurkin, 1967; Mengel, 1963; Pitman and Saddler, 1967).

Nevertheless, we would expect so fundamental a process as ion transport across cellular membranes to show basic biochemical similarities in a variety of systems. There is indeed evidence for a linkage between ATPases in plant tissues and ion transport, although these enzymes may differ a good deal from their counterparts in the animal kingdom.

Histochemical investigations have shown ATPase activity associated with the surfaces of root cells (Chang and Bandurski, 1964; Hall, 1969), and specifically, with the plasmalemma (Hall, 1970). Enzyme activity was associated with plasmalemma vesicles. Hall speculated, in keeping with earlier suggestions (Sutcliffe, 1962), that the plasmalemma forms invaginations which then close to make vesicles entrapping ions from the medium. Subsequently these vesicles dissolve and release the ions into the cytoplasm, this process of "pinocytosis" being energized by an ATP–ATPase system in the plasmalemma. But how would such a scheme account for selectivity in ion uptake? It will in any event be interesting to follow the research on pinocytosis, currently very active (Cocking, 1970), and its implications for the study of ion transport in plant cells.

The most convincing evidence for transport-related ATPases in plant tissue comes from the laboratory of T. K. Hodges in Illinois (now at Purdue University, Indiana). Fisher and Hodges (1969) isolated ATPases from the particulate fraction of homogenates from oat roots, *Avena sativa*. The enzymes were activated by magnesium and further activated by monovalent salts, KCl being the most effective one. However, there was no synergistic effect from the simultaneous presence of both KCl and NaCl, nor was there inhibition by ouabain.

So far, these findings are not different from others mentioned above. However, the authors found that at various concentrations of potassium, the ratio, potassium absorbed by entire roots/potassium-stimulated ATPase, was not far from 1, a correlation which is at least suggestive of a connection between the two. That is, there is enough potassium-activated ATPase to account for the rates of potassium absorption that were measured.

Fisher *et al.* (1970) have extended this investigation to include work with roots of wheat, *Triticum vulgare,* corn, *Zea mays,* and barley, *Hordeum vulgare.* For roots of all species and at all concentrations of salt used (1–50 mM) the ratio between ion transport and ion-stimulated ATPase was approximately 0.85. The findings are consistent with, though they do not prove, the hypothesis of a causal connection between transport and the ion-activated ATPase. There is need for experiments done at lower concentrations (in the range of the type 1 mechanisms).

In conclusion, the problem of the linkage between respiratory metabolism and cellular ion transport in non-photosynthetic tissue has not been solved. The weight of the evidence favors a phosphorylation hypothesis, but no hypothesis now current compels general acceptance. Further evidence which points to ATP as the energy source for ion transport is presented in the next section.

3. ION TRANSPORT IN GREEN TISSUES

Higher Plants.   We have already seen (Chapter 6) that in both algae and green tissues of higher plants, ion uptake is often promoted by light. The early investigators of this phenomenon used aquatic plants, Hoagland and Davis pioneering in the investigation of large-celled algae and Arisz in work with aquatic higher plants. More recently, development of the leaf-slice technique of Smith and Epstein (1964a) has made it possible to do precise experiments with excised leaf tissue of terrestrial higher plants. We shall discuss work on these groups of photosynthetic plants (algae and higher plants) under separate headings because there seem to exist far-reaching differences between algae and green tissues of higher plants in this regard (Steward and Sutcliffe, 1959). But all photosynthetic cells or tissues have in common the advantage that they provide an additional means for studying the connection between ion transport and metabolism, namely experiments on the energy input from chloroplasts in the light and its utilization for ion transport.

Taking up first the matter of ion transport in green tissues of higher plants, we should ask at the outset whether the mechanisms of transport themselves in these tissues are different from those in non-green tissues, quite apart from the problem of their coupling to energy yielding reactions of metabolism. The answer to this question seems, in general, to be no. So many specific features of ion transport in green tissues so closely resemble those in non-green tissues that we are justified in concluding that the basic transport mechanisms are identical.

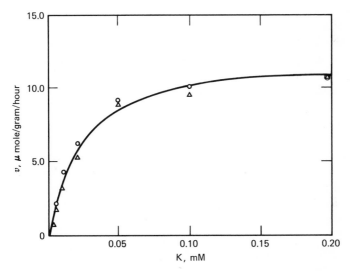

Figure 9-5. Rate, v, of potassium absorption by barley roots and corn leaf tissue as a function of the concentration of KCl. ○, barley roots; △, corn leaf. The curve is a plot of the Michaelis–Menten equation: $K_m = 0.022$ mM; $V_{max} = 12.0$ μmole/g fresh weight/hour. Data from Epstein *et al.* (1963) and Smith and Epstein (1964b).

Figure 9-5 shows the results of two experiments on the rate of absorption of potassium over the range of low concentrations, the range of mechanism 1 (the high-affinity mechanism) of potassium transport. One of these experiments was done with excised barley roots, the other with excised leaf tissue of corn. No statistical normalizing of any kind has been done on these data, which simply report rates of absorption (μmoles per gram fresh weight of tissue per hour) as a function of the external concentration of potassium (mM). The two sets of data agree almost as well as replicates would if they were done with the same tissue. It is unlikely that this result is due to mere coincidence.

Extending the comparison, Smith and Epstein (1964b) showed that the potassium absorption mechanism 1 in corn leaf tissue closely resembles that of barley roots in regard to kinetics, selectivity, effect of the anion, and effect of low temperature. The conclusion was inescapable that the biochemical entities responsible for potassium absorption (the carriers) are identical in roots of barley and leaves of corn—different organs of different species.

Later research by several authors has extended this conclusion. When the concentrations tested include the high range (1 mM and above), green tissues exhibit the dual pattern already familiar from work with non-green tissues (Chapter 6). This applies to the absorption of potassium and other ions as well (Table 6-2).

The problem now before us is therefore not complicated by the operation of different ion transport mechanisms in non-green and green tissues. Rather, we have two clear-cut alternatives to consider. They are whether the transport mechanisms in green tissue depend for their energy supply on the mitochondria, these being supplied by substrates synthesized in the chloroplasts, or whether in green tissue transport can be driven via a direct energy supply from the chloroplasts, by-passing the mitochondria. Anticipating the results briefly summarized below, there is much evidence that green tissue in the light may absorb ions through a direct linkage between metabolites from chloroplasts and the ion absorption mechanisms, without a necessary detour through mitochondrial energy yielding reactions.

Three main lines of evidence support this conclusion: first, the effect of light on the absorption of ions by green tissue in the absence of carbon dioxide, i.e., when the process of photosynthesis cannot go all the way to the fixation of carbon and the synthesis of carbohydrate; second, the action spectrum of photosynthesis and of ion absorption; and third, the effects of inhibitors.

Following earlier work by Arisz, van Lookeren Campagne (1957) studied the absorption of chloride by leaf tissue of the aquatic plant, *Vallisneria spiralis* (tape grass), the classical experimental material in Arisz's laboratory. He found little effect from carbon dioxide on light-induced chloride uptake by the tissue, and concluded that the effect of light energy is not via the process of carbon dioxide reduction and the photosynthetic formation of respiratory substrates. Since such substrates are involved in any scheme implicating the mitochondria, which cannot respire in their absence, the implication of van Lookeren Campagne's finding was that the mitochondria were by-passed under his conditions.

Equally impressive was his evidence on the action spectrum of light-dependent chloride uptake and of photosynthesis. The action spectrum is the relationship between the wavelength of light and the rate of a light-dependent process, care being taken to keep constant the total energy input (quanta) at each wavelength tested. Figure 9-6, taken from van Lookeren Campagne (1957), shows that the two processes, photosynthesis and chloride absorption, responded in parallel fashion to differences in the wave-

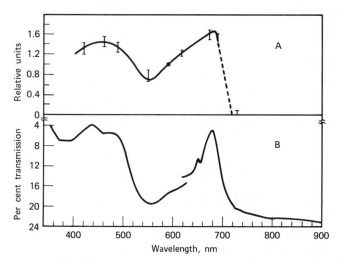

Figure 9-6. (A) Action spectra of chloride absorption and (B) photosynthesis by the leaf of *Vallisneria*. After van Lookeren Campagne (1957).

length of the light used. As before, carbon dioxide was virtually excluded from the reaction vessel; that is, products of the light reactions of photosynthesis were effective in promoting the absorption of chloride, although normal carbon fixation and synthesis of respiratory substrates could not take place.

A more detailed analysis of what takes place was achieved by the third approach, the judicious use of selective inhibitors. Van Lookeren Campagne (1957) found that respiratory inhibitors inhibited only that portion of the chloride absorption by *Vallisneria* leaves that was dependent on respiration, but had no effect on photosynthesis nor on light-dependent absorption of chloride. A tempting conclusion from all this work was that ATP from photophosphorylation was utilized in chloride absorption by this tissue, although at that time the evidence was not compelling.

More definite evidence to the same effect has since been obtained by several investigators (Jeschke, 1967; Jeschke and Simonis, 1969; Nobel, 1969b; Rains, 1967, 1968). They had the benefit of the knowledge of the distinction between cyclic and noncyclic photophosphorylation, and the ability to inhibit one without interfering with the other, or with respiration. Figure 9-7 shows the effects of light and of inhibitors on the rate of absorp-

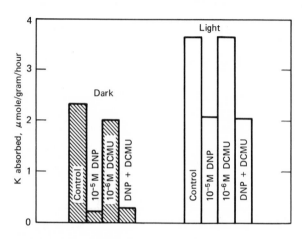

Figure 9-7. Effect of antimetabolites on the rate of absorption of potassium by leaf tissue of corn. Concentration of KCl, 0.1 mM, and of CaSO₄, 0.5 mM. Concentrations of inhibitors indicated on the figure. After Rains (1968).

tion of potassium by corn leaf tissue. Dinitrophenol, the classical uncoupler of mitochondrial (oxidative) phosphorylation, inhibited the rate of absorption of potassium very severely in the dark, much less so in the light. Dichlorophenyl-1,1-dimethylurea (DCMU), which inhibits the photosynthetic evolution of oxygen without interference with cyclic photophosphorylation, was without effect on the absorption of potassium. These and other results by Rains and the other authors mentioned above strongly implicate ATP from photophosphorylation as a source of energy for cellular ion absorption by green tissue in the light. In the absence of carbon dioxide, cyclic phosphorylation alone is involved; in its presence, both cyclic and noncyclic photophosphorylation can donate ATP to drive the ion absorption mechanisms.

In Rains' experiments the kinetics of the absorption of potassium over the low (mechanism 1) range of concentrations strongly suggested that the absorption mechanism was the same in the light and in the dark; only the supply of energy differed. Since the only energy-yielding metabolite generated by both mitochondria and chloroplasts is ATP, his findings and those of the other investigators are strong evidence that mitochondria and chloroplasts perform identical functions in energizing cellular ion transport: the function of supplying ATP.

Green Algae. In this book devoted mainly to the nutrition of higher plants, work on the energy relations of ion transport in algae will be mentioned but briefly. There are several reasons for this, all of them having to do with what seem to be profound differences in this regard between algae and higher plants. These differences induce caution in applying conclusions from work with algae to higher plants. What is more, there seem to be significant differences in ion transport even among the algae, making generalizations and extrapolations even more dubious (Jennings, 1963). It is well to recall W. O. James' (1963) remark: "And in passing I would like to hazard the generalisation that if you wish to concentrate on fundamental questions of form and function it is better to ask them of the higher plant cell, which has been ground down to essentials in the mills of selective evolution, rather than to go to the algae still experimenting with complicated variations."

It seems likely that the differences in mineral ion transport between algae and higher plants have to do with the relations between these plants and their mineral substrates. Algae are wholly immersed in their mineral medium and, their cells once having grown to full size, exist in a state of quasi-equilibrium with the ionic milieu. In higher plants, the root is the prime organ for mineral absorption, but absorption of ions and their retention by the cells of the root amount to only a small fraction of the total ion transport which roots accomplish. Much more is forwarded to the shoot than is retained in the root; if ionic concentrations in root and shoot tissues are about equal, the amount of mineral ions transferred to the shoots is 4 or 5 times as much as the amount retained in the roots of many species, on the basis of the dry weights of roots and shoots (Bray, 1963). For this reason, and the additional one that the concentrations of many ions in the solution bathing them are often exceedingly low, young roots (those most active in absorption) are less likely than algae to be in a state approaching equilibrium with the external solution. The situation is less clear for the cells of leaves. In any event, evidence already reviewed is to the effect that the ion transport mechanisms of the cells of leaves closely resemble those of roots.

Let us examine briefly some outstanding features of ion transport in algae wherein they differ from higher plants. Scott and Hayward (1953a) studied the movements of potassium and sodium by the marine alga, *Ulva lactuca* (sea lettuce). In the dark, the content of potassium in tissue kept in sea water slowly declined below its level in the control tissue kept in continuous light, while that of sodium increased by roughly the same amount. On illumination the tissue in a few hours absorbed potassium to a level well above that of the initial value; after that, the potassium content dropped to that of the (continuous light) control tissue. On the other hand, the effect of light on the sodium content of the tissue was to return it from its elevated value to that of the (continuous light) control. However, it did not seem that these opposing movements of potassium and sodium were tightly coupled, as they are in many animal tissues.

In another experiment with *Ulva lactuca,* Scott and Hayward (1953b) studied the exchange of cellular potassium with (labeled) potassium in the sea water in which the tissue was kept. In the light at 30°C, equilibration was complete in less than one hour; both darkness and a lower temperature (20°C) slowed down the rate of exchange between cellular and external potassium (there was no change in the actual potassium content). In green tissues of higher plants, such exchange is exceedingly slow (Arisz, 1964; Rains, 1968; Rains and Epstein, 1967; see Chapter 6).

Still other characteristic differences between algae and higher plants have to do with ionic selection and mutual ion effects. As we have seen in Chapter 6, the rubidium ion is such a close analog of the potassium ion that the potassium transport mechanisms of higher plant cells fail to distinguish between the two, treating them roughly as though they were isotopes of the same element. In at least some algae, this is not so. For example, West and Pitman (1967) reported that two marine algae discriminate sharply between these two elements, taking up much less rubidium than potassium.

Effects of counterions also vary. In the fresh water alga, *Tolypella intricata,* uptake of potassium was found to depend upon transport of chloride (Smith, 1968); in *Chlorella pyrenoidosa,* also a fresh water alga, it was independent of chloride transport (Schaedle and Jacobson, 1965), and so was absorption of rubidium (Springer–Lederer and Rosenfeld, 1968). In these experiments and most others with algae, there has been no exploration of a wide range of concentrations of the ions investigated, such as has been found revealing in experiments with tissues of higher plants. It may well be that in algae, too, such experiments might turn up different responses at widely different concentrations.

All this said, it nevertheless seems that in spite of the "complicated variations" found in the algae the same basic linkages between metabolism and ion transport must be considered here which we have encountered in higher plants. Specifically, the question whether ion transport in green algae is linked directly to electron transport or is powered by ATP faces us again (MacRobbie, 1966). She concluded from light and dark experiments with the green alga, *Nitella translucens,* in which she used various inhibitors of electron flow and of phosphorylation, that chloride transport was coupled to electron flow but that the transport of potassium depended on ATP. As for the latter conclusion, it is in accord with evidence that in *Chlorella pyrenoidosa,* absorption of rubidium in the light is mainly driven by photophosphorylation (Springer–Lederer and Rosenfeld, 1968).

As for chloride, however, Smith and West (1969) did not find evidence that in another green alga, *Chara corallina* (= *Chara australis* in the older literature), absorption of chloride was coupled directly to electron flow rather than to phosphorylation. They found various inhibitors to have different effects, depending on whether they were applied to entire cells *in vivo* or to isolated chloroplasts *in vitro.* Evidently, the accessibility of inhibitors to the sites at which they produce their effects is an important factor, and one too often overlooked. Furthermore, different plant systems may respond differently to the same inhibitor, and generalizations are therefore notoriously unreliable (see Ellis and MacDonald, 1970). Other recent evidence also favors ATP as the energy source for the absorption of ions by algae (Lilly and Hope, 1971; Penth and Weigl, 1971; Shieh and Barber, 1971).

As with tissues of higher green plants, research on algae has not produced clear proof of potassium–sodium-activated transport ATPases. It may be that such ATPases are unique to systems in which there are tightly coupled, reciprocal movements of potassium and sodium across the cell membrane (Bonting and Caravaggio, 1966; Epstein, 1965). Such movements are common in cells of animals and some microorganisms but not in higher plants.

Organelles. The connection between the cytochrome system and ion absorption by the cells of roots shown by the work of Lundegårdh and by Robertson and his co-workers suggested research on the ionic relations of the mitochondria themselves once these had been recognized as the seat of the respiratory electron transport chain. Such investigations were first carried out in Robertson's laboratory (Robertson *et al.,* 1955). It has become apparent that certain ions are actively transported across mitochondrial membranes and a large body of data has quickly developed. For detailed discussions of this and related work the review by Hanson and Hodges (1967) and the short, well-written book by Robertson (1968) are useful. Energy dependent movements of ions across the membranes of chloroplasts have also been discovered; discussions are given in Robertson's book, by Nobel (1967, 1969a), and by Packer and Crofts (1967). These ionic movements in organelles are by no means incidental but are closely involved in the phosphorylative, energy conserving reactions which both mitochondria and chloroplasts accomplish. However, the mode of this involvement is as yet unknown, and lively research and often controversial discussions are characteristic of this aspect of the study of subcellular metabolism (Packer *et al.*, 1970; Schwartz, 1971).

REFERENCES

Arisz, W. H. 1964. Influx and efflux of electrolytes. Part II. Leakage out of cells and tissues. Acta Bot. Neerl. 13:1–58.

Arnon, D. I. 1955. The chloroplast as a complete photosynthetic unit. Science 122:9–16.

Arnon, D. I. 1967. Photosynthetic activity of isolated chloroplasts. Physiol. Rev. 47:317–358.

Arnon, D. I., M. B. Allen and F. R. Whatley. 1954. Photosynthesis by isolated chloroplasts. Nature 174:394–396.

Atkinson, M. R. and G. M. Polya. 1967. Salt-stimulated adenosine triphosphatases from carrot, beet, and *Chara australis*. Austral. J. Biol. Sci. 20:1069–1086.

Bledsoe, C., C. V. Cole and C. Ross. 1969. Oligomycin inhibition of phosphate uptake and ATP labeling in excised maize roots. Plant Physiol. 44:1040–1044.

Bonting, S. L. and L. L. Caravaggio. 1966. Studies on Na^+–K^+-activated adenosine triphosphatase. XVI. Its absence from the cation transport system of *Ulva lactuca*. Biochim. Biophys. Acta 112:519–523.

Bray, J. R. 1963. Root production and the estimation of net productivity. Can. J. Bot. 41:65–72.

Brown, H. D. and A. M. Altschul. 1964. Glycoside-sensitive ATPase from *Arachis hypogaea*. Biochem. Biophys. Res. Comm. 15:479–483.

Calvin, M. and J. A. Bassham. 1962. The Photosynthesis of Carbon Compounds. W. A. Benjamin, Inc., New York.

Chang, C. W. and R. S. Bandurski. 1964. Exocellular enzymes of corn roots. Plant Physiol. 39:60–64.

Cocking, E. C. 1970. Virus uptake, cell wall regeneration, and virus multiplication in isolated plant protoplasts. Internat. Rev. Cytol. 28:89–124.

Conway, E. J. 1953. A redox pump for the biological performance of osmotic work and its relation to the kinetics of free ion diffusion across membranes. Internat. Rev. Cytol. 2:419–445.

Cram, W. J. 1968. The effects of ouabain on sodium and potassium fluxes in excised root tissue of carrot. J. Expt. Bot. 19:611–616.

Cram, W. J. 1969. Respiration and energy-dependent movements of chloride at plasmalemma and tonoplast of carrot root cells. Biochim. Biophys. Acta 173:213–222.

Dodds, J. J. A. and R. J. Ellis. 1966. Cation-stimulated adenosine triphosphatase activity in plant cell walls. Biochem. J. 101:31P–32P.

Ellis, R. J. and I. R. MacDonald. 1970. Specificity of cycloheximide in higher plant systems. Plant Physiol. 46:227–232.

Epstein, E. 1965. Mineral metabolism. In: Plant Biochemistry. J. Bonner and J. E. Varner, eds. Academic Press, New York. Pp. 438–466.

Epstein, E., D. W. Rains and O. E. Elzam. 1963. Resolution of dual mechanisms of potassium absorption by barley roots. Proc. Nat. Acad. Sci. 49:684–692.

Fisher, J. D., D. Hansen and T. K. Hodges. 1970. Correlation between ion fluxes and ion-stimulated adenosine triphosphatase activity of plant roots. Plant Physiol. 46:812–814.

Fisher, J. and T. K. Hodges. 1969. Monovalent ion stimulated adenosine triphosphatase from oat roots. Plant Physiol. 44:385–395.

Gruener, N. and J. Neumann. 1966. An ion-stimulated adenosine triphosphatase from bean roots. Physiol. Plantarum 19:678–682.

Hackett, D. P. 1955. Recent studies on plant mitochondria. Internat. Rev. Cytol. 4:143–196.

Hackett, D. P., D. W. Haas, S. K. Griffiths and D. J. Niederpruem. 1960. Studies on development of cyanide-resistant respiration in potato tuber slices. Plant Physiol. 35:8–19.

Hall, D. O. and J. M. Palmer. 1969. Mitochondrial research today. Nature 221:717–723.

Hall, J. L. 1969. A histochemical study of adenosine triphosphatase and other nucleotide phosphatases in young root tips. Planta 89:254–265.

Hall, J. L. 1970. Pinocytotic vesicles and ion transport in plant cells. Nature 226:1253–1254.

Hanson, J. B. 1960. Impairment of respiration, ion accumulation, and ion retention in root tissue treated with ribonuclease and ethylenediamine tetraacetic acid. Plant Physiol. 35:372–379.

Hanson, J. B. 1965. Metabolic aspects of ion transport. In: Genes to Genus. F. A. Greer and A. T. Army, eds. International Minerals and Chemical Corporation, Skokie. Pp. 63–74.

Hanson, J. B. and T. K. Hodges. 1967. Energy-linked reactions of plant mitochondria. In: Current Topics in Bioenergetics. D. R. Sanadi, ed. Vol. 2, pp. 65–98.

Hansson, G. and A. Kylin. 1969. ATP-ase activities in homogenates from sugar-beet roots, relation to Mg^{2+} and $(Na^+ + K^+)$-stimulation. Z. Pflanzenphysiol. 60:270–275.

Hatch, M. D. and C. R. Slack. 1970. Photosynthetic CO_2-fixation pathways. Ann. Rev. Plant Physiol. 21:141–162.

Higinbotham, N. 1959. The possible role of adenosine triphosphate in rubidium absorption as revealed by the influence of external phosphate, dinitrophenol and arsenate. Plant Physiol. 34:645–650.

Hill, R. and C. P. Whittingham. 1955. Photosynthesis. Methuen and Company, Ltd., London.

Hodges, T. K. 1966. Oligomycin inhibition of ion transport in plant roots. Nature 209:425–426.

Jacoby, B. and O. E. Plessner. 1970. Oligomycin effect on ion absorption by excised barley roots and on their ATP content. Planta 90:215–221.

James, W. O. 1963. Plant respiration and the microstructure of plant cells. The Advancement of Science 19:375–382.

Jennings, D. H. 1963. The Absorption of Solutes by Plant Cells. Iowa State University Press, Ames.

Jeschke, W. D. 1967. Die cyclische und die nichtcyclische Phosphorylierung als Energiequellen der lichtabhängigen Chloridionenaufnahme bei *Elodea*. Planta 73:161–174.

Jeschke, W. D. and W. Simonis. 1969. Über die Wirkung von CO_2 auf die lichtabhängige Cl⁻-Aufnahme bei *Elodea densa*: Regulation zwischen nichtcyclischer und cyclischer Photophosphorylierung. Planta 88:157–171.

Knaff, D. B. and D. I. Arnon. 1969. A concept of three light reactions in photosynthesis by green plants. Proc. Nat. Acad. Sci. 64:715–722.

LaHaye, P. A. and E. Epstein. 1971. Calcium and salt toleration by bean plants. Physiol. Plantarum. In press.

Laties, G. G. 1957. Respiration and cellular work and the regulation of the respiration rate in plants. Survey Biol. Progress 3:215–299.

Lehninger, A. L. 1965. Bioenergetics. The Molecular Basis of Biological Energy Transformations. W. A. Benjamin, Inc., New York.

Lilley, R. M. and A. B. Hope, 1971. Chloride transport and photosynthesis in cells of *Griffithsia*. Biochim. Biophys. Acta 226:161–171.

Lundegårdh, H. 1966. Plant Physiology. American Elsevier Publishing Company, Inc., New York.

MacRobbie, E. A. C. 1966. Metabolic effects on ion fluxes in *Nitella translucens*. I. Active influxes. Austral. J. Biol. Sci. 19:363–370.

McClurkin, I. T. and D. C. McClurkin. 1967. Cytochemical demonstration of a sodium-activated and a potassium-activated adenosine triphosphatase in loblolly pine seedling root tips. Plant Physiol. 42:1103–1110.

Mengel, K. 1963. Der Einfluss von ATP-Zugaben und weiterer Stoffwechselagenzien auf die Rb-Aufnahme abgeschnittener Gerstenwurzeln. Physiol. Plantarum 16:767–776.

Nobel, P. S. 1967. Calcium uptake, ATPase and photophosphorylation by chloroplasts *in vitro*. Nature 214:875–877.

Nobel, P. S. 1969a. Light-induced changes in the ionic content of chloroplasts in *Pisum sativum*. Biochim. Biophys. Acta 172:134–143.

Nobel, P. S. 1969b. Light-dependent potassium uptake by *Pisum sativum* leaf fragments. Plant Cell Physiol. 10:597–605.

Ordin, L. and L. Jacobson. 1955. Inhibition of ion absorption and respiration in barley roots. Plant Physiol. 30:21–27.

Osmond, C. B. and G. G. Laties. 1968. Interpretation of the dual isotherm for ion absorption in beet tissue. Plant Physiol. 43:747–755.

Packer, L. and A. R. Crofts. 1967. The energized movement of ions and water by chloroplasts. In: Current Topics in Bioenergetics. D. R. Sanadi, ed. Vol. 2, pp. 23–64.

Packer, L., S. Murakami and C. W. Mehard. 1970. Ion transport in chloroplasts and plant mitochondria. Ann. Rev. Plant Physiol. 21:271–304.

Penth, B. and J. Weigl. 1971. Anionen-Influx, ATP-Spiegel und CO_2-Fixierung in *Limnophila gratioloides* und *Chara foetida*. Planta 96: 212–223.

Pitman, M. G. and H. D. W. Saddler. 1967. Active sodium and potassium transport in cells of barley roots. Proc. Nat. Acad. Sci. 57:44–49.

Polya, G. M. and M. R. Atkinson. 1969. Evidence for a direct involvement of electron transport in the high-affinity ion accumulation system of aged beet parenchyma. Austral. J. Biol. Sci. 22:573–584.

Rabinowitch, E. and Govindjee. 1969. Photosynthesis. John Wiley and Sons, Inc., New York.

Rains, D. W. 1967. Light-enhanced potassium absorption by corn leaf tissue. Science 156:1382–1383.

Rains, D. W. 1968. Kinetics and energetics of light-enhanced potassium absorption by corn leaf tissue. Plant Physiol. 43:394–400.

Rains, D. W. and E. Epstein. 1967. Preferential absorption of potassium by leaf tissue of the mangrove, *Avicennia marina*: an aspect of halophytic competence in coping with salt. Austral. J. Biol. Sci. 20:847–857.

Robertson, R. N. 1960. Ion transport and respiration. Biol. Rev. 35:231–264.

Robertson, R. N. 1968. Protons, Electrons, Phosphorylation and Active Transport. Cambridge University Press, Cambridge.

Robertson, R. N. and M. J. Wilkins. 1948a. Quantitative relation between salt accumulation and salt respiration in plant cells. Nature 161:101.

Robertson, R. N. and M. J. Wilkins. 1948b. Studies in the metabolism of plant cells. VII. The quantitative relation between salt accumulation and salt respiration. Austral. J. Sci. Res. B1:17–37.

Robertson, R. N., M. J. Wilkins, A. B. Hope and L. Nestel. 1955. Studies in the metabolism of plant cells. X. Respiratory activity and ionic relations of plant mitochondria. Austral. J. Biol. Sci. 8:164–185.

Robertson, R. N., M. J. Wilkins and D. C. Weeks. 1951. Studies in the metabolism of plant cells. IX. The effects of 2,4-dinitrophenol on salt accumulation and salt respiration. Austral. J. Sci. Res. B4:248–264.

San Pietro, A., F. A. Greer and T. J. Army, eds. 1967. Harvesting the Sun. Photosynthesis in Plant Life. Academic Press, New York.

Schaedle, M. and L. Jacobson. 1965. Ion absorption and retention by *Chlorella pyrenoidosa*. I. Absorption of potassium. Plant Physiol. 40:214–220.

Schwartz, M. 1971. The relation of ion transport to phosphorylation. Ann. Rev. Plant Physiol. 22:469–484.

Scott, G. T. and H. R. Hayward. 1953a. Metabolic factors influencing the sodium and potassium distribution in *Ulva lactuca*. J. Gen. Physiol. 36:659–671.

Scott, G. T. and H. R. Hayward. 1953b. The influence of temperature and illumination on the exchange of potassium ion in *Ulva lactuca*. Biochim. Biophys. Acta 12:401–404.

Shealtiel, M. and G. Ducet. 1959. Considérations sur les relations entre l'absorption des sels minéraux et la respiration. Compt. rend. 249:757–759.

Shieh, Y. J. and J. Barber. 1971. Intracellular sodium and potassium concentrations and net cation movements in *Chlorella pyrenoidosa*. Biochim. Biophys. Acta 233:594–603.

Skou, J. C. 1965. Enzymatic basis for active transport of Na^+ and K^+ across cell membrane. Physiol. Rev. 45:596–617.

Smith, F. A. 1968. Metabolic effects on ion fluxes in *Tolypella intricata*. J. Expt. Bot. 19:442–451.

Smith, F. A. and K. R. West. 1969. A comparison of the effects of metabolic inhibitors on chloride uptake and photosynthesis in *Chara corallina*. Austral. J. Biol. Sci. 22:351–363.

Smith, R. C. and E. Epstein. 1964a. Ion absorption by shoot tissue: technique and first findings with excised leaf tissue of corn. Plant Physiol. 39:338–341.

Smith, R. C. and E. Epstein. 1964b. Ion absorption by shoot tissue: kinetics of potassium and rubidium absorption by corn leaf tissue. Plant Physiol. 39:992–996.

Soemarwoto, O. and L. Jacobson. 1965. The role of respiration in salt absorption. Physiol. Plantarum 18:1077–1084.

Springer–Lederer, H. and D. L. Rosenfeld. 1968. Energy sources for the absorption of rubidium by *Chlorella*. Physiol. Plantarum 21:435–444.

Steward, F. C. and J. F. Sutcliffe. 1959. Plants in relation to inorganic salts. In: Plant Physiology—A Treatise. F. C. Steward, ed. Academic Press, New York and London. Vol. 2, pp. 253–478.

Sutcliffe, J. F. 1962. Mineral Salts Absorption in Plants. Pergamon Press, New York.

Sutcliffe, J. F. and D. P. Hackett. 1957. Efficiency of ion transport in biological systems. Nature 180:95–96.

van Lookeren Campagne, R. N. 1957. Light-dependent chloride absorption in *Vallisneria leaves*. Acta Bot. Neerl. 6:543–582.

Weigl, J. 1963. Die Bedeutung der energiereichen Phosphate bei der Ionenaufnahme durch Wurzeln. Planta 60:307–321.

Weigl, J. 1964. Zur Hemmung der aktiven Ionenaufnahme durch Arsenat. Z. Naturforsch. 19b:646–648.

West, K. R. and M. G. Pitman. 1967. Rubidium as a tracer for potassium in the marine algae *Ulva lactuca* L. and *Chaetomorpha darwinii* (Hooker) Kuetzing. Nature 214:1262–1263.

10

THE ACQUISITION
OF NITROGEN

THE IMPORTANCE OF MICROORGANISMS

We call green plants "autotrophic" (self-growing) because they are able to live and grow by using purely inorganic substrates, that is, without depending on organic foods synthesized by other organisms. Nevertheless the impression of nutritional self-sufficiency which the term autotrophic conveys is misleading when applied to higher green plants. That is because they need nitrogen, the fourth most abundant element in a living plant in terms of the number of atoms it contains (see Table 4-3, p. 63).

Plants can absorb what nitrogen is available in the soil in the form of nitrate or ammonium ions. But the amounts of these ions in most soils at any one time are small, compared with the on-going requirements of the plant cover. Furthermore, nitrate is readily leached by rain or irrigation water into deep layers of soil where roots cannot penetrate, since it is a negatively charged ion which is not attracted by the negatively charged exchange surfaces of soil materials. Ammonium ions are held in the soil by electrostatic attraction because of their positive charge, but are readily converted to nitrate in most soil situations and then become subject to loss through leaching. And finally, fixed nitrogen is reconverted to gaseous nitrogen by denitrifying microorganisms (Broadbent and Clark, 1965; Delwiche, 1965; McKee, 1962).

Therefore, soil nitrogen in a form available for absorption by plants must be constantly replenished from the large reservoir of gaseous nitrogen in the atmosphere. This replenishment takes place to a minor extent through fixation of nitrogen in lightning and other electrical discharges in the atmosphere, which result in the formation of various oxides of nitrogen. These are eventually carried into the soil by rain and snow. An increasingly important contribution to the available supply of fixed nitrogen is made by man through industrial fixation of nitrogen and the manufacture of nitrogenous fertilizers.

All other fixation of atmospheric nitrogen, and that means by far the greatest fraction of the total, worldwide, is biological, and all biological nitrogen fixation involves the activities of microorganisms. Without these, the primary productivity of the soils of the world would be a minute fraction of what it is. Thus, although higher green plants need no organic substrate for their growth, they depend on other organisms for their supply of a key inorganic nutrient, nitrogen, in a form that they can absorb and assimilate.

Nitrate and ammonium ions present in the soil from whatever source—biological nitrogen fixation, fertilizers supplied by man, decay of organic matter, or fixation by atmospheric electricity—are absorbed by the roots of plants. In essence, this process of absorption of nitrogen, once it is fixed, does not differ from other ion absorption processes, for example, the absorption of potassium. This aspect of nitrogen nutrition is discussed in the next section.

However, the initial biological fixation of gaseous nitrogen is a unique process. The microorganisms which accomplish it are free-living or associated in either loose or truly symbiotic relationships with plants. The distinction between these categories is not always sharp (Delwiche, 1965). For example, even free-living nitrogen fixing microorganisms are often most abundant in the vicinity of roots and probably benefit by substances released by the roots into the soil. The biochemical events of nitrogen fixation are probably similar in the free-living and the symbiotic systems.

2. ABSORPTION OF NITRATE AND AMMONIUM IONS

Experimental and Biological Considerations. Experiments on the mechanisms of the absorption of nitrate and ammonium ions have been less popular than experiments on most other macronutrient ions. This might seem surprising in view of the great importance of nitrogen in the fertility of soils and the nutrition of plants. But the reason is simple. For the solution of many problems concerning ion absorption by plant tissues, radioisotopes have been exceedingly useful because of the ease and extreme sensitivity of radioassay methods, and because they provide a means of distinguishing among different pools or fluxes of the same element. But there is no very useful radioisotope of nitrogen, ^{13}N having a half-life of 10.05 minutes.

As a result, experimenters either had to use the stable isotope, ^{15}N, whose assay by mass spectrometry is more difficult and less sensitive than most radioassay procedures, or rely on chemical methods of analysis which also do not approach radioassay in ease and sensitivity, especially for short-term experiments.

Another special aspect of the absorption of nitrate and ammonium ions is inherent in the biology of the process and its metabolic consequences. There are other mineral nutrient elements which are massively incorporated into organic metabolites, for example, phosphorus and sulfur, but none equal nitrogen in the extent to which this takes place, nor in the rapid, far-reaching effects which it has on metabolism. For this reason, absorption of nitrate or ammonium ions does not necessarily lead to an accumulation of these ions within the cells. Experimentally, reduction of nitrate may be inhibited by means of tungstate, and nitrate absorption may then be studied without the complicating effects of its subsequent reduction and assimilation (Heimer *et al.,* 1969; Heimer and Filner, 1971). The uptake system is induced by its substrate, nitrate (see Nitrogen in section 3 of Chapter 11).

Mechanisms of Absorption. The Dutch investigator Becking (1956), using an experimental set-up initially designed by van den Honert for his pioneering studies on phosphate nutrition (Chapter 6), exposed the roots of intact corn plants to flowing solutions containing ammonium ions at concentrations ranging from very low values (approximately 0.005 mM) to 1 mM. He measured the (fairly small) difference in the ammonium concentration of the solutions before and after they had passed through the vessel from which the roots absorbed the ions, and calculated the amounts taken up. The relationship between the concentration of ammonium and its rate of absorption was hyperbolic, with a plateau (maximal uptake) between 0.5 and 1 mM. Half the maximal uptake was obtained at 0.013 mM. In other words, the response of the plants to a range of concentrations of ammonium, in terms of its absorption, much resembled the relation between the concentration of potassium and its rate of absorption in many systems (see, for example, Figures 6-12 and 6-14). This, despite the fact that Becking experimented with intact, growing plants, whereas the experiments with potassium referred to were done with excised tissues.

Using the same experimental set-up, van den Honert and Hooymans

(1955) studied the absorption of nitrate by corn plants. The concentration of nitrate ranged from very low values to about 0.24 mM. Again the relation between concentration and uptake was hyperbolic, with a plateau (maximum uptake) between 0.12 and 0.24 mM nitrate in the solution and half-maximal uptake (at pH 6) at a nitrate concentration of 0.023 mM. That is, absorption of nitrate by growing corn plants showed characteristics very much like those of the absorption of another monovalent anion, chloride, by excised barley roots in short-term experiments (Elzam and Epstein, 1965; and see Table 6-1). As with chloride and other ions, the presence of calcium in the medium is essential for unimpaired absorption of nitrate (Burström, 1954; Rao, 1971), and respiratory poisons, anaerobic conditions, and low temperatures inhibit it (Rao, 1971). Accumulation of nitrate by fungal cells (the mold *Scopulariopsis brevicaulis*), is also inhibited by the respiratory poisons, cyanide and azide (MacMillan, 1956).

In sea water, the concentration of nitrate and ammonium ions is low in the extreme. We might therefore suspect that the plants in the ocean are able to absorb and accumulate these ions with extraordinary avidity. Eppley and Thomas (1969) and Eppley *et al.* (1969) did kinetic experiments on absorption of nitrate and ammonium ions by several species of marine phytoplankton. They found a simple hyperbolic relationship between the external concentration of the ions and their rate of absorption. Interestingly, and in keeping with the expectation, the concentrations giving half the maximal rate of uptake (the "Michaelis constants," K_m) were very low, on the order of a fraction of a μmole per liter for many species. Figure 10-1 shows results for nitrate uptake by the coastal diatom, *Skeletonema costatum*.

In the experiments discussed so far in this section, a single hyperbolic function described the relationship between the concentration of each ion in the solution and its absorption by the plants. Several investigators have used the stable isotope of nitrogen, ^{15}N, in short-term experiments on the absorption of nitrogen by excised roots. They extended the range of concentrations to 10 mM. Fried and co-workers (1965) were the first to use this technique; they studied the absorption of ammonium and nitrate ions by excised roots of rice, *Oryza sativa*. Absorption of ammonium ions showed a typical dual pattern (see Chapter 6). Results for nitrate were less clear-cut but a dual pattern was at least suggested.

Picciurro *et al.* (1967) and Berlier *et al.* (1969) did similar experiments with ^{15}N-labeled ammonium ions; the former used excised roots of wheat, *Triticum durum* and *T. vulgare,* the latter excised roots of corn, *Zea mays.* Typical dual absorption patterns were obtained.

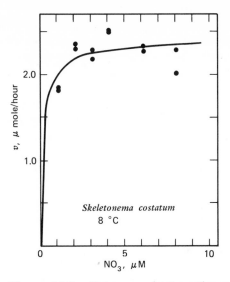

Figure 10-1. Rate, v, of absorption of nitrate by the coastal diatom, *Skeletonema costatum*. The curve is a plot of the Michaelis—Menten equation. $K_m = 0.4$ μM. After Eppley *et al.* (1969).

The evidence we have considered is to the effect that absorption of ammonium and nitrate ions by plants and plant tissues is via mechanisms of ion transport of the same type and having the same basic characteristics as those which mediate the absorption of other ions. In the case of ammonium, the coincidence between the kinetics of its absorption and that of potassium is due to more than just similarity in basic pattern: it is more likely that the carrier mechanisms are identical.

This conclusion is based on experiments on mutual competition among alkali cations and ammonium ions. There is much evidence, already discussed in Chapter 6, that in the low (mechanism 1) range of concentrations, the alkali cations potassium, rubidium, and cesium compete for identical carrier sites in their absorption by plant cells. The affinities of potassium and rubidium for these sites are roughly the same, but that of cesium is less. Bange and Overstreet (1960) studied the absorption of cesium by excised barley roots and concluded that potassium and rubidium were about equally effective competitors for the cesium transporting sites. Ammonium ions also competed, although with considerably less affinity. A more strictly quantitative study is that by Smith and Epstein (1964) on absorption of potassium and rubidium by leaf tissue of corn. Kinetic analysis showed ammonium to be strictly competitive with potassium–rubidium, but its affinity for the common transport mechanism was only about one tenth that of the alkali cations.

Absorption of nitrate is followed by reduction of the nitrogen and its incorporation into organic nitrogen compounds. Going from its oxidation number of $+5$ in NO_3^- to that of -3 in NH_3 requires eight electrons. The nitrate-reducing enzyme system performs a sequential, step-wise reduction. Nitrate reductase is a substrate inducible enzyme, i.e., it is synthesized or activated by the presence of its substrate, nitrate. The subject of nitrate reduction has been reviewed by Bandurski (1965) and by Beevers and Hageman (1969). Metabolic effects of nitrate vs. ammonium nutrition are discussed in Chapter 11.

3. NITROGEN FIXATION

Biological Systems which Fix Nitrogen.   The early history of the discovery of the role of bacteria in symbiotic nitrogen fixation has been briefly given in Chapter 1, and the place of various nitrogen-fixing agents in the nitrogen economy of the biosphere has been indicated in Chapter 2. Table 10-1, an expanded version of one by Burris (1965), lists the biological agents that fix nitrogen. We shall briefly discuss them in the following paragraphs.

Quantitatively, nitrogen fixation by the legume-*Rhizobium* symbiotic system is by far the most important in the temperate regions of the earth. Imshenetskii (1962) has discussed the evolution of biological nitrogen fixation, and Dilworth and Parker (1969) have dealt specifically with the evolution of the nitrogen-fixing system in legumes. There are about seven hundred genera with fourteen thousand species of plants in the family *Leguminosae* (Stewart, 1966) and most of them can be partners with *Rhizobium* in nitrogen-fixing associations. In the United States, leguminous crops fix an estimated 5.5 million tons of nitrogen per year (Burris, 1965). This roughly equals the amount of nitrogen now applied yearly in the form of nitrogenous fertilizers (Harre, 1969). Worldwide, the percentage of nitrogen from synthetic fertilizers is much less, but it is rising fast.

Second to the legumes in quantitative importance among the biological nitrogen-fixing systems are the non-leguminous plants which form root nodules. The symbionts in this case seem to be of various kinds, but their isolation and positive identification has been difficult (Burris, 1966; Stewart, 1966; Thimann, 1963). Becking (1970a) places them in a new family, *Frankiaceae,* of the bacterial order, *Actinomycetales.*

TABLE 10-1.   Biological Nitrogen Fixing Systems

A. Symbiotic associations
  1. Root nodules
     a. Legumes with root nodule bacteria (*Rhizobium*). Peas, beans, clover, soybean, etc.
     b. Non-legumes with root nodule organisms. Alder (*Alnus*), *Ceanothus*, *Casuarina*, etc.
     c. Non-legumes with root nodule blue-green algae. Tropical trees of the order *Cycadales*.
  2. Leaf nodules
     a. Non-legume angiosperms with leaf nodule bacteria. Tropical trees, *Psychotria*, *Ardisia*, etc.
     b. Non-legume angiosperms with leaf gland blue-green algae. Herbaceous species in moist tropical habitats, *Gunnera* spp.
B. Less intimate associations
  1. Bacteria on the leaf surfaces of tropical rainforest trees.
  2. Bacteria on the leaf surfaces of temperate-zone trees.
  3. Blue-green algae associated with ferns, liverworts, and fungi.
C. Free-living
  1. Blue-green algae (*Cyanophyta*), *Nostoc*, etc.
  2. Fungi, yeasts, actinomycetes.
  3. Bacteria
     a. Aerobic. *Azotobacter vinelandii*, *Beijerinckia indica*, etc.
     b. Facultative. *Aerobacter aerogenes*, etc.
     c. Anaerobic
        (1) Nonphotosynthetic. *Clostridium pasteurianum*, etc.
        (2) Photosynthetic. *Rhodospirillum rubrum*, etc.

Compared with the hundreds of genera and thousands of species of legumes, the number of non-leguminous genera and species of angiosperms known to include root nodule nitrogen-fixing symbiotic associations is small—13 genera with a total of 100 species (Bond, 1968). All non-leguminous angiosperms bearing root nodules are woody plants and unlike the legumes, they are not made use of in agriculture. However, they play an important role in the nitrogen economy of wildlands including forests. Most of these plants have temperate or arctic habitats. Bond (1963, 1967, 1968), Allen and Allen (1965), and most recently, Becking (1970b) have interesting reviews of non-leguminous nitrogen-fixing symbiotic systems—a subject sure to be of increasing scientific and practical importance in the future.

Certain tropical, non-leguminous trees and shrubs have leaf nodules inhabited by bacterial endophytes which fix nitrogen (Silver *et al.,* 1963). The moist surfaces of the leaves of many plants in tropical rainforests form a habitat (the "phyllosphere") for bacteria among which there are nitrogen fixers, and it is likely that the host plants acquire some of the nitrogen fixed (Ruinen, 1956, 1961, 1970). Nitrogen fixation in the tropics is a little known topic. But it is scientifically exciting and will certainly be economically rewarding once more research effort gets devoted to it than has so far been the case. According to a report by Jones (1970) such an association of nitrogen-fixing bacteria with the leaves of trees may occur in the temperate zone also. He found this to be so in the phyllosphere of Douglas fir, *Pseudotsuga douglasii,* in the English Lake District.

Blue-green algae are the most important photosynthetic nitrogen-fixing organisms. Not all *Cyanophyta* fix nitrogen but all nitrogen-fixing algae belong to this group (Fogg, 1962). In 39 species belonging to 14 genera, nitrogen fixation has been shown to occur in pure culture (Stewart, 1966). Blue-green algae live both free and in symbiotic as well as less intimate associations. Such associations are formed with certain higher plants, ferns, liverworts, and fungi (Bond, 1968).

The higher plants forming symbiotic root nodule associations with nitrogen-fixing blue-green algae are small, tropical trees belonging to the order *Cycadales,* "living fossils," the lowest form of seed bearing plants. There are nine genera with about 90 species, and in about 25 per cent of these root nodules have been observed. The microorganisms involved seem to be exclusively blue-green algae (Allen and Allen, 1965). The nitrogen-fixing, blue-green alga *Nostoc puntiforme* invades the bases of leaves of herbaceous plants belonging to the genus *Gunnera* which is common in moist habitats in the southern hemisphere. The symbiosis contributes to the nitrogen economy of the host (Silvester and Smith, 1969).

Associations of blue-green algae of the genus *Anabaena* with the floating fern, *Azolla,* seem to be important to the nitrogen economy of rice fields in the tropics (Bond, 1968). Blue-green algae of the genus *Nostoc* and others symbiotically associated with liverworts, and with algae in lichens, fix nitrogen and do so in the free-living state as well (Watanabe and Kiyohara, 1963).

Free-living, blue-green algae which fix nitrogen are of great importance for the nitrogen economy of rice culture in the Orient (Lund, 1967; Singh, 1961). All these algae are filamentous. Interspersed among the normal, vegetative cells are "heterocysts"—cells which are different in appearance from the rest, especially in that they have lost the photosynthetic lamellae of the normal cells. Nitrogen fixation is confined to these specialized cells, according to Stewart *et al.* (1969), but Smith and Evans (1970) demur.

Free-living, nitrogen-fixing, blue-green algae, along with the nitrogen-fixing photosynthetic bacteria (see below), are among the most completely autotrophic organisms known, being capable of acquiring both carbon and nitrogen in a self-sufficient manner. Stewart (1970) has reviewed nitrogen fixation by blue-green algae.

Certain fungi, such as yeasts of the genus *Rhodotorula* and some free-living actinomycetes (*Nocardia*) fix nitrogen, but they are of little quantitative significance in the nitrogen economy of nature and their nitrogen-fixing capabilities have not so far been studied a great deal (Stewart, 1966).

There remain for discussion the free-living nitrogen-fixing bacteria. They are quantitatively less important in the nitrogen balance of nature than are the symbiotic systems and the free-living blue-green algae (Campbell and Lees, 1967). However, their contribution may have been underestimated (Postgate, 1970). They have been important objects of research on the mechanism of nitrogen fixation ever since Winogradsky isolated the anaerobic *Clostridium pasteurianum* in 1893 and showed that it fixed nitrogen. Beijerinck in 1901 discovered nitrogen fixation by the aerobic *Azotobacter chroococcum* and *A. agilis* (Virtanen and Miettinen, 1963). Facultative organisms (*Aerobacter aerogenes* and others) are better nitrogen fixers under anaerobic than aerobic conditions, and it is likely that the aerobes (*Azotobacter, Beijerinckia*) are anomalous and manage nitrogen fixation (a reductive process) only by achieving local intracellular reducing conditions (Burris, 1965).

The anaerobic photosynthetic bacteria that have been examined in this regard all have been found to be capable of nitrogen fixation, but recognition of this capability was slow in coming. Kamen and Gest (1949) made the discovery in experiments with the photosynthetic anaerobe, *Rhodospirillum rubrum*. Their discovery was quickly confirmed and extended to other species of photosynthetic bacteria by University of Wisconsin bacteriologists (Lindstrom *et al.,* 1949, 1950, 1952).

The Physiology of the Formation of Root Nodules. A large body of reliable information is available only for the legume-*Rhizobium* root-nodule associations. They differ in several distinctive ways from the non-legume angiosperm associations of the alder type, and even more widely from the *Cycadales-Cyanophyta* (blue-green algae) root-nodule systems. The former two types have been systematically compared by Allen and Allen (1965). Our discussion will mainly deal with the physiology of the legume-*Rhizobium* system.

The formation of nodules effective in nitrogen fixation is an elaborate process subject to many influences, both endogenous ones supplied by the plant and external factors in the root medium. There is first of all a genetic component governing the compatibility between the particular species and strain of *Rhizobium* and the host legume. Incompatibility between a given host and a given strain of *Rhizobium* may result either in a failure of inoculation and nodule formation, or in the development of nodules more or less ineffective in nitrogen fixation. Accounts of these specificities are by Burton (1965), Vincent (1967), and Nutman (1965).

Given compatible strains of host and symbiont, the sequence of events may be summarized as follows (Munns, 1968a). (1) Development of root hairs. (2) Development of a population of *Rhizobium* bacteria near the surface of the roots, in the rhizosphere. (3) Curling of root hairs and their infection by the bacteria. (4) Development of infection threads. (5) Formation of root-nodules.

The process of infection has been described by several authors; good accounts are by Thimann (1963), Nutman (1963), Stewart (1966), and Dixon (1969). The bacteria congregate in large numbers around the root, often in a layer many cells deep. This proliferation is stimulated by the presence of organic metabolites excreted by the roots (Rovira, 1969). The close association between bacteria and root surface results in the elaboration or release of material which causes the root hairs to become curled and assume a crooked appearance. There is ample evidence that auxin, indole-3-acetic acid, plays a role in this process (Dixon, 1969). Soon after this the first bacteria appear within the epidermal cells of the root. The bacteria now aggregate into threads which are surrounded by a membranous structure containing cellulose and pectin. This is called the infection thread which is continuous with the host cell wall and plasmalemma. It makes its way from the initial site of the infection, the root hair, into the underlying cortical cells, probably traversing the cell walls via the primary pit areas. Eventually, the bacterial cells are released into the cells of the root through dissolution of the wall of the infection thread. In this condition they are many times larger than *Rhizobium* cells in the soil and differ from them in shape; the rhizobia within the root cells are called "bacteroids" and it is they which stimulate the host cell to enlarge. The bacteroids are not in immediate contact with the cytoplasm of the host cells but remain enclosed in the plasmalemma derived from the infection thread (Dixon, 1969).

Not all cells of the root are equally responsive to the stimulus provided by the bacterial infection. All roots contain, in addition to the normal diploid cells, a small number of tetraploid cells, and it is these that upon infection become the center of renewed meristematic activity which results in the formation of the nodule. Most of its cells are therefore tetraploid.

Chemical factors influencing nodulation come from cotyledons, leaves, lateral root initials and from other nodules, but the chemicals involved and their mode of action in nodule morphogenesis are even less completely known than are other aspects of morphogenesis. One fascinating finding is that by Lie (1964, 1969) to the effect that the phytochrome system which governs many plant responses (flowering, seed germination, etc.) plays a role in the control of nodulation.

External factors, including mineral nutrition, which govern nodulation are somewhat better known than the endogenous ones (Vincent, 1965). Chief among them are a high calcium requirement, pH, and the inhibitory effect of nitrate. The process of infection or initiation of the development of the nodule depends on a concentration of calcium in the medium higher than that necessary either for the subsequent development of the nodule or the growth of the host plant when it does not depend on symbiotic nitrogen fixation, i.e., when fixed nitrogen is supplied in the medium. This is shown in Figure 10-2 for subterranean clover, *Trifolium subterraneum* (Lowther and Loneragan, 1968). Going from 246 μM calcium to 720, growth of the plants was not affected (fixed nitrogen was supplied so that

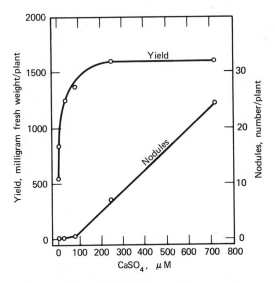

Figure 10-2. Effect of the concentration of calcium in solution on the yield and number of nodules of subterranean clover, *Trifolium subterraneum*. After Lowther and Loneragan (1968).

nitrogen was not a limiting factor), but the number of nodules was trebled. Their distribution along the roots was also influenced. At the 246 μM level of calcium and below, nodules were restricted to near the base of the roots; at 720 μM calcium, their distribution was more even.

The process of infection is sensitive to low pH. Munns (1968a) showed that when the pH of the nutrient solution was kept at 4.4, root hairs of

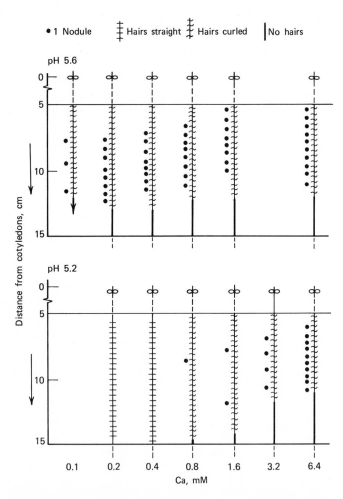

Figure 10-3. Interaction between the concentration of calcium and the pH of the solution in affecting the nodulation of roots of alfalfa, *Medicago sativa*. After Munns (1970).

alfalfa, *Medicago sativa,* failed to curl and become infected even when the cultures were heavily inoculated with the appropriate strain of *Rhizobium*. When the pH was raised to 5.4 the root hairs promptly curled and became infected. Once curling had taken place the pH could be lowered again to 4.4 without adverse effect on subsequent infection and normal development of nodules. The acid-sensitive step occupied only about 12 hours of the four to seven days required for visible nodules to appear. The calcium and pH factors interact: the lower the pH, within the range 4 to 6, the higher the calcium requirement (Loneragan and Dowling, 1958; Munns, 1970). Figure 10-3 shows this interaction for alfalfa, *Medicago sativa.*

Nitrate present in the medium has an inhibitory effect on nodulation; occasional failures to observe this seem to be due to inadequate control of experimental conditions (Munns, 1968b, c). Prior nodulation was as effective as the presence of nitrate in the medium in inhibiting subsequent nodulation in his experiments. The auxin, indole-3-acetate, mitigated but did not eliminate the adverse effects of acidity and nitrate on nodulation (Munns, 1968d).

The mature nodule consists of several well-defined regions. There is an inner core of bacteroid cells. These cells contain hemoglobin which gives this region a pink color. A meristematic region or growing point is sometimes in evidence, but it is not always present or well defined. The vascular tissue is peripheral to the central core. Each vascular bundle has xylem elements oriented toward the center and phloem toward the periphery. These tissues are surrounded by specialized cells of a pericycle and outside this lies an endodermis with a typical Casparian strip. The outermost region of the nodule consists of cortical tissue. Figure 10-4, after Stewart (1966), shows several stages in the development of the pea nodule and its mature configuration. Figure 10-5 shows nodules on the roots of the field bean after infection by *Rhizobium* bacteria. Pate *et al.* (1969) have given a detailed description of pea and clover nodules, with micrographs.

Physiology of Symbiotic Nitrogen Fixation.   Most information is on the legume-*Rhizobium* symbiosis. The chief net result of the symbiosis is the fixation of gaseous nitrogen and the export of the products of the fixation to the host plant and, to a lesser extent and not invariably, to the medium. The main contribution of the host plant to the bacterial symbionts is in the substrates, mainly carbohydrates, furnished by the shoots. The best evidence for this is the quick cessation of nitrogen fixation by nodules once they have been detached from the plant, and the stimulating effect on nitrogen fixation by such nodules of supplying them with

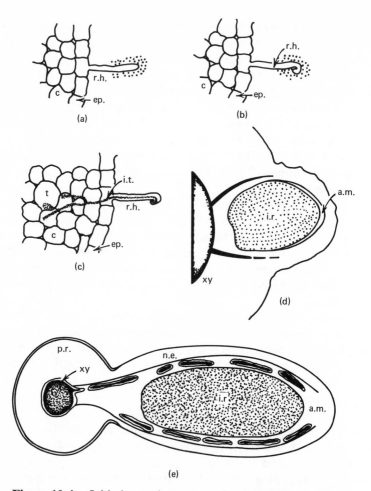

Figure 10-4. Initiation and structure of nodules of pea, *Pisum sativum*. (a) Rhizobia aggregate around root hairs. (b) Root hairs curl. (c) Rhizobia infect root hair and move through the root hair and inner cortex until they enter a tetraploid cell; this stimulates meristematic activity. (d) A central infected region and apical meristem become distinguished. (e) Longitudinal section through nodule showing central infected region, apical meristem and nodule endodermis. (a.m., apical meristem; c., cortex; ep., epidermis; i.r., infected region; i.t., infection thread; n.e., nodule endodermis; p.r., primary root; r.h., root hair; t., tetraploid cell; xy., xylem.) After Stewart (1966).

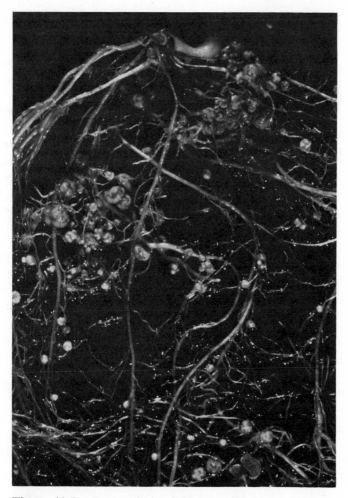

Figure 10-5. Roots of the field bean, *Phaseolus vulgaris,* grown for four weeks in nutrient solution inoculated with *Rhizobium phaseoli* which has formed many large nodules. Photograph by R. Avila–Lozano, courtesy of D. N. Munns.

carbohydrate (Bach *et al.,* 1958). The same workers observed that when soybeans, *Glycine max*, were allowed to photosynthesize CO_2 labeled with ^{14}C for several hours, followed by some hours of continued photosynthesis of unlabeled CO_2, the nodules contained several per cent of the ^{14}C fixed, with a specific activity higher than that of the roots.

Not only the process of nodulation, but also nitrogen fixation by roots already nodulated, is inhibited when fixed nitrogen is available, the more so the higher the concentration of fixed nitrogen furnished the plants (Stewart and Bond, 1961). However, fixation of nitrogen, though diminished, continues at an appreciable rate even when ammonium is furnished. In the field, where levels of fixed nitrogen are seldom high, fixation by nodulated plants probably proceeds apace.

After nitrogen fixation, the organic compounds of nitrogen resulting are transferred to the vascular system and become available to the host plant. Apparently there is no need for the bacteroids to lyse (disintegrate) to liberate the fixed nitrogen. One or the other of two alternative mechanisms seems more likely. Either the nitrogenous compounds are excreted from the bacteroids into the host cells, or the site of nitrogen fixation is not within the bacteroids at all but external to them, obviating the necessity for either lysis or secretion. The details of the export of nitrogenous metabolites from the sites of their synthesis into the vascular tissue remain to be worked out (Pate et al., 1969). It is known in any event through the work of Bond (1956, 1964) that the long-distance transport of the nitrogenous assimilates is mainly in the xylem, via the transpiration stream. This was found in ringing experiments similar to those of Stout and Hoagland discussed in Chapter 7.

What are the organic compounds of nitrogen which are eventually delivered into the xylem and transported into the tissues of the host plant? Studies by Pate and co-workers have shown that they are mainly amino acids and amides (Pate, 1968; Pate et al., 1965). As much as 80 to 90 per cent of the total nitrogen assimilated may be carried by the two amides, asparagine and glutamine, the proportions of these two and of several amino acids varying at different stages of the development of the plant. Figure 10-6 shows the traffic pattern of photosynthate and products of nitrogen fixation in the field pea, *Pisum arvense*.

Biochemistry of Nitrogen Fixation. The nitrogen gas molecule, $N{\equiv}N$, is exceedingly inert. Industrial fixation of nitrogen is accomplished only through application of very high temperatures and pressures (450°C, 250–1000 atmospheres N_2). It is only recently that in the laboratory, complexes of nitrogen have been obtained under relatively mild conditions (Chatt and Leigh, 1968; Hardy and Knight, 1968), and they are at best models or analogs of what nature has achieved under physiological conditions of temperature and pressure (30–35°C, 0.2–1 atmosphere N_2). The fixation of nitrogen by microorganisms therefore ranks with the assimilation of CO_2 by green plants and their accumulation of potassium from micromolar solutions as one of the most remarkable instances of the chem-

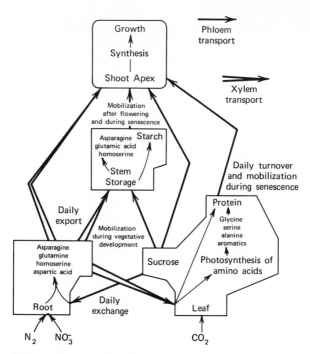

Figure 10-6. The traffic of photosynthate and nitrogen compounds in the field pea, *Pisum arvense,* when nodulated and fixing nitrogen or uninoculated but growing in the presence of a moderately low concentration of nitrate. After Pate (1968).

ical acquisitiveness of the biosphere—its ability to mine an inert environment for chemically elusive nutrient elements.

The study of the biochemistry of this remarkable process had to await the development of methods whereby nitrogen fixation could be achieved in cell-free systems. There were numerous reports of such experiments before 1960, and no doubt real fixation was achieved in a number of these, but results were inconsistent and lead to no sustained advances in the biochemical study of nitrogen fixation (Burris, 1966).

The cell-free period was ushered in by the reports of Carnahan *et al.* (1960a, b) of Du Pont de Nemours on a method of extracting dried cells of the free-living anaerobe, *Clostridium pasteurianum,* resulting in soluble preparations which fixed nitrogen, and did so vigorously and consistently. At the same time, Schneider *et al.* (1960) at the University of Wisconsin obtained consistent and fairly vigorous fixation by extracts from blue-green algae. Cell-free preparations from several other microorganisms have since been prepared which fix nitrogen (Carnahan and Castle, 1963; Hardy and Burns, 1968). For the legume-*Rhizobium* system, the breakthrough to cell-free proceedings was achieved by Evans' group at Oregon State University, using extracts of bacteroids from nodules of the soybean, *Glycine max* (Klucas *et al.,* 1968; Koch *et al.,* 1967a, b).

The sequence of events leading from free nitrogen gas, N_2, to the "key" intermediate is as yet incompletely known. The "key" intermediate was defined by Wilson and Burris (1953) as "the compound which represents the end of the fixation reaction and the start of assimilation of the fixed nitrogen into an organic carbon skeleton." Although other compounds have been cast in the role of the key intermediate, there is no longer any doubt that it is ammonia. The best pieces of evidence for this conclusion are the following. Ammonia and the amides which are early products of its assimilation incorporate ^{15}N with a minimum time lag and at higher concentrations than do other fractions (Burris, 1965). In the experiments of Carnahan *et al.* (1960b) with cell-free extracts of *Clostridium,* ammonia was the first recoverable product of nitrogen fixation. The same is true for nitrogen fixation by cell-free extracts of bacteroids from soybean nodules (Koch *et al.,* 1967a, b).

The nitrogen-fixing enzyme is called nitrogenase. The term refers really to a complex of at least two enzymes, one a molybdenum–iron protein, the other an iron protein (Burris, 1969; Hardy and Knight, 1968; Klucas *et al.,* 1968). The evidence is that once nitrogen is bound to the enzyme it is not released until the sequence of reductions leading to the formation of the key intermediate, NH_3, is over. Evidence for this is the failure to detect sizeable levels of free intermediates between nitrogen gas and ammonia (Burris, 1966). In nitrogen gas, the oxidation number of nitrogen is zero; in NH_3, it is −3. The reduction undoubtedly occurs stepwise.

Nitrogenase is not specific for nitrogen; several other gases, notably hydrogen, H_2, nitrous oxide, N_2O, and carbon monoxide, CO, are competitive inhibitors of nitrogen fixation; they seem to be bound by the same sites on the enzyme which bind nitrogen. Of great interest and utility has been the discovery that nitrogenase binds acetylene, C_2H_2, and reduces it to ethylene, C_2H_4 (Dilworth, 1966). The method provides a sensitive and convenient indirect index for nitrogen fixation, because assay of ethylene by gas chromatography is rapid and sensitive (Stewart *et al.,* 1967). In addition to its capabilities of fixing and reducing nitrogen and some other gases the same versatile enzyme system, nitrogenase, is also able to evolve hydrogen gas, H_2, and such evolution has been observed frequently in nitrogen fixing systems (Hardy and Knight, 1968).

Research with cell-free systems has established that nitrogen fixation requires anaerobic conditions, ATP as an energy source, and a reductant, i.e., a source of electrons. The latter two requirements recall the "assimilatory power" (ATP and reductant) essential to photosynthesis (see Chapter 9). In the cell-free *Clostridium* systems, pyruvic acid can be the source of both. It undergoes a phosphoroclastic split into acetyl phosphate, CO_2, and reducing power:

$$CH_3 \cdot CO \cdot COOH + P_i \rightarrow CH_3 \cdot CO \sim P + CO_2 + 2[H]$$

The acetyl phosphate phosphorylates adenosine diphosphate, ADP, to adenosine triphosphate, ATP,

$$CH_3 \cdot CO \sim P + ADP \rightarrow CH_3 \cdot COOH + ATP$$

and the reducing power, [H], reduces the nitrogen bound to nitrogenase via the intermediary electron carrier, ferredoxin. Figure 10-7, after Burris (1965), schematically shows these reactions. As indicated, H_2 can be the source of reducing power. In the aerobic *Azotobacter,* a type of ferredoxin different from that of plants and the anaerobic *Clostridium,* and possibly a flavoprotein, "azotoflavin," are implicated in transferring electrons to nitrogenase (Benemann *et al.,* 1969; Yoch *et al.,* 1969).

The source of reducing power and ATP for nitrogen fixation by cell-free

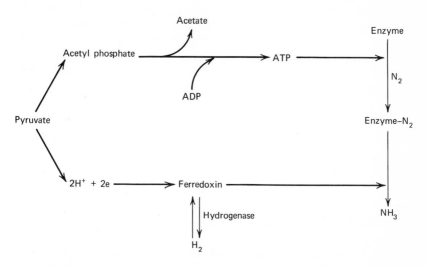

Figure 10-7. Simplified scheme of electron and energy transfer in nitrogen fixation. After Burris (1965). Copyright 1965 by Academic Press, New York and London.

extracts of bacteroids from soybean root nodules is probably the respiratory electron transport chain, but in the experiments done so far, exogenous reductant and ATP had to be added (Klucas and Evans, 1968). Another moot point regarding the biochemistry of nitrogen fixation by the root-nodule system of legumes is the role of hemoglobin, or "leghemoglobin" as it has been called (see Virtanen and Miettinen, 1963). It is essential to nitrogen fixation in this system but absent from free-living nitrogen fixers. The legume, not the bacterium, is the genetic determinant of leghemoglobin production in the nodules (Dilworth, 1969) but the heme is synthesized in the bacteroids (Cutting and Schulman, 1969). Now that the biochemistry of nitrogen fixation by the legume-*Rhizobium* system has become amenable to study the role of leghemoglobin may become clarified.

The following heavy metals are specifically required by nitrogen-fixing systems: iron, molybdenum, and cobalt. Iron and molybdenum, as we have already mentioned, are constituents of nitrogenase. The function of iron seems to be the initial binding of the nitrogen diatomic molecule by the enzyme, and the function of molybdenum, the weakening of the bond between the two nitrogen atoms which renders them susceptible to reduction by the reducing agent (Chatt *et al.,* 1969). In addition, iron is required for synthesis of leghemoglobin in the legume root-nodule system.

A requirement for cobalt in nitrogen fixation was first reported for the symbiotic nitrogen-fixing systems of legumes. Ahmed and Evans (1959, 1960, 1961) worked with the soybean, *Glycine max,* and Reisenauer (1960) and Delwiche *et al.* (1961) with alfalfa, *Medicago sativa.* As low a concentration of cobalt as 2×10^{-8} M in the culture solution was shown to be effective by the latter authors. Free-living, nitrogen-fixing organisms also require cobalt (Evans *et al.,* 1965; Hardy and Knight, 1968; Nicholas *et al.,* 1964). Burris (1969) and Postgate (1970) have very useful summaries of current research on biological nitrogen fixation. For a general account of nitrogen metabolism in plants in its more classical aspects, see the comprehensive work by McKee (1962). The biochemistry of symbiotic nitrogen fixation has been reviewed by Bergersen (1971).

REFERENCES

Ahmed, S. and H. J. Evans. 1959. Effect of cobalt on the growth of soybeans in the absence of supplied nitrogen. Biochem. Biophys. Res. Comm. 1:271–275.

Ahmed, S. and H. J. Evans. 1960. Cobalt: a micronutrient element for the growth of soybean plants under symbiotic conditions. Soil Sci. 90:205–210.

Ahmed, S. and H. J. Evans. 1961. The essentiality of cobalt for soybean plants grown under symbiotic conditions. Proc. Nat. Acad. Sci. 47:24–36.

Allen, E. K. and O. N. Allen. 1965. Nonleguminous plant symbiosis. In: Microbiology and Soil Fertility. C. M. Gilmour and O. N. Allen, eds. Oregon State University Press, Corvallis. Pp. 77–106.

Bach, M. K., W. E. Magee and R. H. Burris. 1958. Translocation of photosynthetic products to soybean nodules and their role in nitrogen fixation. Plant Physiol. 33:118–124.

Bandurski, R. S. 1965. Biological reduction of sulfate and nitrate. In: Plant Biochemistry. J. Bonner and J. E. Varner, eds. Academic Press, New York and London. Pp. 467–490.

Bange, G. G. J. and R. Overstreet. 1960. Some observations on absorption of cesium by excised barley roots. Plant Physiol. 35:605–608.

Becking, J. H. 1956. On the mechanism of ammonium ion uptake by maize roots. Acta Bot. Neerl. 5:1–79.

Becking, J. H. 1970a. *Frankiaceae* fam. nov. (*Actinomycetales*) with one new combination and six new species of the genus *Frankia* Brunchorst 1886, 174. Internat. J. System. Bact. 20:201–220.

Becking, J. H. 1970b. Plant-endophyte symbiosis in non-leguminous plants. Plant and Soil 32:611–654.

Beevers, L. and R. H. Hageman. 1969. Nitrate reduction in higher plants. Ann. Rev. Plant Physiol. 20:495–522.

Benemann, J. R., D. C. Yoch, R. C. Valentine and D. I. Arnon. 1969. The electron transport system in nitrogen fixation by *Azotobacter*, I. Azotoflavin as an electron carrier. Proc. Nat. Acad. Sci. 64:1079–1086.

Bergersen, F. J. 1971. Biochemistry of symbiotic nitrogen fixation in legumes. Ann. Rev. Plant Physiol. 22:121–140.

Berlier, Y., G. Guiraud and Y. Sauvaire. 1969. Etude avec l'azote 15 de l'absorption et du métabolisme de l'ammonium fourni à concentration croissante à des racines excisées de maïs. Agrochimica 13:250–260.

Bond, G. 1956. Some aspects of translocation in root nodule plants. J. Expt. Bot. 7:387–394.

Bond, G. 1963. The root nodules of non-leguminous angiosperms. In: Symbiotic Associations. P. S. Nutman and B. Mosse, eds. Cambridge University Press. Pp. 72–91.

Bond, G. 1964. Isotopic investigations of nitrogen fixation in non-legume root nodules. Nature 204:600–601.

Bond, G. 1967. Fixation of nitrogen by higher plants other than legumes. Ann. Rev. Plant Physiol. 18:107–126.

Bond, G. 1968. Some biological aspects of nitrogen fixation. In: Recent Aspects of Nitrogen Metabolism in Plants. E. J. Hewitt and C. V. Cutting, eds. Academic Press, London and New York. Pp. 15–25.

Broadbent, F. E. and F. Clark. 1965. Denitrification. In: Soil Nitrogen. W. V. Bartholomew and F. E. Clark, eds. American Society of Agronomy, Inc., Madison. Pp. 344–359.

Burris, R. H. 1965. Nitrogen fixation. In: Plant Biochemistry. J. Bonner and J. E. Varner, eds. Academic Press, New York and London. Pp. 961–979.

Burris, R. H. 1966. Biological nitrogen fixation. Ann. Rev. Plant Physiol. 17:155–184.

Burris, R. H. 1969. Progress in the biochemistry of nitrogen fixation. Proc. Roy. Soc. B172:339–354.

Burström, H. 1954. Studies on growth and metabolism of roots. X. Investigations of the calcium effect. Physiol. Plantarum 7:332–342.

Burton, J. C. 1965. The *Rhizobium*-legume association. In: Microbiology and Soil Fertility. C. M. Gilmour and O. N. Allen, eds. Oregon State University Press, Corvallis. Pp. 107–134.

Campbell, N. E. R. and H. Lees. 1967. The nitrogen cycle. In: Soil Biochemistry. A. D. McLaren and G. H. Petersen, eds. Marcel Dekker, Inc., New York. Pp. 194–215.

Carnahan, J. E. and J. E. Castle. 1963. Nitrogen fixation. Ann. Rev. Plant Physiol. 14:125–136.

Carnaham, J. E., L. E. Mortenson, H. F. Mower and J. E. Castle. 1960a. Nitrogen fixation in cell-free extracts of *Clostridium pasteurianum*. Biochim. Biophys. Acta 38:188–189.

Carnahan, J. E., L. E. Mortenson, H. F. Mower and J. E. Castle. 1960b. Nitrogen fixation in cell-free extracts of *Clostridium pasteurianum*. Biochim. Biophys. Acta 44:520–535.

Chatt, J., J. R. Dilworth, R. L. Richards and J. R. Sanders. 1969. Chemical evidence concerning the function of molybdenum in nitrogenase. Nature 224:1201–1202.

Chatt, J. and G. J. Leigh. 1968. The inactivity and activation of nitrogen. In: Recent Aspects of Nitrogen Metabolism in Plants. E. J. Hewitt and C. V. Cutting, eds. Academic Press, London and New York. Pp. 3–12.

Cutting, J. A. and H. M. Schulman. 1969. The site of heme synthesis in soybean root nodules. Biochim. Biophys. Acta 192:486–493.

Delwiche, C. C. 1965. The cycling of carbon and nitrogen in the biosphere. In: Microbiology and Soil Fertility. C. M. Gilmour and O. N. Allen, eds. Oregon State University Press, Corvallis. Pp. 29–58.

Delwiche, C. C., C. M. Johnson and H. M. Reisenauer. 1961. Influence of cobalt on nitrogen fixation by *Medicago*. Plant Physiol. 36:73–78.

Dilworth, M. J. 1966. Acetylene reduction by nitrogen-fixing preparations from *Clostridium pasteurianum*. Biochim. Biophys. Acta 127:285–294.

Dilworth, M. J. 1969. The plant as the genetic determinant of leghaemoglobin production in the legume root nodule. Biochim. Biophys. Acta 184:432–441.

Dilworth, M. J. and C. A. Parker. 1969. Development of the nitrogen-fixing system in legumes. J. Theor. Biol. 25:208–218.

Dixon, R. O. D. 1969. Rhizobia (With particular reference to relationships with host plants). Ann. Rev. Microbiol. 23:137–158.

Elzam, O. E. and E. Epstein. 1965. Absorption of chloride by barley roots: kinetics and selectivity. Plant Physiol. 40:620–624.

Eppley, R. W., J. N. Rogers and J. J. McCarthy. 1969. Half-saturation constants for uptake of nitrate and ammonium by marine phytoplankton. Limnol. and Oceanog. 14:912–920.

Eppley, R. W. and W. H. Thomas. 1969. Comparison of half-saturation constants for growth and nitrate uptake of marine phytoplankton. J. Phycol. 5:375–379.

Evans, H. J., S. A. Russell and G. V. Johnson. 1965. Further consideration of the role of cobalt in organisms that fix atmospheric nitrogen. In: Non-Heme Iron Proteins: Role in Energy Conversion. A. San Pietro, ed. The Antioch Press, Yellow Springs. Pp. 303–313.

Fogg, G. E. 1962. Nitrogen fixation. In: Physiology and Biochemistry of the Algae. R. A. Lewin, ed. Academic Press, New York and London. Pp. 161–170.

Fried, M., F. Zsoldos, P. B. Vose and I. L. Shatokhin. 1965. Characterizing the NO_3 and NH_4 uptake process of rice roots by use of ^{15}N labelled NH_4NO_3. Physiol. Plantarum 18:313–320.

Hardy, R. W. F. and R. C. Burns. 1968. Biological nitrogen fixation. Ann. Rev. Biochem. 37:331–358.

Hardy, R. W. F. and E. Knight, Jr. 1968. Biochemistry and postulated mechanisms of nitrogen fixation. In: Progress in Phytochemistry. L. Reinhold and Y. Liwschitz, eds. Interscience Publishers, John Wiley and Sons, London. Vol. 1, pp. 407–489.

Harre, E. A. 1969. Fertilizer Trends—1969. National Fertilizer Development Center, Tennessee Valley Authority, Muscle Shoals, Alabama.

Heimer, Y. M. and P. Filner. 1971. Regulation of the nitrate assimilation pathway in cultured tobacco cells. III. The nitrate uptake system. Biochim. Biophys. Acta 230:362–372.

Heimer, Y. M., J. L. Wray and P. Filner. 1969. The effect of tungstate on nitrate assimilation in higher plant tissues. Plant Physiol. 44:1197–1199.

Imshenetskii, A. A. 1962. Evolution of the biological fixation of nitrogen. Compar. Biochem. Physiol. 4:353–361.

Jones, K. 1970. Nitrogen fixation in the phyllosphere of the Douglas fir, *Pseudotsuga douglasii*. Ann. Bot. N.S. 34:239–244.

Kamen, M. D. and H. Gest. 1949. Evidence for a nitrogenase system in the photosynthetic bacterium *Rhodospirillum rubrum*. Science 109:560.

Klucas, R. V. and H. J. Evans. 1968. An electron donor system for nitrogenase-dependent acetylene reduction by extracts of soybean nodules. Plant Physiol. 43:1458–1460.

Klucas, R. V., B. Koch, S. A. Russell and H. J. Evans. 1968. Purification and some properties of the nitrogenase from soybean (*Glycine max* Merr.) nodules. Plant Physiol. 43:1906–1912.

Koch, B., H. J. Evans and S. Russell. 1967a. Reduction of acetylene and nitrogen gas by breis and cell-free extracts of soybean root nodules. Plant Physiol. 42:466–468.

Koch, B., H. J. Evans and S. Russell. 1967b. Properties of the nitrogenase system in cell-free extracts of bacteroids from soybean nodules. Proc. Nat. Acad. Sci. 58:1343–1350.

Lie, T. A. 1964. Nodulation of Leguminous Plants as Affected by Root Secretions and Red Light. Ph.D. Thesis, H. Veenman en Zonen N. V., Wageningen.

Lie, T. A. 1969. Non-photosynthetic effects of red and far-red light on root-nodule formation by leguminous plants. Plant and Soil 30:391–404.

Lindstrom, E. S., R. H. Burris and P. W. Wilson. 1949. Nitrogen fixation by photosynthetic bacteria. J. Bact. 58:313–316.

Lindstrom, E. S., J. W. Newton and P. W. Wilson. 1952. The relationship between photosynthesis and nitrogen fixation. Proc. Nat. Acad. Sci. 38:392–396.

Lindstrom, E. S., S. R. Tove and P. W. Wilson. 1950. Nitrogen fixation by the green and purple sulfur bacteria. Science 112:197–198.

Loneragan, J. F. and E. J. Dowling. 1958. The interaction of calcium and hydrogen ions in the nodulation of subterranean clover. Austral. J. Agric. Res. 9:464–472.

Lowther, W. L. and J. F. Loneragan. 1968. Calcium and nodulation in subterranean clover (*Trifolium subterraneum* L.). Plant Physiol. 43:1362–1366.

Lund, J. W. G. 1967. Soil algae. In: Soil Biology. A. Burges and F. Raw, eds. Academic Press, London and New York. Pp. 129–147.

MacMillan, A. 1956. The entry of nitrate into fungal cells. Physiol. Plantarum 9:470–481.

McKee, H. S. 1962. Nitrogen Metabolism in Plants. Clarendon Press, Oxford.

Munns, D. N. 1968a. Nodulation of *Medicago sativa* in solution culture. I. Acid-sensitive steps. Plant and Soil 28:129–146.

Munns, D. N. 1968b. Nodulation of *Medicago sativa* in solution culture. II. Compensating effects of nitrate and of prior nodulation. Plant and Soil 28:246–257.

Munns, D. N. 1968c. Nodulation of *Medicago sativa* in solution culture. III. Effects of nitrate on root hairs and infection. Plant and Soil 29:33–47.

Munns, D. N. 1968d. Nodulation of *Medicago sativa* in solution culture. IV. Effects of indole-3-acetate in relation to acidity and nitrate. Plant and Soil 29:257–262.

Munns, D. N. 1970. Nodulation of *Medicago sativa* in solution culture. V. Calcium and pH requirements during infection. Plant and Soil 32:90–102.

Nicholas, D. J. D., D. J. Fisher, W. J. Redmond and M. Osborne. 1964. A cobalt requirement for nitrogen fixation, hydrogenase, nitrite and hydroxylamine reductases in *Clostridium pasteurianum*. Nature 201:793–795.

Nutman, P. S. 1963. Factors influencing the balance of mutual advantage in legume symbiosis. In: Symbiotic Associations. P. S. Nutman and B. Mosse, eds. Cambridge University Press. Pp. 51–71.

Nutman, P. S. 1965. Symbiotic nitrogen fixation. In: Soil Nitrogen. W. V. Bartholomew and F. E. Clark, eds. American Society of Agronomy, Inc., Madison. Pp. 360–383.

Pate, J. S. 1968. Physiological aspects of inorganic and intermediate nitrogen metabolism (with special reference to the legume, *Pisum arvense* L.). In: Recent Aspects of Nitrogen Metabolism in Plants. E. J. Hewitt and C. V. Cutting, eds. Academic Press, London and New York. Pp. 219–240.

Pate, J. S., B. E. S. Gunning and L. G. Briarty. 1969. Ultrastructure and functioning of the transport system of the leguminous root nodule. Planta 85:11–34.

Pate, J. S., J. Walker and W. Wallace. 1965. Nitrogen-containing compounds in the shoot system of *Pisum arvense* L. II. The significance of amino-acids and amides released from nodulated roots. Ann. Bot. N. S. 29:475–493.

Picciurro, G., L. Ferrandi, R. Boniforti and G. Bracciocurti. 1967. Uptake of ^{15}N-labelled NH_4^+ in excised roots of a *Durum* wheat mutant line compared with *Durum* and bread wheat. In: Isotopes in Plant Nutrition and Physiology. International Atomic Energy Agency, Vienna. Pp. 511–526.

Postgate, J. R. 1970. Biological nitrogen fixation. Nature 226:25–27.

Rao, K. M. P. 1971. Kinetics and Energetics of Nitrate Absorption by Barley Seedlings. M.S. Thesis, University of California, Davis.

Reisenauer, H. M. 1960. Cobalt in nitrogen fixation by a legume. Nature 186:375–376.

Rovira, A. D. 1969. Plant root exudates. Bot. Rev. 35:35–57.

Ruinen, J. 1956. Occurrence of *Beijerinckia* species in the 'phyllosphere.' Nature 177:220–221.

Ruinen, J. 1961. The phyllosphere. I. An ecologically neglected milieu. Plant and Soil 15:81–109.

Ruinen, J. 1970. The phyllosphere. V. The grass sheath, a habitat for nitrogen-fixing micro-organisms. Plant and Soil 33:661–671.

Schneider, K. C., C. Bradbeer, R. N. Singh, L. C. Wang, P. W. Wilson and R. H. Burris. 1960. Nitrogen fixation by cell-free preparations from microorganisms. Proc. Nat. Acad. Sci. 46:726–733.

Silver, W. S., Y. M. Centifanto and D. J. D. Nicholas. 1963. Nitrogen fixation by the leaf-nodule endophyte of *Psychotria bacteriophila*. Nature 199:396–397.

Silvester, W. B. and D. R. Smith. 1969. Nitrogen fixation by *Gunnera-Nostoc* symbiosis. Nature 224:1231.

Singh, R. N. 1961. Role of blue-green algae in nitrogen economy of Indian agriculture. Indian Council of Agricultural Research, New Delhi.

Smith, R. C. and E. Epstein. 1964. Ion absorption by shoot tissue: kinetics of potassium and rubidium absorption by corn leaf tissue. Plant Physiol. 39:992–996.

Smith, R. V. and M. C. W. Evans. 1970. Soluble nitrogenase from vegetative cells of the blue-green alga *Anabaena cylindrica*. Nature 225:1253–1254.

Stewart, W. D. P. 1966. Nitrogen Fixation in Plants. The Athlone Press, University of London.

Stewart, W. D. P. 1970. Algal fixation of atmospheric nitrogen. Plant and Soil 32:555–588.

Stewart, W. D. P. and G. Bond. 1961. The effect of ammonium nitrogen on fixation of elemental nitrogen in *Alnus* and *Myrica*. Plant and Soil 14:347–359.

Stewart, W. D. P., G. P. Fitzgerald and R. H. Burris. 1967. In situ studies on N_2 fixation using the acetylene reduction technique. Proc. Nat. Acad. Sci. 58:2071–2078.

Stewart, W. D. P., A. Haystead and H. W. Pearson. 1969. Nitrogenase activity in heterocysts of blue-green algae. Nature 224:226–228.

Thimann, K. V. 1963. The Life of Bacteria. Their Growth, Metabolism, and Relationships. 2nd ed. The Macmillan Company, New York.

van den Honert, T. H. and J. J. M. Hooymans. 1955. On the absorption of nitrate by maize in water culture. Acta Bot. Neerl. 4:376–384.

Vincent, J. M. 1965. Environmental factors in the fixation of nitrogen by the legume. In: Soil Nitrogen. W. V. Bartholomew and F. E. Clark, eds. American Society of Agronomy, Inc., Madison. Pp. 384–435.

Vincent, J. M. 1967. Symbiotic specificity. Austral. J. Sci. 29:192–197.

Virtanen, A. I. and J. K. Miettinen. 1963. Biological nitrogen fixation. In: Plant Physiology—A Treatise. F. C. Steward, ed. Academic Press, New York and London. Vol. 3, pp. 539–668.

Watanabe, A. and T. Kiyohara. 1963. Symbiotic blue-green algae of lichens, liverworts and cycads. In: Studies on Microalgae and Photosynthetic Bacteria. Japanese Society of Plant Physiologists. The University of Tokyo Press. Pp. 189–196.

Wilson, P. W. and R. H. Burris. 1953. Biological nitrogen fixation—a reappraisal. Ann. Rev. Microbiol. 7:415–432.

Yoch, D. C., J. R. Benemann, R. C. Valentine and D. I. Arnon. 1969. The electron transport system in nitrogen fixation by *Azotobacter*. II. Isolation and function of a new type of ferredoxin. Proc. Nat. Acad. Sci. 64:1404–1410.

11

MINERAL METABOLISM

1. THE FUNCTIONS OF NUTRIENTS

We have seen in several previous chapters that metabolic conditions profoundly affect the absorption, transport, and distribution of mineral elements in plants. The mechanisms of mineral transport evolved in response to physiological, metabolic, or biochemical needs. "Need" is not used here in a naive teleological sense; it is a shorthand notation for any situation in which, in the course of evolution, the presence and functional involvement of a chemical element in a given physico-chemical constellation bestowed a selective advantage upon the organism. On the other hand, an element may inhibit or disrupt a biological system or component so that its absence becomes a "need," i.e., a factor conferring a selective advantage. Thus mechanisms evolved for exclusion or extrusion of specific elements; this, also, is an aspect of transport.

Our knowledge of the functional roles of essential elements in plants varies greatly from element to element. We may set up categories or degrees of knowledge, in ascending order, as follows.

(1) The element is essential. If, in a medium suitably purged of the element in question, the plant fails to grow normally, while in its presence it grows well, then we say that the element is essential, as already discussed in Chapter 4. This finding alone does not imply that we know anything about the specific role or roles that the element plays in the physiology or metabolism of the plant; the finding is only proof that there is at least one such role or function.

(2) The element plays an essential role in a physiological process. If a specific, essential process demonstrably depends upon the presence of the element we have pinpointed at least this one role with some degree of specificity, although we may still not know, in chemical terms, in what manner the element is involved in the process.

(3) The element activates an enzyme or regulates the rate of an enzyme-mediated process, and the evidence is to the effect that the same element normally performs this "cofactor" function in the living plant. Here, we move from physiological criteria to biochemical ones.

(4) The element is an integral constituent of an essential metabolite, complex, or macromolecular assembly. This is definitive evidence for an essential role, and such a finding is itself adequate evidence for the essentiality of the element, as already discussed in Chapter 4.

The importance of many metallic elements as integral constituents of enzymes and electron carriers warrants special mention. A metal may either be a constituent of a prosthetic group or coenzyme which, when attached to the appropriate protein (apoenzyme), results in a catalytically active entity, or it may be an integral part of the protein itself. Enzymes of either kind are called metalloenzymes. While in enzyme activation, ions of several elements may often substitute for one another, such substitution does not normally occur with metals which are constituents of enzyme molecules.

Such a hierarchy of degrees of knowledge, or, if you will, of ignorance, is by the nature of research impermanent and with time, more and more specific biochemical roles (3 and 4) will be discovered for all the essential elements. Meanwhile, such a scheme makes us face up squarely to what we know and do not know about the functions of each element, and serves to guard us from the somnolent effects of such semantic vagaries as, "Element X is functionally or structurally involved in the unique organization of the protoplasmic matrix."

The functions discussed above are those of specific elements. There are certain other functions, some well defined, others less so, which depend to a smaller extent or even not at all on the particular element. Paramount among these is the maintenance of internal osmotic pressure. All ions in solution within the cell, simple or complex, organic or inorganic, and all nonelectrolytes play this role. There are, however, some differences in effectiveness among ions. It depends, for example, on the propensity of ions to form ion pairs or otherwise to depart from ideality (Lee, 1959), and to form complexes with organic ligands (Martell and Calvin, 1952).

Also included among the more or less unspecific functions of ions is their role in the maintenance of electric neutrality. Mineral cations and anions function as counterions of anions and cations, respectively, whether the latter be free in solution or fixed charges, organic or inorganic. These functions are usually called "electrochemical functions." Buffering within cells and organelles and effects on the physical state or conformation of colloids, polymers, and other macromolecular entities including proteins and nucleic acids are also frequently mentioned but as yet poorly understood roles, largely of the "electrochemical" kind.

The methods of investigating essential functions of mineral elements vary depending upon the level in the above hierarchy which our knowledge of the particular element has attained. They range all the way from experiments with nutrient solutions designed to obtain evidence about the essentiality of an element to purely chemical analysis of a metal element in a metalloenzyme. In our discussion of functional roles of mineral nutrients we shall begin with those elements known to be parts of essential compounds and macromolecules and proceed to elements about whose function less is known. The fact that a specific role is known for an element does not, of course, imply that it may not have one or more additional functions, either known or unknown to us.

We shall largely disregard the distinction between "macronutrient" and "micronutrient" elements. The distinction is an arbitrary one, convenient in some contexts but of little use in the present one. As already discussed in Chapter 4, an element which is conventionally called a "macronutrient," such as calcium, may be a "micronutrient" for certain plants, as calcium is for algae. Even in a given plant the quantitative requirement for a given element may vary greatly, depending upon the concentrations of other elements in the medium. Wallace and his co-workers have discussed this in connection with the calcium requirement in higher plants (see section 4 of this chapter). Only carbon, hydrogen, oxygen, nitrogen, phosphorus, and sulfur are universal macronutrients, so that no more than three mineral nutrients invariably belong in this category.

Discussions like this sometimes disregard those elements not known to be essential. But we should keep in mind that some elements, though not essential by the criteria of plant physiological experimentation, may play functional roles in the life of plants in nature. For example, sodium, not known to be essential for sugar beet plants, *Beta vulgaris,* nevertheless has a beneficial effect on their growth under conditions when the level of potassium is suboptimal (Ulrich and Ohki, 1956). Sodium has a similar "potassium sparing" effect in cotton, *Gossypium hirsutum* (Joham and Amin, 1965). The conclusion is inescapable that in at least one process, probably the activation of an enzyme, sodium can substitute for potassium. Specific information of this kind is available for many enzymes which require divalent metal ions for activation. Often several such ions, including some of elements not known to be essential, can activate a given enzyme *in vitro* and hence, probably, *in vivo* as well (Dixon and Webb, 1964; Epstein, 1965).

"Antagonistic" actions of mineral elements also need to be considered. For example, Williams and Vlamis (1957) found that a concentration of manganese in mature leaves of barley plants corresponding to 300–400 ppm on the dry weight basis was toxic when the nutrient solution contained no silicon but harmless when it did. If the concentration of manganese in the plants is kept at a lower level by maintaining a lower concentration of it in the medium, no manganese toxicity develops even when silicon is not provided. Soil-grown plants invariably contain silicon, and it no doubt plays functional roles in many situations (Jones and Handreck, 1967; Lewin and Reimann, 1969).

2. NUTRIENT ELEMENTS AS CONSTITUENTS OF METABOLITES AND COMPLEXES

Nitrogen.   Nitrogen is commonly the fourth most abundant element in plants, following carbon and the elements of water. Proteins contain approximately 18 per cent nitrogen (Table 4-1). Absorbed mainly as nitrate (see Chapter 10), it is reduced and incorporated into organic compounds (Bandurski, 1965; Beevers and Hageman, 1969). It is a constituent of amino acids, nucleotides, and coenzymes. As much as 70 per cent of the total leaf nitrogen may be in chloroplasts (Stocking and Ongun, 1962), suggesting that in a leafy plant roughly half the total nitrogen may be in the plastids. A study of the role of nitrogen is tantamount to a study of plant biochemistry. Unless the supply of nitrogen in the medium is very ample or a deficiency of some other element or of carbohydrate interferes with its assimilation into organic compounds the fraction of the nitrogen in the plant present as free nitrate or ammonium ions is small, 10–20 per cent or less being common (Dijkshoorn *et al.,* 1960; Minotti and Jackson, 1970; Stewart and Porter, 1969).

Phosphorus.   Phosphorus is absorbed in the main as the dihydrogen phosphate ion, $H_2PO_4^-$. It is one of three quantitatively prominent nutrient elements which are absorbed as complex anions, the other two being nitrogen (nitrate) and sulfur (sulfate). But unlike the nitrogen and sulfur of nitrate and sulfate the phosphorus atom of phosphate is not reduced in the cell to a lower oxidation state.

Phosphate plays a key role in energy metabolism, as discussed in Chapter 9. Incorporated into adenosine triphosphate, ATP, it is part and parcel of the universal "energy currency" of all living cells of whatever species. Phosphate occurs in phospholipids including those of membranes, in sugar phosphates, various nucleotides and coenzymes. Phytic acid, the hexaphosphate ester of *myo*-inositol, or its calcium or magnesium salts (phytin), serve as a storage form of phosphate in seeds:

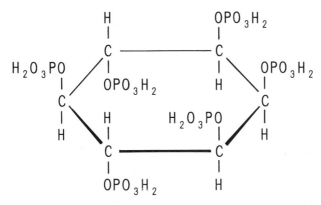

Sulfur. Sulfur, absorbed mainly in the form of sulfate ions, is reduced in plants and incorporated in organic compounds (Bandurski, 1965; Thompson, 1967; Thompson *et al.,* 1970). It is a constituent of the amino acids cystine, cysteine, and methionine, and hence of the proteins containing these amino acids. Thiamine, biotin, and coenzyme A are sulfur-containing, low-molecular weight coenzymes essential in metabolism when attached to appropriate apoenzymes (proteins) which require these coenzymes or prosthetic groups for catalytic activity. The ferredoxins, non-heme iron proteins involved in photosynthesis and other electron transfers (see Iron, p. 291) contain sulfur in amounts equivalent to the iron present (Hall and Evans, 1969). Volatile compounds containing sulfur contribute to the characteristic odors given off by onions, mustards and other plants and plant materials. The functions of these compounds in the life of plants are not understood.

Proteins are the compounds in which most of the nitrogen and most of the sulfur of plant tissues are incorporated. Dijkshoorn *et al.* (1960) grew perennial ryegrass, *Lolium perenne,* in nutrient solutions at varying

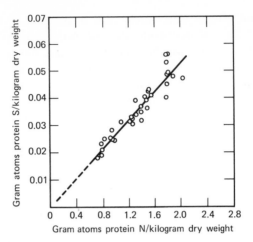

Figure 11-1. The relation between gram atoms of protein–sulfur and gram atoms of protein–nitrogen in perennial ryegrass, *Lolium perenne,* grown in nutrient solutions at various concentrations of sulfate and nitrate. After Dijkshoorn *et al.* (1960).

concentrations of nitrate and sulfate and determined protein–nitrogen and protein–sulfur in the plants. The results (Figure 11-1) showed a fairly fixed stoichiometry between the two, indicating that on the average there are 36 atoms of nitrogen for each atom of sulfur in the proteins of this plant. The ratio of total nitrogen to total sulfur is close to this value in plants generally, under conditions where there is no "luxury consumption" of either element, as shown in the survey of published data by Dijkshoorn and van Wijk (1967). The percentages of nitrogen (1.5) and sulfur (0.1) in plant materials generally considered adequate (Table 4-3) give an atomic ratio, N/S, of 34/1.

The absorption of sulfate, when its supply is ample, may be faster than its reduction and the assimilation of its sulfur atoms into organic compounds (Kylin, 1960). An appreciable fraction of the total sulfur in plants may therefore be in the form of sulfate—a much larger proportion, commonly, than that represented by nitrate as a fraction of the total nitrogen (Dijkshoorn *et al.,* 1960; Eaton, 1966; Stewart and Porter, 1969).

Magnesium.   Chlorophylls are the only major stable compounds of plants which contain an atom of magnesium as a fixed (non-dissociable) constituent. They are magnesium porphyrins. Magnesium represents 2.7 per cent of the molecular weight, but the magnesium of chlorophyll represents only about 10 per cent of the total magnesium in the leaf. Since half or more of the leaf magnesium may be in the chloroplasts (Stocking and Ongun, 1962) the plastids evidently contain much magnesium in addition to that which is part of the chlorophyll. This is not unexpected; energy conversion and conservation are the major functions of chloroplasts, and as we shall see in the next section, magnesium, in addition to its role in chlorophyll, is the most common activator of enzymes concerned with energy metabolism. Marks (1969) has written a monograph on the biochemistry of heme and chlorophyll.

Iron. There are many metabolites containing atoms of iron as fixed (not readily dissociable) constituents of the molecule. The iron is either part of a low molecular-weight prosthetic group or it is an integral part of the protein itself.

The most intensively studied iron containing prosthetic groups are the iron porphyrins (hemes). They function prominently in electron transport, as indicated in the discussion of the roles of the cytochromes in respiration and photosynthesis (Chapter 9). The role of hemoglobin in symbiotic nitrogen fixation, none too well understood, was mentioned in Chapter 10. Other iron porphyrin enzymes are catalase, peroxidases, and certain dehydrogenases. The biochemistry of hemes is discussed in the authoritative monograph by Marks (1969).

There are also non-heme iron proteins including ferredoxins and mitochondrial iron enzymes which play roles in electron transport, although the latter have not yet been studied in plants to any extent. The volume edited by San Pietro (1965), the record of a symposium held in 1965, has numerous reviews by experts on non-heme iron proteins and their role in energy conversion, including photosynthesis, respiration, and nitrogen fixation. A review is by Malkin and Rabinowitz (1967).

Manganese. Manganese acts as an activator of many enzymes, but only one manganoprotein, manganin, has been isolated from a higher plant (Dieckert and Rozacky, 1969). The authors obtained it from seeds of peanut, *Arachis hypogaea,* and partially characterized it.

Zinc. Zinc is the metal component of a number of metalloenzymes, including several dehydrogenases, alcohol dehydrogenase and lactic dehydrogenase among them.

Copper. Like zinc, copper.is a component of several enzymes, including ascorbic acid oxidase, phenolases, laccase, and several others. It is also a constituent of cytochrome oxidase.

Calcium.   Only one metalloenzyme, amylase, has calcium as the metal. In the enzyme obtained from animal and microbial sources, the calcium is quite firmly bound to the molecule; in the enzyme from barley, *Hordeum vulgare,* the binding is relatively weak. Calcium is commonly the major cation of the middle lamella of cell walls, of which calcium pectate is a principal constituent. Calcium has therefore an important bearing on the mechanical strength of tissues (Cleland, 1960; Ito and Fujiwara, 1967; Rasmussen, 1967; Tagawa and Bonner, 1957). In algae, other calcium and magnesium polysaccharides are important components of the cell wall. In some plants including many algae, calcium salts occur in solid form. Calcium oxalate and calcium carbonate are common, calcium phosphate and sulfate less so.

Molybdenum.   Molybdenum is the metal of several metalloenzymes. Prominent among these are enzymes involved in nitrogen fixation and in nitrate reduction.

Cobalt.   Cobalt is a constituent of vitamin B_{12} coenzymes. All nitrogen-fixing systems seem to require these. Cobalt is thus essential to higher plants depending for their nitrogen on symbiotic nitrogen fixation.

Potassium, Chlorine, and Boron.   No enzymes or other essential organic compounds have been isolated which contain these elements as constituents. For boron, however, a specific regulatory step in carbohydrate metabolism has been identified, as discussed in the next section.

3.   NUTRIENT ELEMENTS AS ACTIVATORS, COFACTORS, OR REGULATORS OF ENZYMES

The importance of this aspect of the role of mineral elements in metabolism was well put by Mahler (1961): ". . . there probably does not exist a single enzyme-catalyzed reaction in which either substrate, product, enzyme, or some combination within this triad is not influenced in a very direct and highly specific manner by the precise nature of the inorganic ions which surround and 'modify' it."

Nitrogen.   Certain enzyme activities are not demonstrable if the organism has grown in the absence of the substrate of that enzyme but become evident upon exposure of the organism to the substrate. Factors other than the substrate may also be instrumental in "inducing" an enzyme (Filner *et al.,* 1969). Nitrate reductase is such an inducible enzyme in plants; it is induced by nitrate (Beevers and Hageman, 1969). This process of induction represents enzyme synthesis; it should be clearly distinguished from the activation of preexisting enzymes with which this section mainly deals. Ammonium ions activate certain enzymes for which potassium is normally the activating cation.

Phosphorus.   Phosphate regulates many enzymic processes. The phosphorylation of ADP to ATP and its dependence on the concentration of phosphate have been mentioned in Chapter 9. Phosphate also acts as an activator of some enzymes.

Sulfur.   Sulfate activates some enzymes, sometimes being able to function in lieu of phosphate.

Magnesium.   Magnesium is an activator of more enzymes than any other element. It is a cofactor of nearly all enzymes which act on phosphorylated substrates and is therefore of paramount importance in energy metabolism. The activation is not highly specific; manganese, especially, can frequently substitute, and so, to a lesser extent, can certain other divalent cations. Magnesium also activates certain enzymes not concerned with phosphate transfers.

Manganese.   As pointed out above, manganese can often substitute for magnesium as an activator of phosphate-transferring enzymes. Arginase is specifically activated by manganese. Manganese is prominent as an activator of enzymes mediating reactions of the Krebs (tricarboxylic acid) cycle.

Calcium, Iron, Zinc, Copper, and Cobalt.   Ions of these metals are cofactors of several enzymes but rarely if ever with a high degree of specificity. Calcium may competitively inhibit the activating effect of magnesium, evidently by displacing it from its functional sites.

Molybdenum.   Molybdenum functions mainly as a component of metalloenzymes rather than as an enzyme activator. It plays a role in the induction of nitrate reductase (Beevers and Hageman, 1969).

Chlorine.   Chloride acts in conjunction with an enzyme or enzymes of photosystem II of photosynthesis but the specific reactions and the enzymes catalyzing them are unknown.

Boron.   Boron plays a regulatory role in carbohydrate metabolism. The predominant pathway for degradation of glucose is glycolysis, but an alternative pathway, the "pentose shunt pathway," is significant in many tissues (see Bonner and Varner, 1965b). Lee and Aronoff (1967) have shown that boron combines with 6-phosphogluconate, the first intermediate in this pathway, to form a 6-phosphogluconate–borate complex. This cannot be further metabolized and the pentose shunt pathway is thereby inhibited. In boron deficiency, its operation becomes prominent, and is often accompanied by excessive synthesis of phenolic acids—a recognized biochemical syndrome of boron deficiency (Dear and Aronoff, 1965; Spurr, 1952; Watanabe et al., 1961, 1964).

Potassium.   Potassium is the only monovalent cation essential for all higher plants and indeed, for all living things except a few microorganisms in which rubidium can substitute for it. The principal role of potassium is that of an activator of numerous enzymes. Evans and Sorger (1966) have listed no less than 46 enzymes from animals, higher plants, and microorganisms which require monovalent cations for maximal activity. Potassium is the activating cation in the majority of cases. Other monovalent cations can often substitute, especially rubidium and ammonium ions, but they are usually less effective than potassium. No doubt potassium is the normally functional monovalent cationic cofactor in the living plant. Sodium and lithium ions are often inhibitory.

With the exception of potassium and magnesium, all mineral cations which act as physiological activators or cofactors of plant enzymes are micronutrients, and even magnesium is usually present in plant tissue at only a fraction of the concentration of potassium. Why, for a catalytic role like the activation of enzymes, does the element have to be present at such high concentrations? The answer has to do with the chemistry of potassium and specifically, with the fact that it does not have a high affinity for organic ligands, including the enzymes for which it is a cofactor. For maximal activity of many of these enzymes, concentrations of potassium as high as 50 or 100 mM are common and the Michaelis constants, the concentrations giving half the maximal rates, are on the order of 5–10 mM—still fairly high concentrations. Evidently it is only at concentrations of this magnitude that the reaction between potassium and enzyme or enzyme–substrate proceeds appreciably in the direction of formation of the potassium complex (Hiatt and Evans, 1960).

Such loose binding between potassium and the complexes activated by it may not have been a limiting factor in the early marine phases of the evolution of life because of the fairly high concentration of potassium in sea water which is 12 mM now and probably was similarly high during the early evolution of living things (Redfield, 1958; Rubey, 1951). However, upon emerging onto land, plants had to adapt to a mineral medium with much lower potassium concentrations. This they did by evolving the high-affinity (type 1) mechanism of potassium accumulation (Michaelis constant 0.02 mM) discussed in Chapter 6. Through this mechanism they can accumulate potassium from very dilute solutions such as those encountered in well-leached soils and in bodies of fresh water, and maintain within their cytoplasm the high concentrations of potassium required for complexing of the cation by the low-affinity binding sites of potassium-activated enzymes or enzyme–substrate complexes. As discussed in Chapter 6 there are those who believe that at concentrations of potassium on the order of 10 mM, the ion readily diffuses across the plasmalemma. It is not clear how that idea is compatible with the maintenance within the cytoplasm of the high concentrations of potassium known to be required for appreciable activation of certain cytoplasmic enzymes, under the common conditions when the external potassium concentration is very low.

. NUTRIENT ELEMENTS IN PHYSIOLOGICAL PROCESSES

We have seen that numerous enzymic processes are governed by nutrient elements. As substrates, components of metabolites, activators, and inhibitors they affect the rates of a multitude of enzyme-catalyzed reactions. In the living cell, these reactions do not proceed in isolation. Rather, each enzymic reaction is part of the intricate network of delicately interdependent reactions, each influenced by, and in turn influencing, all the others, some directly, many more indirectly. Were it not for this finely adjusted interplay among all the individual components of metabolism, the machinery of life would become deranged as evidenced by the catastrophic consequences of withholding a single essential element, or supplying a poison known to inhibit a single enzymic step.

Highly integrated sequences of enzymic reactions are recognized as metabolic or physiological processes. For example, the series of enzymic steps which begins when light impinges on a chloroplast is recognized as "photosynthesis." The effects of various factors, including mineral nutrients, on such processes can often be studied to good advantage even when it is not yet feasible to pinpoint the particular enzymic step or steps involved. Other complex physiological processes, not amenable to analysis in terms only of individual enzymic reactions, are often influenced in specific ways by certain mineral nutrients. An example is the opening and closing of stomates.

Thus we see that in governing the rates of individual enzyme-catalyzed reactions, in regulating sequences and networks of such reactions at numerous control points, and in influencing complex physiological functions, mineral nutrients play key roles. In the present section we shall explore these roles at the level of physiological processes, i.e., in situations where our knowledge is insufficient to pinpoint the specific metabolite or reaction involved.

A common method of investigating the physiological functions of a given nutrient is to study biochemical effects of its deficiency. Plants sufficient and deficient in a certain nutrient element may show greater differences in their contents of certain organic metabolites than in their contents of the element itself (Rendig and McComb, 1959). Much valuable information has been obtained through studies of deficiencies, and this approach will loom large in the present discussion of the physiological roles of nutrient elements. Nevertheless, what was said above concerning the intricate interdependence of metabolic steps and sequences urges caution in this regard. A deficiency may have a highly specific primary effect, but this in turn will be followed by catenary secondary, tertiary, and further consequences until eventually the initial derangement caused by the deficiency will have reverberations throughout the metabolic network of the organism. Therefore, a spectacular biochemical abnormality showing up as a result of a nutrient deficiency may be many steps removed from the primary lesion and permit no direct deductions as to the mechanism of action of the deficient element.

These considerations prompt us to ask specific questions about biochemical and metabolic effects of mineral deficiencies. To what extent is the effect specific to a deficiency of the element in question? In other words, can the effect, or very similar effects, be provoked by deficiencies of other elements as well? Is the effect an early consequence of the deficiency? Compared with other results of the particular deficiency, is the effect quantitatively pronounced? If the concentrations or amounts of various metabolites are affected, which of these are known precursors or regulators of the others and therefore perhaps nearer the primary lesion? When the deficient element is resupplied, which of the metabolites whose levels are abnormal as a result of the deficiency respond most quickly and dramatically? To what extent do the manifestations of the deficiency differ at different stages of development of the plant? How consistently does the effect appear in different plants subject to the deficiency? Are there mutants or races of a given species which respond in different ways to the deficiency, and if so, what are the differences?

To ask these questions is to drive home the extent of our ignorance of the metabolic effects of nutrient deficiencies. To provide the answers would require nothing less than the systematic work of a considerable institute devoting many years to the project. Until such a concerted effort is made our knowledge will remain fragmented, haphazard, almost anecdotal. An exceptionally systematic and valuable investigation is that by Bottrill *et al.* (1970). They grew spinach plants, *Spinacea oleracea*, in nutrient solutions and examined the effects of deficiencies of all the essential mineral nutrients, except chlorine, on the rates of photosynthesis and respiration. Deficiencies of most elements reduced the rates of both processes.

Visible symptoms resulting from nutrient deficiencies have been described in Chapter 4. They are, of course, late manifestations of metabolic derangements occurring long before gross, visible effects become apparent. Certain internal symptoms, observable with the light or electron microscope and other techniques of visualization, will be discussed in the present section, along with other physiological effects.

When any essential element is deficient, growth is inhibited. Resupplying the deficient element will have these two most obvious consequences: the content of that element in the plant will increase, because of renewed absorption of it, and growth will resume if no other factor is inhibitory. These two responses—renewed absorption of the deficient element and resumption of growth—are the first and the final results, respectively, of resupplying a deficient nutrient. Many of the specific metabolic effects through which absorption brings about renewed growth will be discussed below.

Nitrogen.   As is to be expected from its major role as a constituent of proteins, purines, pyrimidines, and many coenzymes, an interference with protein synthesis and hence, with growth, is the major biochemical effect of nitrogen deficiency. An early and dramatic symptom of nitrogen deficiency is a general yellowing of leaves, or chlorosis, due to an inhibition of chlorophyll synthesis. The internal appearance of plastids is altered considerably (Thomson and Weier, 1962; Vesk *et al.,* 1966). The resulting slow-down of photosynthesis causes a nitrogen deficient plant to lack not only essential amino acids but also the machinery for synthesis of necessary carbohydrates and carbon skeletons for all manner of organic syntheses. Before chlorosis sets in, however, carbohydrates including starch may accumulate, since, they are not utilized for protein synthesis on account of the deficiency in amino acids.

Plants can absorb both the cationic ammonium ion, NH_4^+, and the anion nitrate, NO_3^- (Chapter 10). A great deal of research effort has gone into comparisons of the metabolic effects of these ions as sources of nitrogen for plants (Coïc *et al.,* 1961; McKee, 1962; Street and Sheat, 1958). When either ammonium or nitrate is available as the sole source of nitrogen, the ion in question will commonly be the most rapidly absorbed ion present in the medium. Unless the medium is buffered or constantly renewed, absorption of ammonium tends to result in a lowering of the pH of the medium (see Chapter 6), while absorption of nitrate has the opposite effect. Control of the pH of the medium frequently eliminates physiological responses which have been attributed to the absorption and metabolism of ammonium or nitrate ions per se (Barker et al., 1966; Karim and Vlamis, 1962).

In most well-aerated soils, nitrate is the principal form of available nitrogen, and plants adapted to such soils grow well with nitrate as the sole source of nitrogen. Many such plants can also utilize ammonium, but suffer various impairments when only ammonium ions furnish nitrogen (Barker et al., 1966; Street and Sheat, 1958). The structure of chloroplasts is affected under conditions of ammonium toxicity (Puritch and Barker, 1967).

Ammonium, once absorbed, can immediately serve in the synthesis of amino acids and other compounds containing reduced nitrogen. Absorption of ammonium therefore causes a great demand for carbon skeletons and, if photosynthesis fails to keep pace, a depletion of carbohydrates. Assimilation of ammonium is usually rapid and there is seldom much accumulation of free ammonium in tissue. Nitrate nitrogen must be reduced before it is assimilated. The immediate demand for carbohydrate is therefore less than in the case of ammonium absorption, and there may be an appreciable buildup of organic acids. Free nitrate may accumulate. Climate, the stage of growth, the species, and the availability of other nutrients all have a bearing on these metabolic adjustments to the absorption of ammonium and nitrate ions.

Phosphorus.   The key role which phosphate plays in the energetics of metabolism and biosynthetic reactions suggests that its deficiency would be scarcely less disastrous than a deficiency of nitrogen. This is so. Phosphate is required for the synthesis of adenosine triphosphate and numerous other phosphorylated compounds; its deficiency, therefore, causes immediate and severe disruptions of metabolism and development. Generalizations are virtually impossible to make and unreliable when attempted in the case of a systemic disorder like phosphorus deficiency.

Phosphate spectacularly promotes the absorption of molybdate by plants (Stout et al., 1951). The effect is puzzling since chemical considerations would suggest the likelihood of competition between the two ions. Perhaps this is one of those situations, discussed by Epstein (1962), in which competition leads to accelerated movement. If there are binding sites along the path of ions with affinity for two ionic species, the movement of one of these ions present at low concentration may be impeded by its becoming attached to these sites. The presence of the second ionic species may accelerate the movement of the first by competing with it for the binding sites and thus keeping it from being immobilized (see Chapter 7, p. 174).

The chloroplasts of phosphorus deficient plants show various abnormalities but these do not appear to be uniform in different plants (Marinos, 1963; Thomson and Weier, 1962; Vesk *et al.,* 1966).

Sulfur.   Much of what was said about deficiencies of nitrogen and phosphorus applies also to sulfur deficiency. The restriction in the synthesis of sulfur containing amino acids and hence, of proteins, causes these similarities, due in the main to shortages of essential proteins including enzymes. The biochemical syndrome of sulfur deficiency often includes a low level of carbohydrates and a buildup of soluble nitrogen fractions including nitrate—the former resulting from diminished photosynthesis, and the latter resulting from the failure of nitrogenous substrates to be used in the synthesis of proteins (Chen, 1967; Ergle and Eaton, 1951; Rendig and McComb, 1961). There are fewer cytoplasmic inclusions, and their appearance is abnormal (Marinos, 1963).

Magnesium.   As a constituent of chlorophyll and activator of numerous enzymes effecting phosphate transfers, magnesium is an element whose deficiency soon affects every facet of the metabolism of the plant. Chlorosis is an early symptom, followed by diminished photosynthesis. Biosynthetic pathways are deranged, as a result of the inhibition of essential enzymic transphosphorylations, and soluble nitrogenous compounds are present at elevated concentrations.

The magnesium content of normal chloroplasts is high (Larkum, 1968; Stocking and Ongun, 1962). Thomson and Weier (1962) and Vesk *et al.* (1966) have shown abnormalities in the fine structure of chloroplasts of magnesium deficient plants. Mitochondria are also affected (Marinos, 1963).

The literature on plant nutrition contains many references to "antagonisms" between magnesium and calcium, on the one hand, and magnesium and potassium, on the other. What is meant is that applications, and increased absorption, of one element may result in diminished absorption of the other. Observations of such interactions are particularly common in work with fruit trees (Emmert, 1961). Such "antagonistic" effects may be so severe that heavy absorption of one nutrient may cause a deficiency of another. For example, potassium deficiency may be induced by heavy applications of magnesium (Ulrich and Ohki, 1966). This phenomenon is puzzling in view of the high specificity of potassium absorption discussed in Chapter 6. Perhaps transport via the (unselective) type 2 mechanisms is involved, or the phenomenon may be due to competitive effects in long-distance translocation.

Iron. A major portion of the total iron of leaves is in the chloroplasts (Price, 1968). Iron is essential for chlorophyll synthesis (Bogorad, 1966). When the supply of iron is varied there is often a good correlation between iron content and chlorophyll content (Jacobson and Oertli, 1956). Machold and Scholz (1969) resupplied iron (labeled with ^{59}Fe) to tomato plants suffering from iron deficiency chlorosis. Figure 11-2 shows a photograph (right) and an autoradiograph (left) of a leaf. The area of the leaf which has greened up (right) coincides precisely with

Figure 11-2. Photograph (right) and autoradiograph (left) of a leaf of a tomato plant, *Lycopersicon esculentum*. The plant suffered from iron deficiency chlorosis because iron was withheld. Iron labeled with ^{59}Fe was then added to the nutrient solution. The area of the leaf which greened up (right) coincides with the area reached by the radioiron (left). From Machold and Scholz (1969).

the distribution of the radioiron in the leaf (left). The appearance of chloroplasts is changed in iron deficiency. The number and the size of the grana are much reduced (Vesk et al., 1966).

The cytochromes are iron-containing electron carriers. Energy relations would therefore be expected to suffer disruption under conditions of iron deficiency. This, in conjunction with the role of iron in chlorophyll biosynthesis, causes considerable similarities between magnesium and iron deficiencies. Price (1968) has discussed numerous biochemical and physiological effects of iron deficiency. On calcareous soils, iron deficiency is frequently due not to a lack of iron but to its being immobilized or inactivated by carbonate or bicarbonate ("lime induced chlorosis"). Brown (1961), Machold (1968), and Thorne et al. (1950) have discussed this subject. Another type of induced iron deficiency is discussed below, under the heading, Manganese. Iron deficiency causes marked changes in the ultrastructure of chloroplasts. In particular, the grana are much reduced in number and in size (Vesk et al., 1966).

Manganese. Disruption of metabolism by manganese deficiency is severe and many metabolites are affected. This is to be expected; manganese plays a prominent role in many of the reactions of the Krebs or tricarboxylic acid cycle, and because of the central role of this cycle in aerobic respiration, repercussions of manganese deficiency are transmitted to other metabolic sequences (Bonner and Varner, 1965b; Steward and Margolis, 1962).

Manganese is also a prominent component of chloroplasts and participates in the reactions leading to the evolution of oxygen (Heath and Hind, 1969; Possingham and Spencer, 1962). This is evidently connected with far-reaching changes in the structure of chloroplasts of manganese deficient plants (Possingham et al., 1964). These authors found no visible alterations in other cell organelles.

Manganese and several other heavy metal ions, when present in the medium at high concentrations, may induce iron deficiency in plants. This phenomenon depends on competitive effects in the absorption and translocation of iron and also on competition by the inhibitory heavy metal for functional sites of iron binding. Hewitt (1963) has reviewed this subject.

Zinc. All deficiencies interfere with growth but a deficiency of zinc does so with such dramatic impact that terms like "little leaf" and "rosetting" have been applied to the condition. "Rosetting" refers to the failure of internodes to elongate, causing the leaves of several nodes to lie telescoped together in a plane, rosette-fashion. The marked effect of zinc deficiency on growth is due to the influence of zinc on the auxin level. The concentration of auxin, indoleacetic acid, in zinc deficient tissue drops well before visible symptoms become apparent, and upon resupplying zinc rises, after which growth resumes (Skoog, 1940). Skoog concluded that zinc is required for maintenance of auxin in an active state, not for its synthesis. Tsui (1948), however, working like Skoog with tomato plants, found zinc essential for the synthesis of tryptophan, a precursor of auxin. Tsui's conclusion has been supported by Salami and Kenefick (1970) and contested by Takaki and Kushizaki (1970). The former authors grew corn, *Zea mays*, in nutrient solutions to which no zinc was added. The plants developed symptoms characteristic of zinc deficiency and grew poorly. Addition of either zinc or tryptophan to the solutions eliminated these effects—indirect evidence that zinc is necessary for synthesis of optimal levels of the amino acid. However, Takaki and Kushizaki (1970), who also worked with corn, found much higher levels of tryptophan in zinc deficient plants than in plants where zinc was sufficient. They concluded that zinc plays a role in the metabolic pathway from tryptophan to auxin via tryptamine.

The involvement of zinc as either a constituent or an activator of several enzymes makes it clear that its role in regulating the level of auxin is not the only one it plays in metabolism. The activity of one enzyme, ribonuclease, has been shown to bear an inverse relation to the zinc concentration in the leaves of fruit trees, suggesting that protein synthesis, which is mediated by ribonucleic acid, might be regulated by the concentration of zinc (Kessler, 1961). Changes in anatomy and histology of zinc deficient plants have been summarized by Hewitt (1963). Chloroplasts are severely affected, but not in a uniquely characteristic way (Thomson and Weier, 1962; Vesk *et al.,* 1966).

Copper. Like deficiencies of other catalytically important elements, copper deficiency interferes with protein synthesis and causes an increase in soluble nitrogen compounds. The concentration of reducing sugars remained low in copper deficient wheat plants, in the experiments of Brown *et al.* (1958), while in copper sufficient plants it rose during development, with a simultaneous drop in the level of organic acids.

Molybdenum. The role of molybdenum in the enzyme systems of symbiotic nitrogen fixation (Chapter 10) suggests that plants relying on symbiotically fixed nitrogen, when subject to molybdenum deficiency, would in effect become nitrogen deficient, and they do. The same consideration applies to plants whose nitrogen source is nitrate. Molybdenum is essential in the enzyme system of nitrate reduction, and therefore molybdenum deficient plants, while able to absorb nitrate, do not reduce it effectively and as a result become functionally nitrogen deficient. When plants are given ammonium ions as a source of nitrogen, many of the characteristic symptoms of molybdenum deficiency, both visual and metabolic, fail to develop or develop only slightly; however, even such plants eventually fail. Molybdenum in higher plants has functions in addition to its role in nitrate reduction.[1] Molybdenum deficient plants generally have lower levels of sugars and of ascorbic acid than do molybdenum sufficient controls. Concentrations of various amino acids are often spectacularly low in molybdenum deficient plants.

Chlorine. It is likely that chloride ions have more than a single specific function but the only one that has been identified is the requirement for chloride in oxygen evolution by photosystem II of photosynthesis (Bové et al., 1963). Larkum (1968) found chloride to be concentrated in chloroplasts of the alga, *Tolypella intricata.* Whether its role in photosynthesis is intimately connected with other manifestations of chlorine deficiency is not known. Among these other effects is the tendency of chlorine deficient plants to wilt, which suggests some derangement of transpiration. Freney et al. (1959) described a buildup of free amino acids in chlorine deficient plants, but it is not known whether the effect is directly related to amino acid synthesis or interconversion.

Deficiency of chloride has never been encountered in the field or in nature, because of the prevalence of cyclic salt. However, excess chloride is very much a matter of concern. Salinity and its effects are considered in Chapter 13. In plants adapted to saline conditions chloride plays an important role as one of the solutes contributing to the lowering of the cellular osmotic potential.

Boron. Boron is unique among the micronutrients in the narrow range of concentrations of it in the medium to which many plants are restricted. A fraction of one part per million may be required, and a few parts per million may be toxic. There is, however, much variation in this regard among species.

The regulatory role boron plays in switching the degradation of glucose to either glycolysis or the pentose shunt pathway has been mentioned in the preceding section. There are, however, several other highly character-

---

[1] This may be wrong. See Hewitt and Gundry (1970).

istic effects of boron or boron deficiency which cannot yet be analyzed in terms of specific biochemical steps.

Growing points of both roots and shoots stop elongating when boron is deficient and if severe deficiency continues, become discolored, disorganized, and die. This is apparently related to an effect of boron on ribonucleic acid (RNA) metabolism. Albert (1965) and Johnson and Albert (1967) found elongation of boron deficient tomato root tips to stop before their RNA content decreased. This might suggest that the effect of boron deficiency on RNA metabolism is indirect or secondary. However, supplying the nitrogen bases, thymine, guanine, or cytosine prevented both the stoppage of elongation and the decrease in RNA content which normally follows it in boron deficient tips (Johnson and Albert, 1967). The response to the nitrogen bases implicates RNA in the boron effect.

In boron deficiency, flowering is inhibited and if the deficiency is severe, prevented altogether. Water relations are abnormal and specifically, leaves and stems of herbaceous plants appear desiccated and may have a stiff, woody feel. Cell walls are thin and brittle and may collapse as a result of pressure exerted by adjacent cells. The germination of pollen grains and growth of the pollen tube are severely inhibited by lack of boron (Stanley and Lichtenberg, 1963). Much early literature has been reviewed by Gauch and Dugger (1954), and more recent work is included in Hewitt's (1963) discussion of essential elements.

The role of boron in the translocation of sugars from leaves has been much discussed. Gauch and Dugger (1953) drew attention to the propensity of borate to form complexes with polyhydroxyl compounds including alcohols and sugars. They postulated that sugar-borate complexes would more easily traverse cellular membranes than would the highly polar sugar molecules themselves. They reported that boron greatly accelerated the translocation of sucrose from leaves of tomato plants dipped into solutions containing sucrose and boron, compared with controls in which the solutions contained no boron. However, contradictory results have since been reported. Weiser et al. (1964) concluded that absorption of sucrose by the leaf, rather than translocation per se, was influenced by boron, but that conclusion in turn has been questioned (Lee et al., 1966). The role of boron in translocation of sugar remains a moot point. Whether the accumulation of sugars and starch in leaves of boron deficient plants—a general phenomenon—is related to diminished export from the leaves is also unresolved.

To sum up, it is still true that "it can be stated with little fear of contradiction that the exact role or mode of action of this element has not been unequivocally demonstrated" (McIlrath and Skok, 1964). Two main factors share the blame: the analytical chemistry of boron is difficult, and there is no suitable radioisotope. The former difficulty has recently been largely overcome (Carlson and Paul, 1968); the second one is a fact of nature. In any event, the "minimal cell"—the irreducible machinery of life—can apparently function without boron, since it has not been shown to be essential for many microorganisms, nor for animals.

Calcium. Concentrations of calcium in healthy tissues of many crop plants range from about 0.2 per cent (dry weight basis) to several per cent. It is nevertheless likely that these values are well in excess of minimal metabolic requirements. Provided that other divalent metal ions were kept at low concentrations in the nutrient solutions, tobacco and corn plants have been shown to grow well in solutions in which the concentration of calcium was only 2 ppm, or 0.05 mM (Wallace *et al.*, 1966). Leaves of corn grown in these solutions contained 0.011 per cent calcium in their dry matter, and leaves of tobacco, 0.08 per cent. Since such low values are tolerable only at low concentrations of other divalent cations it seems that the normally higher levels of calcium serve to render innocuous otherwise toxic concentrations of these metals.

Roots of germinating seeds will grow well in moist air for a few days (Marschner and Günther, 1964). Their calcium content is low under these conditions, because the nutrient supply from the seed is via the phloem in which calcium is notably immobile. If roots of germinating seeds are exposed to a nutrient solution lacking calcium they stop growing at once. Evidently the ions of the nutrient solution are toxic when no calcium is included in it. Roots also fail to grow into calcium-free soil (Rios and Pearson, 1964). Again, detoxification of other ions emerges as a role of calcium; when that is not required (when the roots grow in moist air), the low-calcium roots persist for longer. That calcium counteracts the effects of low pH in the medium has already been brought out in the discussion of ion transport (Chapter 6, pp. 117–118). It is equally important when the medium contains sodium at high levels. The greater a demand is made on the selective machinery of the plasmalemma, through the presence in the medium of potentially damaging ions at high concentrations, the more crucial seems to be the role of calcium in maintaining the integrity of the membrane.

Loneragan and his co-workers at the University of Western Australia have conducted carefully controlled experiments on the calcium nutrition of many crop and pasture species (Loneragan *et al.*, 1968; Loneragan and Snowball, 1969a, b). The plants grew in flowing nutrient solutions in which the concentrations of calcium were kept constant at low, predetermined values ranging from 0.0003 to 1 mM (Loneragan and Snowball, 1969a). The plants grew increasingly well in the range 0.0003 to 0.0025 mM calcium, but at higher calcium concentrations in the solutions there was little additional gain in growth, although more calcium was absorbed. Many of the grasses gave maximal or near maximal yields when their tops contained 0.1 per cent calcium (dry weight basis); for legumes, the value was about twice that, and for herbs (*Cryptostemma calendula, Erodium botrys,* and *Lycopersicon esculentum*) it was higher yet (Figure 11-3).

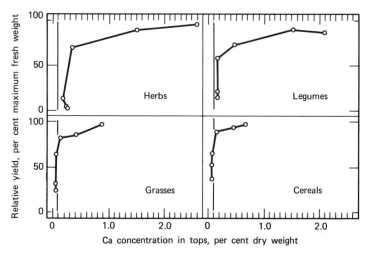

Figure 11-3.   The relation between the calcium concentration in the tops of herbs, legumes, grasses, and cereal crops after 17 to 19 days' growth in solutions containing calcium at concentrations ranging from 0.0003 mM to 1 mM. The vertical line indicates 0.1 per cent calcium in the tops. After Loneragan and Snowball (1969a).

Plants grown in the field or in conventional (high-calcium) nutrient solutions often contain much higher concentrations of calcium in their tissues than the values recorded in these experiments and those of Wallace *et al.* (1966). Much of this uptake must be in the nature of "luxury consumption."

The immobility of calcium in the phloem and its consequent failure to be retranslocated from older leaves have an interesting result, as noted by Loneragan (1968) and Loneragan and Snowball (1969a). When plants grew in solutions maintained at 0.0003 mM calcium, they were severely deficient; at 1 mM calcium, they grew well and there was "luxury consumption" of calcium. When plants were transferred from 1 mM to 0.0003 mM calcium solutions growth slowed down and young leaves and growing points became calcium deficient. Nevertheless, the older leaves of these plants, which had matured while the plants were in the high-calcium solution, still contained ample calcium—much more than healthy plants growing continuously in solutions of low but adequate calcium concentrations (0.002 to 0.01 mM). In other words, plants contained "luxury" quantities of calcium in some tissues and organs, and were calcium deficient to the point of growth inhibition in others. Such uneven distribution may also occur in nature because the availability of elements for absorption often varies from time to time. For phloem-immobile elements like calcium, therefore, metabolic derangements due to deficiency may be very localized. The foregoing discussion points up the fact that calcium is required for normal growth at lower concentrations in the tissues than are often encountered. Whether it "ought to be regarded as a 'micro' nutrient element" (Burström, 1968) is a semantic point we need not belabor.

Although all growing points are sensitive to calcium deficiency, those of the roots are affected most severely. They cease growing, become disorganized and discolored and in severe deficiency, die. Before that happens the normal process of cell division by mitosis may become deranged, and there may be polyploid nuclei, binucleate cells, constricted nuclei, and amitotic cell divisions, as observed in a study of calcium deficiency in the pea, *Pisum sativum* (Sorokin and Sommer, 1940). Deficiencies in flowers and fruits are also spectacular, as indicated by the term "blossom-end rot" (Spurr, 1959).

Brewbaker and Kwack (1963) found calcium to be indispensable for the germination of pollen and growth of the pollen tube in plants from numerous families. They attributed the importance of calcium to its involvement in the synthesis of cell wall materials, but a role in the plasmalemma is equally plausible. Obviously, these are not mutually exclusive functions.

The general disorganization of cells and tissues afflicted with calcium deficiency suggests that the role of calcium in maintaining membranes in a functional state, already discussed in connection with ion transport (Chapter 6), may be one of its key functions. This is so because cellular organization is to a considerable extent a matter of compartmentation and metabolic pools, and cell compartments such as the protoplast, the vacuole, the nucleus, and other organelles are delimited by membranes.

Direct evidence for the paramount role of calcium in the maintenance of the structure of membranes has been obtained through electron microscopy. Marinos (1962) studied cells in the shoot apex of barley as affected by calcium deficiency. In contrast to the clearly organized cytoplasm of normal cells, that of calcium dificient cells had "structureless areas," fragmented membranes, and various vesicles and amorphous inclusions. Magnesium deficiency produced no comparable disintegration of cell structure (Marinos, 1963). Marschner and Günther (1964) and Marschner et al. (1966) have reported similar findings for calcium deficient cells of barley and corn root tips, respectively. Membranous organelles like mitochondria (Florell, 1956; Lindblad, 1959) and chloroplasts (Vesk et al., 1966) are also affected by calcium deficiency.

When Stocking and Ongun (1962) isolated chloroplasts from calcium sufficient tobacco and bean leaves by the nonaqueous method to prevent loss by leaching during the isolation procedure, they found that the chloroplasts contained about 60 per cent of the total leaf calcium. They and Larkum (1968) concluded that chloroplasts act as sites of calcium accumulation. Energy-dependent calcium transport in chloroplasts has since been described (Nobel, 1967, 1969). Mitochondria also effect an energy-driven transport of calcium (Elzam and Hodges, 1968; and see Chapter 9).

Recent reviews on the functions of calcium in plants are by Jones and Lunt (1967) and by Burström (1968), the latter with surprisingly little attention to the role which has emerged as paramount in the present discussion, that of maintaining membranes and hence, intracellular organization.

Potassium. Unlike calcium, potassium is highly mobile in the phloem. Its utilization is therefore efficient in the sense that it is readily redistributed from older leaves to young, growing organs. As a result, symptoms of potassium deficiency usually appear first in the older leaves.

The function of potassium as an activator of many enzymes, discussed in the preceding section and reviewed in detail by Evans and Sorger (1966), causes a deficiency to have pervasive effects on metabolic events. In early stages of potassium deficiency soluble carbohydrates including reducing sugars often accumulate (Eaton, 1952), and the concentrations and ratios of organic acids may vary from the norm of potassium sufficient plants (Jones, 1961). Soluble nitrogen compounds including the amines, agmatine and putrescine, often accumulate, the latter probably being responsible for the necrotic spots which often appear on potassium deficient leaves (Coleman and Richards, 1956; Hackett et al., 1965; Sinclair, 1969). Recent work on the regulation of amine synthesis in potassium deficiency has been discussed by Smith (1968).

The accumulation of carbohydrates and soluble nitrogen compounds point to diminished protein synthesis in potassium deficiency. In keeping with this conclusion is the observation by Ozbun *et al.* (1965) that fully mature leaves of bean plants, *Phaseolus vulgaris,* were much less sensitive to potassium deficiency than were young, growing leaves. A direct role of potassium in protein synthesis, first recognized for microbial systems, has also been suggested in work with higher plants (Evans and Sorger, 1966) but recent evidence does not support the hypothesis (Hsiao *et al.,* 1970). As with other systemic disorders, the patterns of metabolic derangements in potassium deficiency are not uniform in different species, and depend also on the stage of development of the plant and various external factors.

When there is much "luxury consumption" of potassium, due to high concentrations of it in the nutrient medium, the absorption or, more likely, the translocation of other cations may be depressed. Magnesium levels, in particular, may be low in the leaves (Cain, 1955; Emmert, 1961; Shear *et al.,* 1953)—so low that in the field and in nutrient solutions, high concentrations of potassium may bring on or aggravate magnesium deficiency (Embleton, 1966). Obviously, interference with the distribution of magnesium would eventually have a bearing on photosynthesis. Two other roles indirectly linking potassium with photosynthesis have been proposed.

In 1959 and 1960, Fujino of Nagasaki University proposed a mechanism for the opening of stomates (Fujino, 1967). It was based on earlier work by another Japanese scientist, S. Imamura. Previous hypotheses had been mainly in terms of osmosis due to increased concentrations of soluble carbohydrates in the guard cells. Fujino found that upon illumination, closed guard cells accumulated potassium and opened. Darkened, they lost potassium to the surrounding tissue and closed. These movements of potassium showed characteristics of active transport (cf. Chapter 9).

These findings have been confirmed and extended. Fischer and Hsiao (1968) floated epidermal strips of *Vicia faba* leaves on potassium solutions and estimated that the increase in the concentration of potassium in the illuminated guard cells reached 300 mM. Assuming an equal rise in the concentration of a univalent anion (not determined, however), the resulting decrease in the water potential of the guard cells was enough to account for the opening of the stomates.

Humble and Hsiao (1969, 1970) investigated the ion specificity and other aspects of the effect. Potassium and rubidium caused maximal opening at very low concentrations (Figure 11-4), suggesting that absorption was via a high-affinity mechanism (cf. Chapter 6). The other monovalent ions were much less effective.

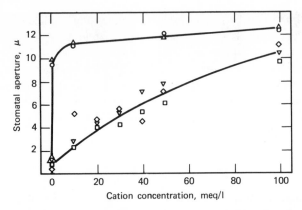

Figure 11-4. The degree of opening of stomates of *Vicia faba* as a function of the concentration of monovalent cations in the solutions on which strips of epidermis were floated, in the light. △, K^+; ○, Rb^+; ◇, Li^+; □, Cs^+; ▽, Na^+. After Humble and Hsiao (1969).

Sawhney and Zelitch (1969) and Humble and Raschke (1971) used the electron probe analyzer to demonstrate potassium accumulation in the guard cells. Figure 11-5, from Humble and Raschke (1971), shows the accumulation of potassium in illuminated guard cells (stomates open) and the paucity of it after darkening (stomates closed). [The opening and closing of the leaflets of the silk-tree, *Albizzia julibrissin,* is also effected through intercellular, osmotic migration of water, and in this case, too, transport of potassium is the direct cause of the water flux (Satter *et al.,* 1970)].

A second role proposed for potassium and indirectly linking it with photosynthesis is that it promotes the translocation of photosynthate from leaves. According to a much debated hypothesis first proposed by Boussingault in 1868 and comprehensively reviewed by Neales and Incoll (1968) a hundred years later, the rate of photosynthesis is slowed down when assimilates accumulate in the leaf. If so, rapid export of photosynthate from the leaf would be important for maintenance of a high net photosynthetic rate. Hartt (1969) has presented evidence that potassium accelerates the movement of assimilates from leaves of sugar cane, *Saccharum officinarum.* Even a mild deficiency of potassium decreased the translocation of photosynthate, before causing a drop in the rate of photosynthesis. However, to what extent the deficiency might have damaged the phloem, thereby slowing transport, was not investigated.

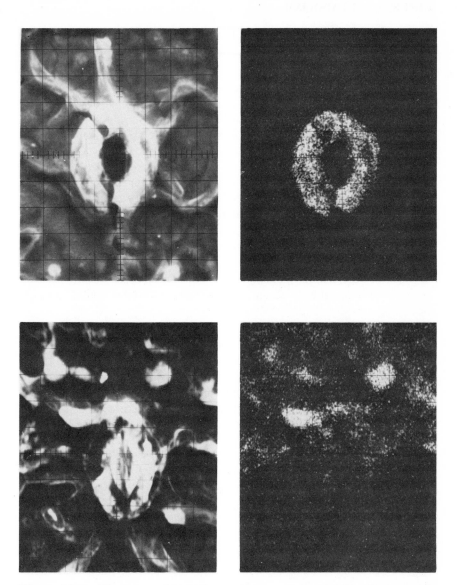

Figure 11-5. Electron probe analyzer pictures of guard cells of *Vicia faba* (left) and of the distribution of potassium (right). The open stomate (top) has potassium concentrated in the guard cells; the closed stomate (bottom) shows no potassium accumulation in the guard cells, the distribution of potassium in the epidermal cells being spotty and diffuse. The pictures of the guard cells (left) are secondary electron images; those of potassium distribution (right) are potassium X-ray images. From Humble and Raschke (1971).

In any event, direct interference with photosynthetic processes from potassium deficiency has long been suspected and is likely on evidence concerning the potassium relations of chloroplasts. Stocking and Ongun (1962) and Larkum (1968), using the nonaqueous method for isolating chloroplasts, found high concentrations of potassium in them. Larkum estimated that the potassium concentration within chloroplasts of the alga, *Tolypella intricata,* was 340 mM—about three times as high as in the cytoplasm (87–97) and the vacuole (110–119). In leaves of tobacco, *Nicotiana rustica,* Stocking and Ongun (1962) found that 55 per cent of the potassium was in the chloroplasts, and in leaves of bean plants, *Phaseolus vulgaris,* 39 per cent. For mesophyll cells, therefore, they concluded that the findings are "at variance with the concept of the vacuole as a major site of ion accumulation." The same conclusion would apply to *Tolypella,* on the basis of Larkum's (1968) results, but not equally to another fresh water alga, *Nitella,* for which Stocking and Ongun (1962) found only 11 per cent of the cellular potassium in the chloroplasts. Nevertheless, in *Nitella,* also, potassium is accumulated in the chloroplasts (Saltman *et al.,* 1963).

That the high concentrations of potassium in chloroplasts are not incidental is evident from the results of potassium deficiency on their structure. Thomson and Weier (1962), Marinos (1963) and Vesk *et al.* (1966) have described various alterations in chloroplasts of potassium deficient plants. Early development of the chloroplasts was not affected, however, and the consequences of this deficiency on the structure of chloroplasts appear to be not as devastating as those of other deficiencies, calcium deficiency in particular.

More direct evidence for the participation of potassium (and other cations) in the metabolic activities of chloroplasts was furnished by experiments in which the experimental treatment of isolated chloroplasts caused a large loss of their cations (Jagendorf and Smith, 1962). Such chloroplasts lost their phosphorylative activity; in effect, they became uncoupled. It appears that light-induced ion movements in both chloroplasts (Packer and Crofts, 1967) and mitochondria (Hanson and Hodges, 1967) are essential features of the energy metabolism of these organelles (Chapter 9). The symposium volume edited by Kilmer *et al.* (1968) contains several chapters on physiological and metabolic roles of potassium.

General Information and Other Elements.   In order not to overburden this chapter with references, generally known and frequently reviewed facts have often been given without literature citations. General references and sources of further information are therefore collected here.

The review by Pirson (1955) on functional aspects comments on much of the older literature. Though no longer up to date, this thoughtful review is still useful. The books by Gilbert (1957) and Schütte (1964) are valuable sources of general information; both discuss the role of mineral elements in animals as well as in plants. Bowen (1966) has collected a great deal of information, much of it in condensed, tabular form, on the geochemistry, cycling, absorption, and biological functions of mineral elements. The chapters by Hewitt (1963) and Nason and McElroy (1963) in the third volume of the treatise of *Plant Physiology* edited by F. C. Steward of Cornell University are comprehensive reviews, the former with a plant physiological, the latter, a biochemical orientation. The book by Price (1970), *Molecular Approaches to Plant Physiology,* contains a great deal of information on the biochemical roles of nutrient elements in plants. Several chapters in *Plant Biochemistry,* edited by Bonner and Varner (1965a) deal with related topics. Concerning elements not known to be essential (and some which are) Bollard and Butler (1966) have compiled a large miscellany of references.

REFERENCES

Albert, L. S. 1965. Ribonucleic acid content, boron deficiency symptoms, and elongation of tomato root tips. Plant Physiol. 40:649–652.

Bandurski, R. S. 1965. Biological reduction of sulfate and nitrate. In: Plant Biochemistry. J. Bonner and J. E. Varner, eds. Academic Press, New York and London. Pp. 467–490.

Barker, A. V., R. J. Volk and W. A. Jackson. 1966. Root environment acidity as a regulatory factor in ammonium assimilation by the bean plant. Plant Physiol. 41:1193–1199.

Beevers, L. and R. H. Hageman. 1969. Nitrate reduction in higher plants. Ann. Rev. Plant Physiol. 20:495–522.

Bogorad, L. 1966. The biosynthesis of chlorophylls. In: The Chlorophylls. L. P. Vernon and G. R. Seely, eds. Academic Press, New York and London. Pp. 481–510.

Bollard, E. G. and G. W. Butler. 1966. Mineral nutrition of plants. Ann. Rev. Plant Physiol. 17:77–112.

Bonner, J. and J. E. Varner, eds. 1965a. Plant Biochemistry. Academic Press, New York and London.

Bonner, J. and J. E. Varner. 1965b. The path of carbon in respiratory metabolism. In: Plant Biochemistry. J. Bonner and J. E. Varner, eds. Academic Press, New York and London. Pp. 213–230.

Bottrill, D. E., J. V. Possingham and P. E. Kriedemann. 1970. The effect of nutrient deficiencies on photosynthesis and respiration in spinach. Plant and Soil 32:424–438.

Bové, J. M., C. Bové, F. R. Whatley and D. I. Arnon. 1963. Chloride requirement for oxygen evolution in photosynthesis. Z. Naturforsch. 18b: 683–688.

Bowen, H. J. M. 1966. Trace Elements in Biochemistry. Academic Press, London and New York.

Brewbaker, J. L. and B. H. Kwack. 1963. The essential role of calcium ion in pollen germination and pollen tube growth. Am. J. Bot. 50:859–865.

Brown, J. C. 1961. Iron chlorosis in plants. Adv. Agron. 13:329–369.

Brown, J. C., L. O. Tiffin and R. S. Holmes. 1958. Carbohydrate and organic acid metabolism with ^{14}C distribution affected by copper in Thatcher wheat. Plant Physiol. 33:38–42.

Burström, H. G. 1968. Calcium and plant growth. Biol. Rev. 43:287–316.

Cain, J. C. 1955. The effect of potassium and magnesium on the absorptions of nutrients by apple trees in sand culture. Proc. Am. Soc. Hort. Sci. 65:25–31.

Carlson, R. M. and J. L. Paul. 1968. Potentiometric determination of boron as tetrafluoroborate. Anal. Chem. 40:1292–1295.

Chen, C.–L. H. 1967. Assimilation of $^{14}CO_2$ by *Medicago sativa* Leaves in Relation to Sulfur Nutrition. Ph.D. Thesis, University of California, Davis.

Cleland, R. 1960. Effect of auxin upon loss of calcium from cell walls. Plant Physiol. 35:581–584.

Coïc, Y., C. Lesaint and F. Le Roux. 1961. Comparaison de l'influence de la nutrition nitrique et ammoniacale combinée ou non avec une déficience en acide phosphorique, sur l'absorption et le métabolism des anions-cations et plus particulièrement des acides organiques chez le maïs. Comparaison du maïs et de la tomate quant à l'effet de la nature de l'alimentation azotée Ann. Physiol. Vég. 3:141–163.

Coleman, R. G. and F. J. Richards. 1956. Physiological studies in plant nutrition. XVIII. Some aspects of nitrogen metabolism in barley and other plants in relation to potassium deficiency. Ann. Bot. N. S. 20:393–409.

Dear, J. and S. Aronoff. 1965. Relative kinetics of chlorogenic and caffeic acids during the onset of boron deficiency in sunflower. Plant Physiol. 40:458–459.

Dieckert, J. W. and E. Rozacky. 1969. Isolation and partial characterization of manganin, a new manganoprotein from peanut seeds. Arch. Biochem. Biophys. 134:473–477.

Dijkshoorn, W., J. E. M. Lampe and P. F. J. van Burg. 1960. A method of diagnosing the sulphur nutrition status of herbage. Plant and Soil 13:227–241.

Dijkshoorn, W. and A. L. van Wijk. 1967. The sulphur requirements of plants as evidenced by the sulphur–nitrogen ratio in the organic matter. A review of published data. Plant and Soil 26:129–157.

Dixon, M. and E. C. Webb. 1964. Enzymes. 2nd ed. Academic Press, New York.

Eaton, F. M. 1966. Sulfur. In: Diagnostic Criteria for Plants and Soils. H. D. Chapman, ed. University of California, Division of Agricultural Sciences. Pp. 444–475.

Eaton, S. V. 1952. Effects of potassium deficiency on growth and metabolism of sunflower plants. Bot. Gaz. 114:165–180.

Elzam, O. E. and T. K. Hodges. 1968. Characterization of energy-dependent Ca^{2+} transport in maize mitochondria. Plant Physiol. 43:1108–1114.

Embleton, T. W. 1966. Magnesium. In: Diagnostic Criteria for Plants and Soils. H. D. Chapman, ed. University of California, Division of Agricultural Sciences. Pp. 225–263.

Emmert, F. H. 1961. The bearing of ion interactions on tissue analysis results. In: Plant Analysis and Fertilizer Problems. W. Reuther, ed. American Institute of Biological Sciences, Washington, D. C. Pp. 231–243.

Epstein, E. 1962. Mutual effects of ions in their absorption by plants. Agrochimica 6:293–322.

Epstein, E. 1965. Mineral metabolism. In: Plant Biochemistry. J. Bonner and J. E. Varner, eds. Academic Press, New York and London. Pp. 438–466.

Ergle, D. R. and F. M. Eaton. 1951. Sulphur nutrition of cotton. Plant Physiol. 26:639–654.

Evans, H. J. and G. J. Sorger. 1966. Role of mineral elements with emphasis on the univalent cations. Ann. Rev. Plant Physiol. 17:47–76.

Filner, P., J. L. Wray and J. E. Varner. 1969. Enzyme induction in higher plants. Science 165:358–367.

Fischer, R. A. and T. C. Hsiao. 1968. Stomatal opening in isolated epidermal strips of *Vicia faba*. II. Responses to KCl concentration and the role of potassium absorption. Plant Physiol. 43:1953–1958.

Florell, C. 1956. The influence of calcium on root mitochondria. Physiol. Plantarum 9:236–242.

Freney, J. R., C. C. Delwiche and C. M. Johnson. 1959. The effect of chloride on the free amino acids of cabbage and cauliflower plants. Austral. J. Biol. Sci. 12:160–166.

Fujino, M. 1967. Role of adenosinetriphosphate and adenosinetriphosphatase in stomatal movement. Sci. Bull. Fac. Educ. Nagasaki Univ. 18:1–47.

Gauch, H. G. and W. M. Dugger, Jr. 1953. The role of boron in the translocation of sucrose. Plant Physiol. 28:457–466.

Gauch, H. G. and W. M. Dugger, Jr. 1954. The Physiological Action of Boron in Higher Plants: A Review and Interpretation. University of Maryland Agricultural Experiment Station Bull. A-80 (Technical).

Gilbert, F. A. 1957. Mineral Nutrition and the Balance of Life. University of Oklahoma Press, Norman.

Hackett, C., C. Sinclair and F. J. Richards. 1965. Balance between potassium and phosphorus in the nutrition of barley. I. The influence on amine content. Ann. Bot. N. S. 29:331–345.

Hall, D. O. and M. C. W. Evans. 1969. Iron–sulphur proteins. Nature 223:1342–1348.

Hanson, J. B. and T. K. Hodges. 1967. Energy-linked reactions of plant mitochondria. In: Current Topics in Bioenergetics. D. R. Sanadi, ed. Vol. 2, pp. 65–98.

Hartt, C. E. 1969. Effect of potassium deficiency upon translocation of ^{14}C in attached blades and entire plants of sugarcane. Plant Physiol. 44:1461–1469.

Heath, R. L. and G. Hind. 1969. On the functional site of manganese in photosynthetic oxygen evolution. Biochim. Biophys. Acta 189:222–233.

Hewitt, E. J. 1963. The essential nutrient elements: requirements and interactions in plants. In: Plant Physiology—A Treatise. F. C. Steward, ed. Academic Press, New York and London. Vol. 3, pp. 137–360.

Hewitt, E. J. and C. S. Gundry. 1970. The molybdenum requirement of plants in relation to nitrogen supply. J. Hort. Sci. 45:351–358.

Hiatt, A. J. and H. J. Evans. 1960. Influence of certain cations on activity of acetic thiokinase from spinach leaves. Plant Physiol. 35:673–677.

Hsiao, T. C., R. H. Hageman and E. H. Tyner. 1970. Effects of potassium nutrition on protein and total free amino acids in Zea mays. Crop Sci. 10:78–82.

Humble, G. D. and T. C. Hsiao. 1969. Specific requirement of potassium for light-activated opening of stomata in epidermal strips. Plant Physiol. 44:230–234.

Humble, G. D. and T. C. Hsiao. 1970. Light-dependent influx and efflux of potassium of guard cells during stomatal opening and closing. Plant Physiol. 46:483–487.

Humble, G. D. and K. Raschke. 1971. Stomatal opening quantitatively related to potassium transport: evidence from electron probe analysis. Plant Physiol. In press.

Ito, A. and A. Fujiwara. 1967. Functions of calcium in the cell wall of rice leaves. Plant Cell Physiol. 8:409–422.

Jacobson, L. and J. J. Oertli. 1956. The relation between iron and chlorophyll contents in chlorotic sunflower leaves. Plant Physiol. 31:199–204.

Jagendorf, A. T. and M. Smith. 1962. Uncoupling phosphorylation in spinach chloroplasts by absence of cations. Plant Physiol. 37:135–141.

Joham, H. E. and J. V. Amin. 1965. Role of sodium in the potassium nutrition of cotton. Soil Sci. 99:220–226.

Johnson, D. L. and L. S. Albert. 1967. Effect of selected nitrogen-bases and boron on the ribonucleic acid content, elongation, and visible deficiency symptoms of tomato root tips. Plant Physiol. 42:1307–1309.

Jones, L. H. 1961. Some effects of potassium deficiency on the metabolism of the tomato plant. Can. J. Bot. 39:593–606.

Jones, L. H. P. and K. A. Handreck. 1967. Silica in soils, plants, and animals. Adv. Agron. 19:107–149.

Jones, R. G. W. and O. R. Lunt. 1967. The function of calcium in plants. Bot. Rev. 33:407–426.

Karim, A. Q. M. B. and J. Vlamis. 1962. Comparative study of the effects of ammonium and nitrate nitrogen in the nutrition of rice. Plant and Soil 16:32–41.

Kessler, B. 1961. Ribonuclease as a guide for the determination of zinc deficiency in orchard trees. In: Plant Analysis and Fertilizer Problems. W. Reuther, ed. American Institute of Biological Sciences, Washington, D. C. Pp. 314–322.

Kilmer, V. J., S. E. Younts and N. C. Brady, eds. 1968. The Role of Potassium in Agriculture. American Society of Agronomy, Crop Science Society of America, Soil Science Society of America, Madison.

Kylin, A. 1960. The incorporation of radio-sulphur from external sulphate into different sulphur fractions of isolated leaves. Physiol. Plantarum 13:366–379.

Larkum, A. W. D. 1968. Ionic relations of chloroplasts in vivo. Nature 218:447–449.

Lee, K.–W., C. M. Whittle and H. J. Dyer. 1966. Boron deficiency and translocation profiles in sunflower. Physiol. Plantarum 19:919–924.

Lee, S. and S. Aronoff. 1967. Boron in plants: a biochemical role. Science 158:798–799.

Lee, T. S. 1959. Chemical equilibrium and the thermodynamics of reactions. In: Treatise on Analytical Chemistry. I. M. Kolthoff and P. J. Elving, eds. The Interscience Encyclopedia, Inc., New York. Part I, Vol. 1, pp. 185–275.

Lewin, J. and B. E. F. Reimann. 1969. Silicon and plant growth. Ann. Rev. Plant Physiol. 20:289–304.

Lindblad, K.-L. 1959. The influence of growth conditions on the amount and ribonucleic acid content of wheat root mitochondria. Physiol. Plantarum 12:400–411.

Loneragan, J. F. 1968. Nutrient requirements of plants. Nature 220:1307–1308.

Loneragan, J. F., J. S. Gladstone and W. J. Simmons. 1968. Mineral elements in temperate crop and pasture plants. II. Calcium. Austral. J. Agric. Res. 19:353–364.

Loneragan, J. F. and K. Snowball. 1969a. Calcium requirements of plants. Austral. J. Agric. Res. 20:465–478.

Loneragan, J. F. and K. Snowball. 1969b. Rate of calcium absorption by plant roots and its relation to growth. Austral. J. Agric. Res. 20:479–490.

Machold, O. 1968. Einfluss der Ernährungsbedingungen auf den Zustand des Eisens in den Blättern, den Chlorophyllgehalt und die Katalase-sowie Peroxydaseaktivität. Flora A159:1–25.

Machold, O. and G. Scholz. 1969. Eisenhaushalt und Chlorophyllbildung bei höheren Pflanzen. Naturwiss. 56:447–452.

Mahler, H. R. 1961. Interrelationships with enzymes. In: Mineral Metabolism. C. L. Comar and F. Bronner, eds. Academic Press, New York and London. Vol. IB, pp. 743–879.

Malkin, R. and J. C. Rabinowitz. 1967. Nonheme iron electron-transfer proteins. Ann. Rev. Biochem. 36:113–148.

Marinos, N. G. 1962. Studies on submicroscopic aspects of mineral deficiencies. I. Calcium deficiency in the shoot apex of barley. Am. J. Bot. 49:834–841.

Marinos, N. G. 1963. Studies on submicroscopic aspects of mineral deficiencies. II. Nitrogen, potassium, sulfur, phosphorus, and magnesium deficiencies in the shoot apex of barley. Am. J. Bot. 50:998–1005.

Marks, G. S. 1969. Heme and Chlorophyll. Chemical, Biochemical and Medical Aspects. D. Van Nostrand Company, Ltd., London.

Marschner, H. and I. Günther. 1964. Ionenaufnahme und Zellstruktur bei Gerstenwurzeln in Abhängigkeit von der Calcium-Versorgung. Z. Pflanzenern. Düng. Bodenk. 107:118–136.

Marschner, H., R. Handley and R. Overstreet. 1966. Potassium loss and changes in the fine structure of corn root tips induced by H-ion. Plant Physiol. 41:1725–1735.

Martell, A. E. and M. Calvin. 1952. Chemistry of the Metal Chelate Compounds. Prentice-Hall, Inc., New York.

McIlrath, W. J. and J. Skok. 1964. Distribution of boron in the tobacco plant. Physiol. Plantarum 17:839–845.

McKee, H. S. 1962. Nitrogen Metabolism in Plants. Clarendon Press, Oxford.

Minotti, P. L. and W. A. Jackson. 1970. Nitrate reduction in the roots and shoots of wheat seedlings. Planta 95:36–44.

Nason, A. and W. D. McElroy. 1963. Modes of action of the essential mineral elements. In: Plant Physiology—A Treatise. F. C. Steward, ed. Academic Press, New York and London. Vol. 3, pp. 451–536.

Neales, T. F. and L. D. Incoll. 1968. The control of leaf photosynthesis rate by the level of assimilate concentration in the leaf: a review of the hypothesis. Bot. Rev. 34:107–125.

Nobel, P. S. 1967. Calcium uptake, ATPase and photophosphorylation by chloroplasts in vitro. Nature 214:875–877.

Nobel, P. S. 1969. Light-induced changes in the ionic content of chloroplasts in Pisum sativum. Biochim. Biophys. Acta 172:134–143.

Ozbun, J. L., R. J. Volk and W. A. Jackson. 1965. Effects of potassium deficiency on photosynthesis, respiration and the utilization of photosynthetic reductant by mature bean leaves. Crop Sci. 5:497–500.

Packer, L. and A. R. Crofts. 1967. The energized movement of ions and water by chloroplasts. In: Current Topics in Bioenergetics. D. R. Sanadi, ed. Vol. 2, pp. 23–64.

Pirson, A. 1955. Functional aspects in mineral nutrition of green plants. Ann. Rev. Plant Physiol. 6:71–114.

Possingham, J. V. and D. Spencer. 1962. Manganese as a functional component of chloroplasts. Austral. J. Biol. Sci. 15:58–68.

Possingham, J. V., M. Vesk and F. V. Mercer. 1964. The fine structure of leaf cells of manganese-deficient spinach. J. Ultrastructure Res. 11:68–83.

Price, C. A. 1968. Iron compounds and plant nutrition. Ann. Rev. Plant Physiol. 19:239–248.

Price, C. A. 1970. Molecular Approaches to Plant Physiology. McGraw-Hill Book Company, New York.

Puritch, G. S. and A. V. Barker. 1967. Structure and function of tomato leaf chloroplasts during ammonium toxicity. Plant Physiol. 42:1229–1238.

Rasmussen, H. P. 1967. Calcium and strength of leaves. I. Anatomy and histochemistry. Bot. Gaz. 128:219–223.

Redfield, A. C. 1958. The biological control of chemical factors in the environment. Am. Scientist 46:205–222.

Rendig, V. V. and E. A. McComb. 1959. Effect of nutritional stress on plant composition: I. The interaction of added nitrogen with varying sulfur supply. Soil Sci. Soc. Am. Proc. 23:377–380.

Rendig, V. V. and E. A. McComb. 1961. Effect of nutritional stress on plant composition. II. Changes in sugar and amide nitrogen content of normal and sulfur-deficient alfalfa during growth. Plant and Soil 14:176–186.

Rios, M. A. and R. W. Pearson. 1964. The effect of some chemical environmental factors on cotton root behavior. Soil Sci. Soc. Am. Proc. 28:232–235.

Rubey, W. W. 1951. Geologic history of sea water. An attempt to state the problem. Bull. Geol. Soc. Am. 62:1111–1147.

Salami, A. U. and D. G. Kenefick. 1970. Stimulation of growth in zinc-deficient corn seedlings by the addition of tryptophan. Crop Sci. 10:291–294.

Saltman, P., J. G. Forte and G. M. Forte. 1963. Permeability studies on chloroplasts from Nitella. Expt. Cell Res. 29:504–514.

San Pietro, A., ed. 1965. Non-Heme Iron Proteins: Role in Energy Conversion. The Antioch Press, Yellow Springs.

Satter, R. L., P. Marinoff and A. W. Galston. 1970. Phytochrome controlled nyctinasty in Albizzia julibrissin. II. Potassium flux as a basis for leaflet movement. Am. J. Bot. 57:916–926.

Sawhney, B. L. and I. Zelitch. 1969. Direct determination of potassium ion accumulation in guard cells in relation to stomatal opening in light. Plant Physiol. 44:1350–1354.

Schütte, K. H. 1964. The Biology of the Trace Elements. Their Role in Nutrition. J. B. Lippincott Company, Philadelphia and Montreal.

Shear, C. B., H. L. Crane and A. T. Myers. 1953. Nutrient element balance: response of Tung trees grown in sand culture to potassium, magnesium, calcium, and their interactions. U. S. Department of Agriculture Tech. Bull. No. 1085.

Sinclair, C. 1969. The level and distribution of amines in barley as affected by potassium nutrition, arginine level, temperature fluctuation and mildew infection. Plant and Soil 30:423–438.

Skoog, F. 1940. Relationships between zinc and auxin in the growth of higher plants. Am. J. Bot. 27:939–951.

Smith, T. A. 1968. The biosynthesis of putrescine in higher plants and its relation to potassium nutrition. In: Recent Aspects of Nitrogen Metabolism in Plants. E. J. Hewitt and C. V. Cutting, eds. Academic Press, London and New York. Pp. 139–146.

Sorokin, H. and A. L. Sommer. 1940. Effects of calcium deficiency upon the roots of Pisum sativum. Am. J. Bot. 27:308–318.

Spurr, A. R. 1952. Fluorescence in ultraviolet light in the study of boron deficiency in celery. Science 116:421–423.

Spurr, A. R. 1959. Anatomical aspects of blossom-end rot in the tomato with special reference to calcium nutrition. Hilgardia 28:269–295.

Stanley, R. G. and E. A. Lichtenberg. 1963. The effect of various boron compounds on in vitro germination of pollen. Physiol. Plantarum 16:337–346.

Steward, F. C. and D. Margolis. 1962. The effects of manganese upon the free amino acids and amides of the tomato plant. Contrib. Boyce Thompson Inst. 21:393–410.

Stewart, B. A. and L. K. Porter. 1969. Nitrogen-sulfur relationships in wheat (Triticum aestivum L.), corn (Zea mays), and beans (Phaseolus vulgaris). Agron. J. 61:267–271.

Stocking, C. R. and A. Ongun. 1962. The intracellular distribution of some metallic elements in leaves. Am. J. Bot. 49:284–289.

Stout, P. R., W. R. Meagher, G. A. Pearson and C. M. Johnson. 1951. Molybdenum nutrition of crop plants. I. The influence of phosphate and sulfate on the absorption of molybdenum from soils and solution cultures. Plant and Soil 3:51–87.

Street, H. E. and D. E. G. Sheat. 1958. The absorption and availability of nitrate and ammonia. In: Encyclopedia of Plant Physiology. W. Ruhland, ed. Springer–Verlag, Berlin. Vol. 8, pp. 150–165.

Tagawa, T. and J. Bonner. 1957. Mechanical properties of the Avena coleoptile as related to auxin and to ionic interactions. Plant Physiol. 32:207–212.

Takaki, H. and M. Kushizaki. 1970. Accumulation of free tryptophan and tryptamine in zinc deficient maize seedlings. Plant and Cell Physiol. 11:793–804.

Thompson, J. F. 1967. Sulfur metabolism in plants. Ann. Rev. Plant Physiol. 18:59–84.

Thompson, J. F., I. K. Smith and D. P. Moore. 1970. Sulfur requirement and metabolism in plants. In: Symposium: Sulfur in Nutrition. O. H. Muth and J. E. Oldfield, eds. The Avi Publishing Company, Inc., Westport. Pp. 80–96.

Thomson, W. W. and T. E. Weier. 1962. The fine structure of chloroplasts from mineral-deficient leaves of *Phaseolus vulgaris*. Am. J. Bot. 49:1047–1055.

Thorne, D. W., F. B. Wann and W. Robinson. 1950. Hypotheses concerning lime-induced chlorosis. Soil Sci. Soc. Am. Proc. 15:254–258.

Tsui, C. 1948. The role of zinc in auxin synthesis in the tomato plant. Am. J. Bot. 35:172–179.

Ulrich, A. and K. Ohki. 1956. Chlorine, bromine and sodium as nutrients for sugar beet plants. Plant Physiol. 31:171–181.

Ulrich, A. and K. Ohki. 1966. Potassium. In: Diagnostic Criteria for Plants and Soils. H. D. Chapman, ed. University of California, Division of Agricultural Sciences. Pp. 362–393.

Vesk, M., J. V. Possingham and F. V. Mercer. 1966. The effect of mineral nutrient deficiencies on the structure of the leaf cells of tomato, spinach, and maize. Austral. J. Bot. 14:1–18.

Wallace, A., E. Frolich and O. R. Lunt. 1966. Calcium requirements of higher plants. Nature 209:634.

Watanabe, R., W. Chorney, J. Skok and S. H. Wender. 1964. Effect of boron deficiency on polyphenol production in the sunflower. Phytochem. 3:391–393.

Watanabe, R., W. J. McIlrath, J. Skok, W. Chorney and S. H. Wender. 1961. Accumulation of scopoletin glucoside in boron-deficient tobacco leaves. Arch. Biochem. Biophys. 94:241–243.

Weiser, C. J., L. T. Blaney and P. Li. 1964. The question of boron and sugar translocation in plants. Physiol. Plantarum 17:589–599.

Williams, D. E. and J. Vlamis. 1957. The effect of silicon on yield and manganese-54 uptake and distribution in the leaves of barley plants grown in culture solutions. Plant Physiol. 32:404–409.

PART IV

HEREDITY AND ENVIRONMENT IN PLANT NUTRITION

12

PHYSIOLOGICAL GENETICS OF PLANT NUTRITION

INTRODUCTION

In 1909 the British physician Archibald E. Garrod published a book, *Inborn Errors of Metabolism,* which stands as a milestone in the history of biology (Garrod, 1909). Garrod had studied four biochemical abnormalities in humans which were genetically controlled. The discoveries of Gregor Mendel of Austria, published in 1866 in his *Experiments on Plant Hybrids* and long neglected, had been rediscovered some years before, in 1900, and it soon became clear that Garrod's "inborn errors" showed the pattern of inheritance to be expected for double recessive Mendelian factors, or genes.

The abnormality which Garrod studied most was alkaptonuria, an inborn condition which causes the subject to pass urine which on contact with air blackens. This is due to its large content of homogentisic acid, a product of the breakdown of the amino acids phenylalanine and tyrosine. Normally homogentisic acid is further metabolized, but in alkaptonuria its degradation is blocked. These findings gave rise to the idea that an enzyme essential to the breakdown of homogentisic acid is lacking in persons with this condition.

There were, then, two main concepts which Garrod correlated and thereby became the father of biochemical genetics. They were, first, that a biochemical abnormality was due to the absence of a single, specific

325

enzyme, and second, that the presence or absence of that enzyme was controlled by a single pair of Mendelian genes—a proposition which later became known as the "one gene—one enzyme" hypothesis.

If a is a metabolite and b the product of its degradation by enzyme e, we may depict the scheme advanced by Garrod as follows:

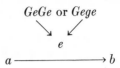

where Ge is the dominant gene specifying the synthesis of the enzyme, e, and ge the recessive. An "error of metabolism" occurs if the organism carries both recessive alleles:

The enzyme, e, is absent, and metabolite a accumulates, instead of being degraded enzymically to b (faint arrows and lettering indicate missing syntheses and products).

Perhaps this statement, through benefit of hindsight, is more clear-cut than these ideas were in Garrod's time. In any event, like Mendel's concepts before, those advanced by Garrod did not attract wide attention until much later. In the early 1940's G. W. Beadle and E. L. Tatum, then at Stanford University, adopted the bread mold, *Neurospora,* as their experimental organism and the deliberate induction of gene mutations as their experimental strategy. Their idea was that rather than hope to discover a fortunate happenstance like alkaptonuria, they would induce mutations by radiation or other mutagenic agents. Then they would look for mutants that fail to grow in the absence of an organic nutrient which normal or "wild type" *Neurospora* does not need to have in its medium. Such an abnormal requirement would be evidence that the mutant lacked the ability to synthesize the nutrient, presumably through absence of an enzyme essential for its synthesis.

The first mutant they discovered in this way was one requiring the presence of vitamin B_6 in the medium. Wild type *Neurospora* are able to synthesize vitamin B_6 and grow well on a medium not containing it. In securing a mutant which required for growth a medium supplemented with vitamin B_6 Beadle and Tatum had found what they were after: a mutant deficient in the ability to synthesize an essential metabolite. It only remained to establish, by appropriate crosses, whether the deficiency wàs the result of a mutation in a single gene. It was.

After this initial success, the investigators soon produced dozens of mutant *Neurospora* strains with a great variety of nutritional deficiencies. (See Beadle, 1959, which is the text of the address he gave in December 1958 in Stockholm on the occasion of receiving the Nobel prize, with Tatum, for their *Neurospora* research.) Since then, innumerable instances have been discovered in many organisms of the dependence of specific biochemical events on gene action (Wagner and Mitchell, 1964).

In this book, we are concerned, not with the synthesis of specific organic metabolites and nutrients, but with physiological processes involving many metabolites as well as complex cellular structures and organelles. These processes are governed by a multitude of enzymes and therefore are less easy to analyze in terms of specific enzymes and the genes required for their synthesis than are single enzyme-governed reactions.

As a result, this genetic approach has had little influence on the study of the mineral nutrition of plants. The genetics of mineral nutrition are not even discussed in the current encyclopedic works on plant physiology. We shall see, nevertheless, that there is ample evidence to the effect that various aspects of the transport and utilization of mineral nutrients in plants are under genetic control. In a few cases, single gene mutations have been shown to be responsible for nutritional abnormalities. In many more, nutritional features do not differ in such a dramatic on-off fashion but quantitatively, suggesting that a number of gene loci is involved. We are dealing here with a potentially exceedingly useful approach to mineral plant nutrition, a method that so few investigators have so far used that it is sure to lead to important discoveries once it is applied with a will.

2. MUTANTS

In this section we shall discuss those regrettably few instances in which aspects of mineral nutrition have been shown to be governed by single gene loci.

Mineral Deficiency Mutants. At the same time that Beadle and Tatum were working on nutritionally defective mutants of *Neurospora,* M. G. Weiss chose for his Ph.D. research at Iowa State College (now Iowa State University) the problem of the iron nutrition of several lines of soybean, *Glycine max,* which had been introduced from the Orient by the Division of Plant Exploration and Introduction of the U. S. Department of Agriculture (Weiss, 1943). The oriental varieties showed iron deficiency chlorosis when grown in a medium low in available iron which, however, did not cause chlorosis in soybeans of several standard pure lines. The former varieties were called inefficient and the latter, efficient in iron utilization.

Genetic analysis revealed that a single pair of alleles governed efficiency and inefficiency, with efficiency dominant. *FeFe* plants and *Fefe* plants were efficient, where *Fe* stands for the dominant and *fe* the recessive allele. Plants with the *fefe* genotype were inefficient.

Brown and his associates in the U. S. Department of Agriculture have done much research on two of these soybean varieties, one (Hawkeye, or HA) efficient, the other (PI-54619-5-1, or PI) inefficient. They grafted scions of the varieties onto rootstocks in all possible combinations and

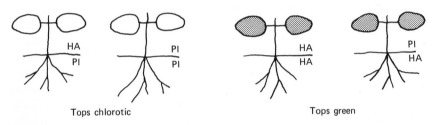

Tops chlorotic   Tops green

Figure 12-1. Diagrammatic representation of the reciprocal grafting experiment of Brown *et al.* (1958).

established that it is in the roots that the control of efficiency of iron utilization resides (Brown *et al.,* 1958). On calcareous soil, HA (efficient) and PI (inefficient) tops were chlorotic on PI roots and green on HA roots (Figure 12-1). Efficient varieties translocate much more iron to the tops than inefficient ones. In some experiments, the difference amounted to a factor of 25 (Brown *et al.,* 1967). The controlling factor appears to be the ability of the roots to reduce iron from the ferric to the ferrous state.

Weiss published his discovery of monogenic control of efficiency of iron utilization in soybeans in 1943. Ten years later, Pope and Munger of Cornell University described two more cases of nutritional disorders subject to control by single gene loci. One concerned a susceptibility to magnesium deficiency in celery, *Apium graveolens* (Pope and Munger, 1953a), the other susceptibility to boron deficiency in the same species (Pope and Munger, 1953b).

The case of magnesium deficiency came to light because certain varieties of celery, for example Utah 10B, regularly showed severe chlorosis when growing on organic (muck) soils in New York while other varieties grew normally beside them. By analysis and other tests the chlorosis was identified as due to magnesium deficiency. Crosses between Utah 10B and varieties not prone to magnesium deficiency revealed simple Mendelian segregation, with the normal condition (*Mg*) dominant and susceptibility (*mg*) recessive. The metabolic lesion causing magnesium deficiency must reside in the roots or stems of the *mgmg* plants because the condition is most efficiently corrected by foliar sprays with magnesium solutions, rather than with soil treatments (Johnson *et al.*, 1961). In a similar investigation, the celery strain S48-54-1 was found to be susceptible to boron deficiency. Genetic analysis revealed that susceptibility to the disorder was conditioned by double recessive alleles at a single locus (Pope and Munger, 1953b).

Before G. W. Beadle, of *Neurospora* fame, elevated that lowly mold to its present preeminence as an organism for research on biochemical genetics he had worked with other plants, the kingly corn, *Zea mays*, among them. He was the first to describe a mutant, which he called yellow stripe (ys), characterized by chlorotic stripes or streaks between the main vascular bundles (Beadle, 1929). Beadle did not examine the physiology or biochemistry of the condition but thought that it might be due to "an inadequate rate of transport of some essential substance to or from the regions between the larger veins."

Bell *et al.* (1958, 1962) have determined that the yellow stripe condition is an iron deficiency, and that the lesion resides in the roots. Factors similar to those investigated by Brown and his co-workers in soybeans appear to govern the unavailability of iron in yellow stripe corn—mainly the inability to reduce ferric to ferrous iron.

In the discussion of boron deficiency (Chapter 4) it was brought out that afflicted tissues are often hard and brittle. Wall and Andrus (1962) have described a mutant called "brittle stem" in the tomato, *Lycopersicon esculentum*. Crosses with strains not prone to this condition showed it to be due to a recessive gene, *btl*. Plants with the *btlbtl* genotype had much less boron in their leaves than normal plants. When ample boron was supplied in the medium the condition did not show up and the plants thrived. In other words, the brittle stem genotype causes an inefficient utilization of boron. The roots of brittle stem plants absorb boron from the medium just as readily as do those of normal plants but do not forward it efficiently to the shoots.

Other Mineral Nutrition Mutants.   Not only mineral nutrient deficiencies but some other conditions related to mineral metabolism have been shown to be governed by single gene loci. One such instance was discovered by J. Langridge of the Commonwealth Scientific and Industrial Research Organization at Canberra, Australia. Langridge (1958) studied a small member of the family Cruciferae, Arabidopsis thaliana. This plant has become a favorite experimental subject for plant physiologists interested in genetic control of physiological processes because it has a short life cycle (28 days), a low chromosome number ($n = 5$), and can be grown to maturity in a test tube under aseptic conditions (Langridge, 1955).

The mutant Langridge (1958) studied had an unusually low internal osmotic pressure when grown at 23°C and therefore failed to grow, although at 28°C it was normal. At 23°C it could be made to grow by adding an absorbable solute to the medium at a substantial concentration. Sucrose, glucose, and potassium sulfate all were effective; mannitol, which is not absorbed appreciably, was not. Apparently the mutant is unable, at 23°C, to absorb and retain solutes at a rate high enough to keep its internal osmotic pressure above that of the medium. Osmotic uptake of water is therefore prevented and growth is stopped.

Quite a different situation was uncovered in the soybean, Glycine max, but here, too, a single gene pair governed the response. Most varieties of soybean grow well even at high concentrations of phosphate in the solution (several millimoles per liter), but some varieties are sensitive to such high levels of phosphate and respond by developing blotches and chlorosis in the leaves and by failing to grow. Bernard and Howell (1964) found on crossing a phosphate-tolerant with a phosphate-sensitive variety and studying their progeny that a pair of alleles at a single locus, which they named Np and np, governed the response. NpNp plants showed little sensitivity to high phosphate levels, the npnp genotype was very sensitive, and the heterozygote Npnp was intermediate.

Abel (1969) studied the tolerance of varieties of this same species, soybean, to saline conditions. Most varieties, bred as they are for areas where soils are not saline, were highly sensitive to salt when grown on the saline soils of the Imperial Valley of southern California; however, a few varieties were tolerant of salt. The roots of both kinds absorbed chloride about equally. However, the salt-tolerant varieties translocated very little chloride to the leaves while the sensitive ones translocated much, causing necrotic leaf burn which eventually killed the plants. Crosses revealed a single locus to determine the difference, Ncl being the dominant chloride excluder and ncl the recessive chloride includer.

Shea et al. (1967, 1968) at the University of Wisconsin investigated the genetic control of the efficiency with which strains of the snap bean, Phaseolus vulgaris, utilize potassium. "Efficient" and "inefficient" strains did not differ consistently in the amounts of potassium absorbed and translocated. The difference lay rather in the efficiency with which the potassium was utilized. Efficient strains showed no evidence of potassium deficiency when the percentage of potassium in the tops was between 0.5 and 1.0 per cent, at which level inefficient strains were apt to be potassium deficient. Genetic analysis revealed the efficient strains to be homozygous for a recessive pair of alleles, named k_e.

This survey of mutations governing aspects of mineral nutrition in higher plants shows that so far, only a few such situations have come to light. Biochemical genetics has made a truly revolutionary impact in the study of microorganisms and we might well ask why this approach has not been equally important in the area of plant physiology in general and plant nutrition in particular. The reasons are fairly obvious. The generation time of higher plants is long, compared with that of microorganisms, and physiologists and biochemists interested in gene effects have preferred bacteria and fungi for that reason. When a mutant is obtained as evidenced by obvious phenotypic differences there is seldom any clue as to the nature of the metabolic lesion causing it. Mutants with severe nutrient deficiencies, especially of macronutrients, would be lethals, and it is noteworthy that of the three deficiencies due to single gene pairs discovered so far (iron, boron, and magnesium), two are micronutrient deficiencies.

Enzyme Deficiency Mutants.   Nitrate and sulfate must be reduced before their nitrogen and sulfur atoms can be incorporated into essential metabolites. When the mechanism of reduction of either element fails, the plants become functionally deficient in the element despite the fact that they may absorb ample quantities of the anion. Such failure occurs when one of the enzymes in the sequence of reductions is lacking. Shafer *et al.* (1961) irradiated a culture of the green alga, *Chlorella,* with ultraviolet light to induce mutations and then grew individual colonies in a medium in which KNO_3 was the sole source of nitrogen. They then studied a mutant which they identified because it made very little growth on this medium. They found that the mutant grew well when nitrogen was furnished in the form of ammonium ions, NH_4^+, and (somewhat less readily) with nitrite, NO_2^-, as the source of nitrogen, but not on nitrate, NO_3^-. The findings suggested that the mutant lacked nitrate reductase, and assay for the enzyme confirmed this. The normal or "wild type" *Chlorella* possesses nitrate reductase and is competent to reduce nitrate to nitrite.

In corn, *Zea mays,* the level of nitrate reductase activity in two inbred lines is controlled by two loci (Warner *et al.,* 1969). It is likely that many enzyme activities in plants are determined by multiple loci but this is the first such instance reported for a higher plant. Nelson (1967) has given a review of biochemical genetics of higher plants but with the single exception of Langridge's osmotic mutant in *Arabidopsis* did not include any cases bearing on mineral nutrition, transport, or metabolism.

Although only a few clear-cut cases are known of mineral nutritional mutants in higher plants there is ample evidence that aspects of absorption, translocation, and utilization of mineral nutrients are under genetic control. In addition to the mutations discussed already, there are two lines of evidence for this conclusion. One concerns quantitative differences among varieties or cultivars of crop species with respect to the absorption, translocation, or utilization of nutrients or their toleration of mineral elements present at high concentrations. These are discussed in the remainder of the present chapter. The other has to do with nutritional ecotypes, a matter treated in the next chapter.

3.   VARIETAL DIFFERENCES

When plants of different varieties of a species, having grown side by side in the same soil under identical conditions, are analyzed for their content of nutrient and other mineral elements, large differences often become apparent. Such differences do not necessarily indicate that the variety with larger amounts of some nutrient in its leaves than another has a more efficient mechanism for absorption or translocation of that nutrient. It may merely mean that the variety has a larger or more finely branched root system and therefore explores its soil more effectively (Lee, 1960; Rabideau *et al.*, 1950).

Nevertheless, there are many instances which suggest that such intraspecific variations in the uptake of nutrients reflect genetically controlled differences in mechanisms of mineral nutrition, especially those concerned with absorption and translocation of a given element. Gross differences in the morphology of the root would be expected to result in parallel variations in the uptake of all or at least many elements. When two varieties differ significantly with respect to the absorption of certain elements but not in regard to that of others we are led to the hypothesis that the differences are due to genetically controlled mechanisms of mineral nutrition. When the differences are quantitative, the process is probably controlled by a number of loci. Important earlier work is discussed by Burkholder and McVeigh (1940) in their paper on genetic effects in the nitrogen nutrition of inbred lines and hybrids of corn, *Zea mays*.

TABLE 12-1. Varietal Effects in Mineral Nutrition

Species	Common name	Elements investi- gated	Remarks	Reference
Avena sativa	Oat	Mn, Mg		Vose and Griffiths (1961)
Avena sativa and *A. byzantina*	Oat	Mn		Munns *et al.* (1963a, b, c)
Beta vulgaris	Sugar beet	K, P, N		Ulrich (1961)
Citrus sp.	Citrus	Cl		Hewitt and Furr (1965)
Dactylis glomerata	Orchardgrass	Ca, P		Crossley and Bradshaw (1968)
Fragaria sp.	Strawberry	K, Ca, Mg, P, Ń		Lineberry and Burkhart (1943)
Hordeum vulgare	Barley	Sr	Excised roots	Pinkas and Smith (1966)
Lolium perenne	Ryegrass	N		Vose and Breese (1964)
Lolium perenne	Ryegrass	Ca, P		Crossley and Bradshaw (1968)
Lolium perenne	Ryegrass	Al, Mn		Vose and Randall (1962)
Lycopersicon esculentum	Tomato	Ca		Greenleaf and Adams (1969)
Lycopersicon esculentum	Tomato	Na	Excised roots	Picciurro and Brunetti (1969)
Malus sp.	Apple	N		Ruck and Bolas (1956)
Medicago sp.	Alfalfa	Mn	Tolerance to toxicity	Dessureaux and Ouellette (1958)
Oryza sativa	Rice	Fe	Excised roots	Shim and Vose (1965)
Pelargonium hortorum	Geranium	Na, Ca, B	Tolerance to toxicities	Kofranek *et al.* (1958)
Zea mays	Corn	N	Inbred lines and hybrids	Burkholder and McVeigh (1940)
Zea mays	Corn	Rb	Excised roots	Millaway and Schmid (1967)

Table 12-1 lists instances of quantitative differences among varieties. They all bear on some aspect of mineral nutrition. Further examples have been listed by Epstein and Jefferies (1964). When plants grow in soil many uncontrolled and unobserved variables affect their nutrition. Some of these factors are themselves influenced by the genotype and may have indirect effects on mineral nutrition. The degree of proliferation of the roots is an example. Quite apart from that there may be competition for light and water and other environmental factors which cause variability (Ulrich, 1961). For these reasons, only experiments in which the plants grew under controlled conditions in nutrient solutions are included in Table 12-1.

Nevertheless, experiments with soil-grown plants have also provided much convincing evidence on genetically controlled nutritional differences among varieties. For example, workers at the Pennsylvania Agricultural Experiment Station have investigated the genetic component in the differential accumulation of numerous mineral elements by genotypes of corn, *Zea mays* (Barber *et al.*, 1967; Gorsline *et al.*, 1964a, b; Baker *et al.*, 1964). Another group, at the University of Minnesota, has paid particular attention to the genetic factor in the accumulation of strontium. Strontium is chemically much like calcium, and its radioisotope, ^{90}Sr, a fallout product of atomic weapons, is absorbed by plants and on being ingested by animals including humans is, like calcium, incorporated into bone. This has been the impetus for research on the inheritance of strontium accumulation in barley, *Hordeum vulgare*, wheat, *Triticum aestivum*, and other species (Rasmusson *et al.*, 1964; Rasmusson and Kleese, 1967; Smith *et al.*, 1963). These investigators found varieties to differ in their accumulation of calcium and strontium. In some varieties, the ratio of strontium-89 to calcium in the grain was lower than in others; in other words, there was discrimination against strontium. Control of these differences is not understood but reciprocal grafting experiments have shown that in soybeans, *Glycine max,* it resides in the shoots rather than the roots (Kleese, 1968). Butler and his colleagues in New Zealand studied the mineral content of ryegrass derived by crossing *Lolium perenne* with *L. multiflorum* (Butler *et al.*, 1962). Of the twelve mineral elements investigated, ten showed significant differences which were attributed to genetic factors.

In acid soils there often are high concentrations of aluminum, manganese, and other ions of heavy metals (Chaputer 13, pp. 357–358). They are an important factor in the infertility of such soils. Several groups have studied varietal differences in the sensitivity of plants to aluminum and other heavy metal ions. For example, Foy *et al.* (1967) of the U. S. Department of Agriculture found varieties of cotton, *Gossypium hirsutum* and *G. barbadense,* to differ in their sensitivity to aluminum. The same observation has repeatedly been made for varieties of other species, for example for ryegrass, *Lolium* sp., by Vose and Randall (1962) and for alfalfa, *Medicago* sp., by Dessureaux and Ouellette (1958).

Much evidence for genetical control of uptake and transport of mineral elements has come from experiments on the mineral composition of the scion as influenced by the rootstock of grafted plants. Fruit trees are often not grown on their own roots but grafted onto rootstocks which are better adapted to soil conditions, more resistant to soil-borne infections, or other-

TABLE 12-2. Effects of Rootstocks on the Mineral
Composition of Scions

Species	Common name	Elements investigated	Culture technique	Reference
Citrus sp.	Sweet and sour orange	Fe	Soil	Wallihan and Garber (1968)
Citrus sp.	Grapefruit	P, K, Mg, Ca, Fe, Mn, Zn, Cu, B, Cl	Soil	Wutscher *et al.* (1970)
Malus sp.	Apple	N, P, K, Ca, Mg, Fe, Mn, Cu, B, Mo, Al	Soil	Awad and Kenworthy (1963)
Malus sp.	Apple	Ca	Soil	Sistrunk and Campbell (1966)
Persea sp.	Avocado	Cl, N, P, K, Ca, Mg	Soil	Embleton *et al.* (1962)
Prunus spp.	Peach, almond, plum, apricot	Na, Cl	Soil	Bernstein *et al.* (1956)
Vitis sp.	Grape	N, P, K, Ca, Mg	Soil	Cook and Lider (1964)
Vitis sp.	Grape	Cl	Sand culture	Bernstein *et al.* (1969)

wise superior to the roots of the scion itself, which is chosen for the quality of the fruit. When different rootstocks are compared for their effects on the mineral composition of the scion, marked differences often show up. Table 12-2 lists a number of examples. Other instances have been given by Epstein and Jefferies (1964), and many more are recorded in the horticultural literature.

STUDIES ON MECHANISMS

Only rarely have genetically controlled quantitative differences in mineral nutrition been examined in detail to determine what aspect of nutrition was affected. The genetic control may govern the initial absorption of an element by the roots, its subsequent translocation into and through the xylem, the degree of its retention in tissues adjacent to the conducting elements, its mobility in the phloem, the efficiency of its metabolic utilization, and others. With some genotypes differing in a single gene locus such findings have been made, as discussed above in the section on mutants, but where inheritance of a character is polygenic and differences are quantitative, detailed investigations are usually lacking. That such investigations can be fruitful is nevertheless clear from some that have been made, and there is no doubt that physiological and biochemical genetics will contribute much to the clarification of problems in mineral plant nutrition.

Foy and Barber (1958) investigated the magnesium nutrition of two inbred lines of corn, *Zea mays*. One of them, Indiana WF9, typically has a much higher content of magnesium in the leaves than does the other, Ohio 40B. The difference is not due to a higher ability of roots of Indiana WF9 to absorb magnesium. Rather, the Ohio 40B variety retains more magnesium in the stem than does Indiana WF9. The mechanism of this immobilization was not examined.

Two varieties of barley, *Hordeum vulgare*, differ consistently in their transport of strontium into the xylem (Pinkas and Smith, 1966). The process was shown to be affected by antimetabolic conditions (low temperature, cyanide, and dinitrophenol), and equally so in both varieties. Whether tissues other than those of the roots of the two varieties also differ in their transport of strontium was not examined. It would also be interesting to know whether the transport of calcium in the two varieties differs in parallel fashion.

Still another investigation demonstrating the utility of the genetic approach to problems of mineral nutrition is that by Munns *et al.* (1963a, b, c) on the uptake and distribution of manganese in oat plants, *Avena sativa and A. byzantina*. Plants of different varieties grown in nutrient solutions differed slightly or not at all in their contents of several divalent cations and iron but some showed significant differences in the manganese content of their shoots.

Kinetic experiments with radiomanganese, [54]Mn, similar in general outline to experiments with other elements discussed in Chapter 7, led to the recognition of three compartments or metabolic pools of manganese in the roots, as shown in Figure 12-2. Although the authors defined these pools or fractions in operational terms it is nonetheless likely that the replaceable fraction represents manganese in the "outer" space and manganese held reversibly by the electronegative sites of the cell wall, and that the labile and non-labile fractions represent manganese in the cytoplasm and vacuole, respectively.

Varieties differed considerably in the size of the pools, especially the vacuolar fraction. For example, roots of the variety Ventura retained very little manganese in the non-labile (presumably vacuolar) pool, while the variety Palestine retained a substantial fraction in that pool (Munns *et al.*, 1963c). Obviously, kinetic experiments of this kind are more revealing than the mere finding that upon analysis of plants grown under the same

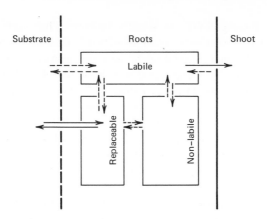

Figure 12-2. Scheme showing pools of manganese in oat roots, *Avena* sp. After Munns *et al.* (1963b).

conditions, those of one variety are found to contain a higher concentration of a certain element than those of another variety.

In still other experiments, excised roots of plants of different varieties have been used to study varietal differences in the kinetics of ion absorption (see Table 12-1). The experiments discussed in this chapter, and others of the same general nature, represent the beginnings of research on the physiological and biochemical genetics of mineral plant nutrition. Further research along these lines is discussed in connection with nutritional or edaphic (soil-related) ecotypes in the next chapter.

IMPLICATIONS FOR PLANT BREEDING

When sexually reproducing plants of a given species differ in genetically controlled features or performances which can be observed or measured, the conditions exist for breeding programs aimed at creating strains in which desired features or performances are optimized (Allard and Hansche, 1964). Genetic variability has been made use of by plant breeders with spectacular results in terms of increased yields, better resistance to disease, more advantageous chemical composition, and other desiderata.

The evidence presented in this chapter, and further evidence discussed in the next, shows that genotypes within a species may differ greatly in various aspects of mineral nutrition: rates of absorption and translocation of specific elements, efficiency of metabolic utilization, tolerance to high concentrations of an element in the medium, and other features. Yet the opportunity which these findings present for plant breeding have not been widely perceived, let alone realized in terms of actual programs of breeding. Why should this be, in view of the success that has attended breeding endeavors aimed at higher yields and other features?

There is no one reason for this state of affairs; there are several. The breeder has to be guided by the phenotype, and features of the phenotype such as yield or degree of infection with a disease are easily observed and measured. Aspects of mineral nutrition are often more subtle and may escape attention. When the condition of concern is a mineral deficiency it is usually easy and economically worthwhile, in agriculturally and technically advanced countries, to supply the deficient nutrient through fertilization and therefore there has not been much impetus for breeding for efficiency of uptake or utilization of nutrients. Except for the matter of salinity, discussed in the next chapter, toxic conditions have not been sufficiently widespread or serious to provide motivation for plant breeding endeavors.

Our slim knowledge of the genetic basis of aspects of mineral nutrition has also been a factor. Plant physiologists, including those dealing with mineral nutrition, have usually worked with pure lines of relatively few species, most of them crop plants. There are good reasons for this, in addition to the obvious one of economic importance: reduced variability, availability, and a body of prior knowledge which serves as a base for further experimentation. Nevertheless, this preoccupation with a few species, and pure lines of such species at that, means that we tend to narrow our field of view. It is one of the reasons that we have been slow in recognizing the immense genotypic wealth that is at our disposal, both for experimentation and as a gene pool from which the plant breeder can select combinations with superior performance.

In the last decade several investigators have recognized both the potential utility of breeding plants with desirable nutritional traits and the existence of the requisite genotypic variability. Vose (1963), Gerloff (1963), Epstein (1963), and Epstein and Jefferies (1964) reviewed then available evidence and pointed out the prospects for breeding, Vose emphasizing breeding for efficient nutrient uptake and utilization, and Epstein and Jefferies, both that and tolerance of toxic conditions, especially salinity. Myers (1960) made the same point with respect to the ability of plants to exclude harmful radionuclides.

Indications are that the need to breed strains of plants with various desirable nutritional traits will be increasingly felt. For example, the new varieties of tomato, *Lycopersicon esculentum,* VF-145 and VF-13L, bred for mechanical harvesting, are prone to potassium deficiency (Lingle and Lorenz, 1969). This is not due to any genetic impairment of their ability to absorb or utilize potassium. Rather, several features combine to make unusually high demands on the capacity of the plants to absorb potassium—demands which the mechanisms of absorption and transport cannot cope with. The features responsible are extremely dense planting, a very heavy fruit load, and a shorter growing season than that of conventional varieties. Along with other features that have been bred into these varieties it would be advantageous to equip them, through breeding, with superior capabilities of taking up potassium.

Forest trees are notable for growth on soils which are infertile by conventional agricultural standards. Trees grow slowly in comparison with agricultural crops, and this partly accounts for their ability to develop well under such poor conditions of mineral nutrition. However, through both breeding and management, artificial manipulation of forest stands will become increasingly frequent (Libby et al., 1969). Inevitably, forest trees will have to extract larger amounts of nutrients per unit time from their soils, a point which forest tree breeders would be well advised to keep in mind, in view of the experience with tomatoes related above.

Breeding can also enhance tolerance of toxic materials. Deficient materials can be supplied, but it is more difficult to remove matter present in excess. On a worldwide scale, by far the most important deleterious component of soils is salt. Hence, breeding for salt tolerance emerges as the greatest challenge to plant breeding aimed at problems of mineral metabolism. This subject is discussed further below and in the next chapter.

What beginnings have been made clearly indicate that plant breeding for desirable traits of mineral metabolism is feasible. Vose and Randall (1962) in just a few generations of selection increased the resistance of ryegrass, *Lolium* sp., to aluminum. They also made selections for resistance to manganese toxicity. Dewey (1962) pointed out that for species in which there is ample heritable variation in regard to salt tolerance, effective selection of salt tolerant strains should be feasible. He demonstrated that the requisite genetic variability is to be found in crested wheatgrass, *Agropyron desertorum,* and outlined a breeding plan for salt tolerance.

Not all such investigations have turned up much genetic variability with regard to some particular trait. For example, Ayers *et al.* (1951) and Bernstein and Ayres (1953) found little variability in the salt tolerance of six varieties of lettuce, *Lactuca sativa,* and five varieties of carrot, *Daucus carota,* respectively. Bernstein (1963) concluded that there is "little likelihood" of genetically improving salt tolerance, because of the relative uniformity in this trait of the commonly available varieties. However, the common varieties are the product of breeding, and a great deal of the original variability has been lost in the process. No pessimistic estimates are justified since, as Dewey (1962) has remarked, no species has yet been adequately surveyed for its heritable variability in respect to salt tolerance. We thus return to a point made before: existing crop species afford too narrow a basis for selection but we have available limitless genotypic wealth on which to draw. It is ours for the asking.

REFERENCES

Abel, G. H. 1969. Inheritance of the capacity for chloride inclusion and chloride exclusion by soybeans. Crop Sci. 9:697–698.

Allard, R. W. and P. E. Hansche. 1964. Some parameters of population variability and their implications in plant breeding. Adv. Agron. 16:281–325.

Awad, M. M. and A. L. Kenworthy. 1963. Clonal rootstock, scion variety and time of sampling influences in apple leaf composition. Proc. Am. Soc. Hort. Sci. 83:68–73.

Ayers, A. D., C. H. Wadleigh and L. Bernstein. 1951. Salt tolerance of six varieties of lettuce. Proc. Am. Soc. Hort. Sci. 57:237–242.

Baker, D. E., W. I. Thomas and G. W. Gorsline. 1964. Differential accumulation of strontium, calcium, and other elements by corn (Zea mays L.) under greenhouse and field conditions. Agron. J. 56:352–355.

Barber, W. D., W. I. Thomas and D. E. Baker. 1967. Inheritance of relative phosphorus accumulation in corn (Zea mays L.). Crop Sci. 7:104–107.

Beadle, G. W. 1929. Yellow stripe—a factor for chlorophyll deficiency in maize located in the Prpr chromosome. Am. Naturalist 63:189–192.

Beadle, G. W. 1959. Genes and chemical reactions in Neurospora. Science 129:1715–1719.

Bell, W. D., L. Bogorad and W. J. McIlrath. 1958. Response of the yellow-stripe mutant (ys_1) to ferrous and ferric iron. Bot. Gaz. 120:36–39.

Bell, W. D., L. Bogorad and W. J. McIlrath. 1962. Yellow-stripe phenotype in maize. I. Effects of ys_1 locus on uptake and utilization of iron. Bot. Gaz. 124:1–8.

Bernard, R. L. and R. W. Howell. 1964. Inheritance of phosphorus sensitivity in soybeans. Crop Sci. 4:298–299.

Bernstein, L. 1963. Salt tolerance of plants and the potential use of saline waters for irrigation. In: Desalination Research Conference. National

Academy of Sciences—National Research Council Publication 942. Pp. 273–283.

Bernstein, L. and A. D. Ayers. 1953. Salt tolerance of five varieties of carrots. Proc. Am. Soc. Hort. Sci. 61:360–366.

Bernstein, L., J. W. Brown and H. E. Hayward. 1956. The influence of rootstock on growth and salt accumulation in stone-fruit trees and almonds. Proc. Am. Soc. Hort. Sci. 68:86–95.

Bernstein, L., C. F. Ehlig and R. A. Clark. 1969. Effect of grape rootstocks on chloride accumulation in leaves. J. Am. Soc. Hort. Sci. 94:584–590.

Brown, J. C., R. S. Holmes and L. O. Tiffin. 1958. Iron chlorosis in soybeans as related to the genotype of rootstalk. Soil Sci. 86:75–82.

Brown, J. C., C. R. Weber and B. E. Caldwell. 1967. Efficient and inefficient use of iron by two soybean genotypes and their isolines. Agron. J. 59:459–462.

Burkholder, P. R. and I. McVeigh. 1940. Growth and differentiation of maize in relation to nitrogen supply. Am. J. Bot. 27:414–424.

Butler, G. W., P. C. Barclay and A. C. Glenday. 1962. Genetic and environmental differences in the mineral composition of ryegrass herbage. Plant and Soil 16:214–228.

Cook, J. A. and L. A. Lider. 1964. Mineral composition of bloomtime grape petiole in relation to rootstock and scion variety behavior. Proc. Am. Soc. Hort. Sci. 84:243–254.

Crossley, G. K. and A. D. Bradshaw. 1968. Differences in response to mineral nutrients of populations of ryegrass, *Lolium perenne* L., and orchard grass, *Dactylis glomerata* L. Crop Sci. 8:383–387.

Dessureaux, L. and G. J. Ouellette. 1958. Tolerance of alfalfa to manganese toxicity in sand culture. Can. J. Soil Sci. 38:8–13.

Dewey, D. R. 1962. Breeding crested wheatgrass for salt tolerance. Crop Sci. 2:403–407.

Embleton, T. W., M. Matsumura, W. B. Storey and M. J. Garber. 1962. Chlorine and other elements in avocado leaves as influenced by rootstock. Proc. Am. Soc. Hort. Sci. 80:230–236.

Epstein, E. 1963. Selective ion transport in plants and its genetic control. In: Desalination Research Conference. National Academy of Sciences—National Research Council Publication 942. Pp. 284–298.

Epstein, E. and R. L. Jefferies. 1964. The genetic basis of selective ion transport in plants. Ann. Rev. Plant Physiol. 15:169–184.

Foy, C. D., W. H. Armiger, A. L. Fleming and C. F. Lewis. 1967. Differential tolerance of cotton varieties to an acid soil high in exchangeable aluminum. Agron. J. 59:415–418.

Foy, C. D. and S. A. Barber. 1958. Magnesium absorption and utilization by two inbred lines of corn. Soil Sci. Soc. Am. Proc. 22:57–62.

Garrod, A. E. 1909. Inborn Errors of Metabolism. Henry Frowde Hodder and Staughton, London. (Also available in a reprint with a supplement by H. Harris, in: Garrod's Inborn Errors of Metabolism. 1963. Oxford University Press, London.)

Gerloff, G. C. 1963. Comparative mineral nutrition of plants. Ann. Rev. Plant Physiol. 14:107–124.

Gorsline, G. W., W. I. Thomas, D. E. Baker and J. L. Ragland. 1964a. Relationships of strontium-calcium accumulation within corn. Crop Sci. 4:154–156.

Gorsline, G. W., W. I. Thomas and D. E. Baker. 1964b. Inheritance of P, K, Mg, Cu, B, Zn, Mn, Al, and Fe concentrations by corn (*Zea mays* L.) leaves and grain. Crop Sci. 4:207–210.

Greenleaf, W. H. and F. Adams. 1969. Genetic control of blossom-end rot disease in tomatoes through calcium metabolism. J. Am. Soc. Hort. Sci. 94:248–250.

Hewitt, A. A. and J. R. Furr. 1965. Uptake and loss of chloride from seedlings of selected *Citrus* rootstock varieties. Proc. Am. Soc. Hort. Sci. 86:194–200.

Johnson, K. E. E., J. F. Davis and E. J. Benne. 1961. Occurrence and control of magnesium-deficiency symptoms in some common varieties of celery. Soil Sci. 91:203–207.

Kleese, R. A. 1968. Scion control of genotypic differences in Sr and Ca accumulation in soybeans under field conditions. Crop Sci. 8:128–129.

Kofranek, A. M., H. C. Kohl, Jr. and O. R. Lunt. 1958. Effects of excess salinity and boron on geraniums. Proc. Am. Soc. Hort. Sci. 71:516–521.

Langridge, J. 1955. Biochemical mutations in the crucifer *Arabidopsis thaliana* (L.) Heynh. Nature 176:260–261.

Langridge, J. 1958. An osmotic mutant of *Arabidopsis thaliana*. Austral. J. Biol. Sci. 11:457–470.

Lee, J. A. 1960. A study of plant competition in relation to development. Evolution 14:18–28.

Libby, W. J., R. F. Stettler and F. W. Seitz. 1969. Forest genetics and forest-tree breeding. Ann. Rev. Genetics 3:469–494.

Lineberry, R. A. and L. Burkhart. 1943. Nutrient deficiencies in the strawberry leaf and fruit. Plant Physiol. 18:324–333.

Lingle, J. C. and O. A. Lorenz. 1969. Potassium nutrition of tomatoes. J. Am. Soc. Hort. Sci. 94:679–683.

Millaway, R. M. and W. E. Schmid. 1967. Genetic control of rubidium absorption by excised corn roots: a preliminary survey using several inbred varieties. Transact. Illinois State Acad. Sci. 60:250–258.

Munns, D. N., C. M. Johnson and L. Jacobson. 1963a. Uptake and distribution of manganese in oat plants. I. Varietal variation. Plant and Soil 19:115–126.

Munns, D. N., L. Jacobson and C. M. Johnson. 1963b. Uptake and distribution of manganese in oat plants. II. A kinetic model. Plant and Soil 19:193–204.

Munns, D. N., C. M. Johnson and L. Jacobson. 1963c. Uptake and distribution of manganese in oat plants. III. An analysis of biotic and environmental effects. Plant and Soil 19:285–295.

Myers, W. M. 1960. Genetic control of physiological processes: consideration of differential ion uptake by plants. In: A Symposium on Radioisotopes in the Biosphere. R. S. Caldecott and L. A. Snyder, eds. University of Minnesota, Minneapolis. Pp. 201–226.

Nelson, O. E., Jr. 1967. Biochemical genetics of higher plants. Ann. Rev. Genetics 1:245–268.

Picciurro, G. and N. Brunetti. 1969. Assorbimento del Sodio (Na^{22}) in radici escisse di alcune varietà di *Lycopersicum esculentum*. Agrochimica 13:347–357.

Pinkas, L. L. H. and L. H. Smith. 1966. Physiological basis of differential strontium accumulation in two barley genotypes. Plant Physiol. 41:1471–1475.

Pope, D. T. and H. M. Munger. 1953a. Heredity and nutrition in relation to magnesium deficiency chlorosis in celery. Proc. Am. Soc. Hort. Sci. 61:472–480.

Pope, D. T. and H. M. Munger. 1953b. The inheritance of susceptibility to boron deficiency in celery. Proc. Am. Soc. Hort. Sci. 61:481–486.

Rabideau, G. S., W. G. Whaley and C. Heimsch. 1950. The absorption and distribution of radioactive phosphorus in two maize inbreds and their hybrid. Am. J. Bot. 37:93–99.

Rasmusson, D. C. and R. A. Kleese. 1967. Isogenic analysis of strontium-89 accumulation in barley. Crop Sci. 7:617–619.

Rasmusson, D. C., L. H. Smith and R. A. Kleese. 1964. Inheritance of Sr-89 accumulation in wheat and barley. Crop Sci. 4:586–589.

Ruck, H. C. and B. D. Bolas. 1956. Studies in the comparative physiology of apple rootstocks. I. The effect of nitrogen on the growth and assimilation of Malling apple rootstocks. Ann. Bot. N. S. 20:57–68.

Shafer, J., Jr., J. E. Baker and J. F. Thompson. 1961. A *Chlorella* mutant lacking nitrate reductase. Am. J. Bot. 48:896–899.

Shea, P. F., W. H. Gabelman and G. C. Gerloff. 1967. The inheritance of efficiency in potassium utilization in snap beans, (*Phaseolus vulgaris* L.). Proc. Am. Soc. Hort. Sci. 91:286–293.

Shea, P. F., G. C. Gerloff and W. H. Gabelman. 1968. Differing efficiencies of potassium utilization in strains of snap beans, *Phaseolus vulgaris* L. Plant and Soil 28:337–346.

Shim, S. C. and P. B. Vose. 1965. Varietal differences in the kinetics of iron uptake by excised rice roots. J. Expt. Bot. 16:216–232.

Sistrunk, J. W. and R. W. Campbell. 1966. Calcium content differences in various apple cultivars as affected by rootstock. Proc. Am. Soc. Hort. Sci. 88:38–40.

Smith, L. H., D. C. Rasmusson and W. M. Myers. 1963. Influence of genotype upon relationship of strontium[89] to calcium in grain of barley and wheat. Crop Sci. 3:386–389.

Ulrich, A. 1961. Variability of sugar beet plants grown in pots without competition for light, water and nutrients. J. Am. Soc. Sugar Beet Technol. 11:595–604.

Vose, P. B. 1963. Varietal differences in plant nutrition. Herbage Abstracts 33:1–13.

Vose, P. B. and E. L. Breese. 1964. Genetic variation in the utilization of nitrogen by ryegrass species *Lolium perenne* and *L. multiflorum*. Ann. Bot. N. S. 28:251–270.

Vose, P. B. and D. J. Griffiths. 1961. Manganese and magnesium in the grey speck syndrome of oats. Nature 191:299–300.

Vose, P. B. and P. J. Randall. 1962. Resistance to aluminium and manganese toxicities in plants related to variety and cation-exchange capacity. Nature 196:85–86.

Wagner, R. P. and H. K. Mitchell. 1964. Genetics and Metabolism. 2nd ed. John Wiley and Sons, Inc., New York.

Wall, J. R. and C. F. Andrus. 1962. The inheritance and physiology of boron response in the tomato. Am. J. Bot. 49:758–762.

Wallihan, E. F. and M. J. Garber. 1968. Iron uptake by two *Citrus* rootstock species in relation to soil moisture and $CaCO_3$. Agron. J. 60: 50–52.

Warner, R. L., R. H. Hageman, J. W. Dudley and R. J. Lambert. 1969. Inheritance of nitrate reductase activity in *Zea mays* L. Proc. Nat. Acad. Sci. 62:785–792.

Weiss, M. G. 1943. Inheritance and physiology of efficiency in iron utilization in soybeans. Genetics 28:253–268.

Wutscher, H. K., E. O. Olson, A. V. Shull and A. Peynado. 1970. Leaf nutrient levels, chlorosis, and growth of young grapefruit trees on 16 rootstocks grown on calcareous soil. J. Am. Soc. Hort. Sci. 95:259–261.

13

ECOLOGICAL ASPECTS OF PLANT NUTRITION

CONCEPTS

Ecology and Evolution. Why do certain kinds of organisms live where they live, and why do they not live in other places? This, in operational terms, is the central question to which the science of ecology addresses itself. But the kinds and distributions of living things we now observe have come about gradually, as the result of long-continued historical processes; and therefore problems of evolution are inextricably bound up with those of ecology.

The central concept in considering these problems is adaptation. The organism is adapted, or fit, for the environment in which it thrives. This adaptedness is a property of the phenotype; it is the phenotype that is directly exposed to, and tested by, a given condition of life. The evolutionary view, first propounded by Darwin (1859) over a century ago and since then greatly elaborated, is today briefly as follows.

The range of potentialities of the phenotype is defined by its genetic constitution, the genotype. Changes in the genotype occurring randomly, i.e., mutations, modify the potentialities of the phenotype, and these innovations are then subjected to the screening action of natural selection. Those mutations that make individuals better adapted, or more fit for their environment, make in successive generations a proportionally greater contribution to the gene pool of the population than mutations that tend to make them less well adapted. Through the evolution of diploidy and the sexual exchange of genetic material there is available in populations of higher organisms an immense pool of latent variability from which, through natural selection, new forms can come to the fore in response to changes in the conditions of life. Geographic, anatomical, and physiological barriers to the interchange of genes cause the genetic isolation of populations and speciation, and make for often highly specialized, even bizarre, adaptations to every kind of environment.

The conditions of life, i.e., the environment, include everything which has a bearing on the success or failure of the organism in contributing offspring to the next generation, thereby influencing its gene pool. They include the obvious physical and chemical features of temperature, light, wind, water, and nutrients, and also other organisms both of the same and of different kinds which exist within the same environment. Such are the potentialities of the genetic material that only environments with the most extremely unfavorable conditions—utter dryness, extreme heat—fail to support living things.

Adaptive success implies that the organism uses the environment advantageously for its physiological needs, and that it copes with those features of the environment which are inimical to it. Considering the nutrition of living things, the most important difference in this regard between terrestrial plants and animals is that animals are mobile but plants are not. The animal moves in search of its food, and it moves not to become the food of others. The plant stays put and hence lacks the whole range of locomotive and behavioral adaptations which animals have at their disposal in coping with the environment.

Roots and Mineral Nutrition.   We are here concerned with the mineral nutrition of plants, and in this regard the versatility which terrestrial plants have had to evolve collectively is astonishing. Being individually incapable of locomotion they must take potluck where they are, and do so successfully in the vast array of contrasting mineral media already discussed in Chapter 3. Soil conditions range from dry to permanently inundated, from acidic to basic, from chemically denuded to saturated with respect to salt, from well aerated to nearly anoxic, from almost sterile to teeming with microbial life. Thus, different soils represent mineral nutritional environments of extraordinary variety. The nature of the soil as the mineral medium of plants is therefore apt to be a potent determinant of plant adaptation and distribution, i.e., a potent ecological factor. We shall see that this is so.

For the success of stationary and long-lived organisms like terrestrial plants, however, more comes into play than genetic adaptation to a specific set of conditions. The roots of the individual plant encounter within the soil a great range of conditions in space and in time. Growing through the soil, the root may encounter here a particle of calcium carbonate, or lime, imparting a pH to that microregion of over 8, and in another place it may extend into a volume containing decayed organic matter causing a pH of 4 in its immediate surroundings. Droppings and drippings from an animal may locally raise the nitrogen content of the soil far above that of a neighboring region not equally favored. At one time, a period of dryness may cause the volume of the soil solution to shrink and the concentration of solutes in it to rise, to be followed by sudden rain which, soaking the soil, dilutes the soil solution.

Crop plants, no less than wild ones, are subject to such variations. In fact, there are added sources of variability under cultivation. An application of fertilizer may suddenly cause very high local concentrations of nutrients. Chemical heterogeneity attends irrigation no less than fertilization. The effect of irrigation is much like that of rain, but the water added is frequently much less uniformly distributed within the soil.

Variability of the kind discussed above is encountered within the living space and within the lifespan of the individual plant, and adaptive genetic changes involving plants in successive generations are not competent to deal with it. To cope with such variability the plant must possess phenotypic plasticity, the physiological versatility to respond to environmental change. Of the three types of environment of living things—bodies of water, the air, and soil—soil is by far the most heterogeneous in time and in space. Roots would therefore be expected to be paradigms of phenotypic plasticity.

Stebbins (1950), in Chapter 3 of his book on *Variation and Evolution in Plants,* points out that vegetative characteristics are more subject to plastic variability than are reproductive ones. Long-continued meristematic activity, followed by gradual enlargement of the cells formed, offers ample opportunity for external factors, including nutrition, to exert their influence. In contrast, the development of reproductive organs is determinate and results, in a relatively short time, in the formation of small structures, seeds, so that the influence of the environment is minimized. The growth habits of roots—their continuous, open-ended development, their lack of nodes and internodes and of determinate structures—render them capable of exhibiting the degree of phenotypic plasticity shown in the following examples.

Phenotypic Plasticity. Duncan and Ohlrogge (1958) divided the root system of young corn plants between adjacent containers of soil with different fertilizer treatments. When part of the root system grew in unfertilized soil, and the other part in the can in which the soil had been fertilized with both nitrogen and phosphorus, the roots in the fertilized soil were more numerous than those in the control (untreated) soil, they were finer and silkier in appearance, and their total weight was nearly twice as much. Even this gives an inadequate impression of the differences, since the numerous fine, hair-like roots in the fertilized soil contributed

Figure 13-1. Proliferation of a root of soybean, *Glycine max*, where it enters a zone of soil fertilized with both nitrogen and phosphorus. The remainder of the root (top of picture) was in infertile soil. From Wilkinson (1961).

little to the weight of the root but must have enormously added to the total root-soil interface and hence, the capacity to absorb nutrients. In a slightly different experiment roots were let to grow into a volume of unfertilized soil containing a band of fertilized soil. Figure 13-1 shows how the roots proliferated in the fertilized volume.

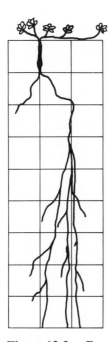

Figure 13-2. *Franseria bipinnatifida* at the ridge garden. Scale: 1 square = 1 foot (0.3 meter). After Martin and Clements (1939). Courtesy of Carnegie Institution of Washington.

For an example of morphological plasticity in native plants we turn to a classic study by Martin and Clements (1939). They grew dune plants of various species in different locations, including a ridge garden and a shelter garden 12 miles west of Santa Barbara, California. The ridge garden was on a small dune rising 20 feet above a lagoon which defined the water table. The shelter garden was in a depression in the dunes, 300 feet from the ridge garden. Figures 13-2 and 13-3 show the growth habits of *Franseria bipinnatifida* in the ridge and shelter gardens, respectively.

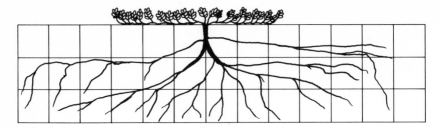

Figure 13-3. *Franseria bipinnatifida* at the shelter garden. Scale: 1 square = 1 foot (0.3 meter). After Martin and Clements (1939). Courtesy of Carnegie Institution of Washington.

Marked phenotypic adaptations to the contrasting conditions of the two habitats are evident, being elicited by differences in exposure, height of the water table, salt, and other factors. Lyr and Hoffman (1967) give other instances of growth responses of the roots of individual plants to localized soil conditions, such as pockets of high fertility in otherwise poor soil.

In a sense, morphological plasticity represents physiological plasticity also, since it is through physiological processes of growth and development that plants or their organs assume the sizes and shapes they do. However, purely physiological responses to different environments do occur, giving evidence of physiological plasticity. The organic acid metabolism of plants is a case in point. Both qualitative and quantitative aspects of the organic acid profile of plants are to a marked extent functions of the mineral medium in which they grow. This was one of the important points to emerge from the classical researches of Vickery and his co-workers of the Connecticut Agricultural Experiment Station (Vickery *et al.,* 1940). They grew tobacco plants, *Nicotiana tabacum,* in nutrient solutions with variable ratios of nitrate to ammonium ions, the total concentration of nitrogen being constant. The form in which nitrogen was supplied markedly affected the levels of many compounds within the plants. Table 13-1 lists some selected items from their Table 13, dealing with plants which had grown in solutions with zero and 60 per cent of their nitrogen in the form of ammonium ions, the balance being furnished as nitrate. The different levels of various organic acids which were brought about by the form in which nitrogen was supplied are no less manifestations of phenotypic plasticity than are visible differences in, say, the size and extent of branching of the root system.

Other instances of physiological versatility in regard to aspects of min-

TABLE 13-1. The Levels of Three Organic Acids of Tobacco
Plants, *Nicotiana tabacum,* as Influenced by the
Form of Nitrogen Supplied[1]

	Percentage of N of the culture solution supplied as ammonium ion			
	Leaves		Stalks	
	0	60	0	60
	Gram per kilogram fresh tissue			
Malic acid	3.70	0.099	1.80	0.67
Citric acid	1.20	0.054	0.227	0.113
Oxalic acid	1.28	0.606	0.676	0.508

[1] After Vickery *et al.* (1940).

eral nutrition readily come to mind. For example, Jefferies *et al.* (1969)
have described phenotypic modifications of the properties of enzymes resulting from changes in the ionic composition of the medium in which the
plants grew.

The Concept of the Ecotype.   What has been said about the physiological versatility of roots should not be interpreted to mean that roots
of all plants are competent to cope with all nutritional media. Within
a given genotype there are limits to the latitude of phenotypic responses
to the environment (Bradshaw, 1965). Plants have become adapted to
the mineral habitats of the earth in all their diversity, as they have become
adapted to all other conditions of life, through the evolution of a vast
range of genotypes. Genotypical response of plants to the habitat was
demonstrated experimentally by the Swedish investigator, Göte Turesson
(1922), who showed that there is evolutionary diversification even within
species, and created the concept of the "ecotype."
    Turesson collected plants belonging to the same species growing in geo-
graphically separated areas. The plants at their several locations showed
characteristic features related to their habitats. For example, plants of
Atriplex litorale (salt bush) in exposed coastal situations were low and
spreading while those in sheltered places were tall and erect. When plants
of this annual species from contrasting locations were grown together in
the same experimental area for several years they retained the features

characterizing the plants in their original habitats. In other words, the plants represented races or varieties of the species adapted to different habitats. Such ecological races Turesson called "ecotypes"; they represent "the genotypical response of the plant species to the habitat." Ecotypical differentiation has since become recognized as a common result of natural selection.

We may note in passing that the plan of Turesson's experiments, though not their execution, was anticipated by Darwin (1859) in Chapter 2 of the *Origin*. Speaking of the term "variation" as used with the implication that it refers to modifications directly due to environmental influences (i.e., to phenotypic responses as discussed under the previous heading), he writes: ". . . but who can say that the dwarfed condition of shells in the brackish waters of the Baltic, or dwarfed plants on Alpine summits, or the thicker fur of an animal from far northwards, would not in some cases be inherited for at least some few generations? and in this case, I presume that the form would be called a variety."

Environmental factors responsible for such differentiation that have been studied most are climatic: temperature and light variations along gradients of altitude and latitude (Hiesey and Milner, 1965). However, evidence is also at hand for edaphic (soil-related) ecotypes (Bradshaw, 1969; Epstein and Jefferies, 1964). For experimental studies, nutritional ecotypes of a given species are apt to be more fortunate choices than different species, because the condition of "other things being equal" is more nearly met when plants of the same species are compared. In what follows we shall encounter examples of work at both levels: species and races or ecotypes.

Nutritional Ecology.   Since the properties of soils are to a large extent conditioned by climatic factors (Jenny, 1941) it is often difficult to single out aspects of nutritional ecology from the array of climatic factors which are also ecological determinants. However, there are many situations where soil properties vary abruptly or gradually within a given climatic region, and in these situations the role of the nutritional factor in ecological adaptation can be studied to good advantage.

Indeed, because of the sharpness of the boundaries that may be encountered between contrasting types of soil, nutritional ecology offers situations ideally suited to ecological studies, but these and other advantages have not drawn to this aspect of ecology the attention it merits. Krause (1958), Mason (1946), Salisbury (1959), and Epstein and Jefferies (1964), among others, drew attention to the nutritional factor in plant ecology. However, they made few converts, and we find that at a recent UNESCO symposium on functioning of terrestrial ecosystems at the primary production level, F. L. Milthorpe of Australia was moved to plead with his colleagues not to forget the functional role of roots "in our enthusiasm for the above-ground system" (Eckardt, 1968, pp. 506–507).

There is nevertheless no doubt that nutritional ecology or more broadly, edaphic (soil-related) plant ecology, is entering an exciting period of development, along with the closely related subject of the genetic factor in plant nutrition (Chapter 12). In addition to the intrinsic scientific interest of these aspects of plant nutrition there are major developments which forcibly draw attention to them, especially the spread of modern agricultural methods in the developing countries, whose soils pose many problems, and the growing concern with the environment and its contamination, as serious in connection with soils as it is in respect to air and water.

Plant Geography.   There is a considerable literature devoted to plant distribution as related to geography (Mason and Stout, 1954). It is mainly European and goes back to the classical works of De Candolle (1855) and Schimper (1935), the first edition of whose book was published in 1898. The climatic factor looms large in the treatments of plant geography, but some attention is also paid to the soil factor including soil chemistry and its bearing on mineral plant nutrition. The books by Walter (1968, 1969) are the most comprehensive recent treatises in this tradition, and the one by Eyre (1968) specifically stresses the relation between the distribution of plants, world-wide, and the world's soils.

To those interested in the ecology of mineral plant nutrition there is much of interest in the work of the plant geographers. Two aspects, in particular, merit mention. Unavoidably, the practicing plant physiologist is a specialist. Viewing the broad canvas of the plant geographer is useful to him in that it places particular and sharply focused investigations in a broad geobiological context. Secondly, plant physiological research including research on plant nutrition favors a few species, mainly economic crop plants, for reasons already discussed in Chapter 12, pp. 337–338. Billings (1957), speaking from the point of view of an ecologist, has rightly remarked that there are all too few studies on the mineral nutrition of wild plants. A few others have echoed this sentiment (Epstein and Jefferies, 1964; Gerloff, 1963; Kinzel, 1963). The work of the plant geographers can give an added impetus to this incipient interest in the mineral nutrition of wild plants because surveying the world's vegetation tends to counteract the tendency to consider the barley root, the *Nitella* cell, and the potato slice as adequately representative of the plant kingdom.

Inspiring as the work of the plant geographers is, it does have the drawback, from the point of view of the physiologist, of being almost wholly descriptive. Although terms like "physiological" (Schimper, 1935) and "eco-physiological" (Walter, 1968, 1969) are applied to such investigations, they are of necessity broad surveys of what can be seen, with occasional references to measurements of pH, salinity, and some other easily measured parameters. But they do not involve incisive experimental work aimed at probing the physiological mechanisms whereby the observed plant adaptations and distributions come about. Thus, at a recent symposium devoted to ecological aspects of the mineral nutrition of plants (Rorison, 1969), not one of the participants ever referred to the plant geographical works mentioned above. In the present context, we shall confine ourselves entirely to experimental investigations, recognizing that even these, in the present state of our knowledge, are often mainly descriptive rather than analytical and mechanistic in their approach.

2. MINERAL ELEMENTS IN PLANT ECOLOGY

Calcium and pH.   We have seen in the discussion of ion transport (Chapter 6) that calcium plays a crucial role in the relationship between roots and their media. For well over a century European plant geographers and ecologists have observed that soils derived from limestone rock, rich in lime ($CaCO_3$) and hence high in pH, bear a flora which differs characteristically from that found on acidic soils poor in calcium.

In his book, *Downs and Dunes: Their Plant Life and its Environment* the botanist Sir Edward Salisbury (1952) gives an account of the "chalk downs" of the British Isles, and references to other literature concerning plants on calcareous soils in Britain, while Kinzel (1963) gives references to the Continental literature. Similar situations occur in North America (Bamberg and Major, 1968; Major and Bamberg, 1963; Whittaker and Niering, 1968; Wright and Mooney, 1965) and Australia (Parsons and Specht, 1967). Evidently, the nature of the parent material and of the soils resulting has had a powerful selective effect in the evolution of these "calcicole" and "calcifuge" floras (as those found on limestone soils and those avoiding them, respectively, have been called in the European literature).

It has been known ever since the work of Iljin in the first half of this century (summarized by Kinzel, 1963) that many of the plants common on calcareous soils contain high concentrations of intracellular calcium and high concentrations of malic acid, while those normally confined to acid, low-calcium soils contain very little soluble calcium because of the presence of intracellular oxalate which precipitates calcium as calcium oxalate. There are thus fairly well marked differences in the calcium metabolism of plants characteristic of these contrasting types of soil (Kinzel, 1963, 1969).

However, the calcium level is by no means the only one governing the ecology of plants on these soils. Because of the number of properties in which calcareous and acidic soils differ a causal analysis of their effects on plants has not yet been possible. The properties most important in the nutrition of plants in which these contrasting types of soils differ are listed in Table 13-2.

TABLE 13-2.   Contrasting Properties of Calcareous and Acid Soils[1]

Calcareous	Acid
High in calcium	Low in calcium
High in pH and carbonate	Low in pH; no carbonate
Rich in nutrients	Poor in nutrients
Solubility of heavy metal ions low	Solubility of heavy metal ions high
Activity of nitrogen fixing and nitrifying bacteria high	Activity of nitrogen fixing and nitrifying bacteria low

[1] There are certain exceptional situations in which statements made here do not apply.

The importance of the calcium concentration of the soil for the growth of white clover, *Trifolium repens,* was studied in an investigation by Snaydon (1962a). He sampled an area of only 10 square meters in Caernarvonshire, Wales, which showed a patchy distribution of white clover. At each sampling point, he determined the degree to which white clover covered a 10 square centimeter area, and the calcium concentration in the soil. Figure 13-4 shows that there was an essentially sigmoidal relationship between the concentration of calcium in the soil and the extent of growth of the clover—much the same kind of relationship obtained with increasing levels of fertilization of a soil very deficient in some nutrient.

The findings were especially interesting in that there was no corresponding correlation between the calcium concentration of the soil and its pH. Such correlation is common, but in this situation the variation in pH over the area (4.7 to 5.8) was not related to the distribution of calcium. Although pH was therefore ruled out as a factor, the findings do not prove

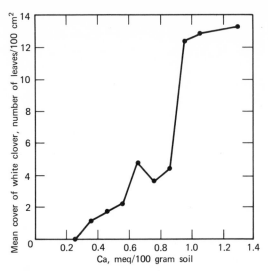

Figure 13-4. Cover of white clover, *Trifolium repens,* as a function of the calcium concentration of the soil at the point sampled. After Snaydon (1962a).

a causal connection between the concentration of calcium and the growth of *T. repens.* Conceivably, for example, soil calcium and the growth of the plants were both governed by a third, unknown factor.

That calcium per se was nevertheless the governing factor in the distribution of *T. repens* observed by Snaydon is made likely by the results of another investigation. Snaydon and Bradshaw (1969) collected specimens of *T. repens* from calcareous and acid soils and grew them in nutrient solutions in which the calcium concentration ranged upwards from 4 ppm (0.1 mM) by steps of factors of 2 (4, 8, etc.). Plants belonging to the populations from the acid (low-calcium) soils grew better at the lowest calcium concentrations (4 and 8 ppm) than did plants of the populations from the calcareous soils. At increasingly higher calcium levels, the calcareous populations responded more markedly than the acid populations. These experiments not only documented the role of calcium but also provided evidence for the existence of nutritional ecotypes in *T. repens* adapted to soils of different calcium status.

The importance of calcium per se was also brought out by Clarkson (1965b) in a comparison of "calcicole" and "calcifuge" species of *Agrostis* (bent grass). When *A. stolonifera,* common on high-calcium soils, and

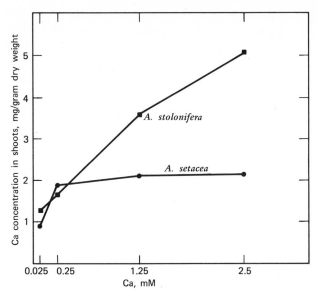

Figure 13-5. Calcium concentration in the shoots of *Agrostis* species as a function of the concentration of calcium in the nutrient solution. After Clarkson (1965b).

A. setacea, which occurs on acid, low-calcium soils in Britain, were grown at pH 4.5 in solutions of increasing calcium concentrations the former accumulated progressively higher concentrations of calcium in its tissues (Figure 13-5) and grew better. The plants of *A. setacea,* native to low-calcium soil, did not respond to increasing concentrations of calcium above 0.25 mM. Apparently in this species, the calcium absorbing mechanism has a high affinity for calcium and operates at its maximal rate at 0.25 mM calcium in the medium. Germination of the seeds and the establishment of the seedlings is often more strictly dependent on the calcium status of the medium than is the subsequent growth of the plants (Jefferies and Willis, 1964).

A factor of great importance in acid soil is the high solubility of heavy metals at low pH values, and aluminum looms large in this connection because of its prevalence in the soil solutions of acid soils and its toxicity to plants. Rorison (1960a, b) showed that the inability of the calcicole, *Scabiosa columbaria,* to grow in acid soils was mainly due to its intolerance of aluminum ions. The plants grew poorly in an acid-sand soil of pH 4.8, but in nutrient solutions grew almost equally well at pH 4.8 and 7.6. Adding calcium sulfate (which does not change the pH) to the acid soil caused no improvement in the growth of the plants, but adding calcium hydroxide did. Addition of aluminum to acid (pH 4.8) solutions in which the plants grew made them respond in much the same way as they did on the acid soil. That is, *Scabiosa columbaria,* unlike *Trifolium repens* in the investigations described above, did not respond to calcium per se. Rather, its intolerance of acid (low-calcium) soils was related to the high aluminum concentrations of these soils. The mechanism of aluminum toxicity is not fully understood. One important feature is that aluminum phosphate is precipitated in the root tissue, causing phosphorus deficiency, as shown earlier for barley, *Hordeum vulgare,* by Vlamis (1953). There is also a direct effect of aluminum and other trivalent cations on cell division (Clarkson, 1965a, 1966).

Just as high concentrations of heavy metal ions in acid soils may limit the growth of certain plants, so their low solubility in calcareous, high pH soils may become a limiting factor. Snaydon (1962b) studied populations of white clover, *Trifolium repens,* from calcareous "chalk" soils and from acid soils in Wales. Plants of each population grew best on their own soil type. Plants of the populations from the acid soils, when grown on calcareous soil, developed lime-induced chlorosis (iron deficiency). Evidently, the populations from the acid and the calcareous sites represent edaphic ecotypes, and the former are not adapted to utilize iron from calcareous soil in which the concentration of iron in solution is exceedingly low. The plants from the calcareous soil are adapted to this condition; they did not develop iron deficiency chlorosis.

Ecotypic differentiation in terms of the ability to utilize iron from calcareous soils has been shown in other species, for example in *Teucrium scorodonia* (Hutchinson, 1967). In this species it was shown specifically that the susceptibility of the ecotypes from acid soils to lime-induced chlorosis is not due to the inability of their roots to absorb iron from calcareous soil (Hutchinson, 1968). Rather, they seem to lack the ability to form iron complexing compounds by means of which the iron pool in the roots can be mobilized for translocation to the shoots in a form which is physiologically effective. The calcareous-soil ecotypes generate such iron-transport ligands (cf. Brown and Tiffin, 1965; Schmid and Gerloff, 1961; Tiffin, 1970). Once chlorotic, leaves are prone to excessive water loss by transpiration and are more sensitive to the resulting water deficits than are green leaves (Hutchinson, 1970).

The significance of calcium and of pH in the legume-*Rhizobium* system has been discussed in Chapter 10. Acid, low-calcium soils are generally poor habitats of legumes depending on symbiotic nitrogen fixation for their nitrogen supply. In addition to factors associated with pH and calcium level already discussed there is the added one that molybdenum, essential for nitrogen fixation, is more likely to be deficient on acid than on high-calcium, high-pH soils (Davies, 1956; Williams and Andrew, 1970). Robson (1969) has discussed soil factors important in the distribution of legumes, with emphasis on alfalfa, *Medicago*.

The above discussion has focused on experimental investigations of chemical features of soils as factors in plant distribution. An important technique we have encountered, and shall encounter again, is that of growing plants from different soil types in their normal and in the contrasting type of soil, and observing their responses. When in such experiments plants of a single type (species or ecotype) are grown alone a factor is missing which, in the natural environment, looms large indeed. That is the factor of competition.

In experimental pure cultures, plants frequently grow reasonably well even on soils from which they are virtually excluded in nature. The reason is that protected from competing, better adapted plants, they suffer some degree of impairment when grown in soil to which they are none too well adapted, but nevertheless manage after a fashion. Under natural conditions, subject to competition by plants better adapted to that soil, the disabilities of the less well adapted plants soon tell, and result eventually in their elimination.

A classical demonstration of the effect of competition was given long ago by Tansley (1917) who studied two bedstraws, *Galium saxatile,* which grows in Britain on soils of low calcium status, and *G. pumilum,* which is confined to calcareous soils. Tansley found that *G. saxatile* in isolation could grow on calcareous soil, and *G. pumilum,* on acid, at least for some years. But in mixed stands, the species which was on "alien" soil was severely handicapped by competition from the better adapted species for which that soil was the normal substrate. A more recent demonstration of the effect of competition has been given by Snaydon (1962b). Although these examples deal with responses of plants to calcareous and acid soils the general conclusion applies to other situations. It is that the effects of slight differences in physiological adaptedness among populations will be amplified by competition, often to the point where the end result is an all-or-nothing situation, the less adapted type being eliminated from the habitat.

Heavy Metals.   Plant geographers and ecologists have long been interested in the often unique floras which are characteristic of soils containing high concentrations of heavy metals. Such sites occur naturally, as the result of geological processes and the agencies of weathering and soil formation. Others are man-made, as when mining operations result in the deposition of slag heaps and tailings which form a substrate with high concentrations of heavy metals and usually, low concentrations of nutrient elements. The man-made habitats have aroused special attention because they are of recent origin. Their history is therefore known, and they make a veritable stage upon which natural selection can be observed to act.

Tailings left in the wake of mining operations seldom remain completely barren for long. As a rule, plants colonize these sites, but the floristic composition of this plant cover differs from that inhabiting the surrounding "normal" soils. The most critical selective agent in these habitats is the high concentration of lead, zinc, copper, or other heavy metals. These metals, when present in high concentrations, are generally toxic to plants, and their presence would tend to keep most plants from invading such sites. An additional factor is the low nutrient status of these soils, and often, their low water holding capacity.

Seeds of most plants alighting on such sites would perish as a result of these unfavorable conditions, especially the toxic levels of heavy metal ions. However, the genetic variability existing in populations of wild plants is such that there will be occasional individuals with the capacity to withstand these conditions. Although they, too, may suffer impairment in comparison with their potential performance on normal soil, they will have the advantage, on the heavy metal site, of much diminished competition. The point has already been made, under the previous heading, that competition greatly magnifies the effects of even slight differences in degree of adaptedness. Conversely, diminished competition brought about by a condition intolerable to most plants amplifies the advantage which accrues to those plants able to cope with it.

One genus which has been much studied is bent grass, *Agrostis,* species of which grow in Wales and Scotland on mine tailings containing lead, zinc, copper, nickel and other heavy metals at concentrations toxic to most plants (Bradshaw, 1952; Bradshaw *et al.,* 1965; Jowett, 1958). An obvious hypothesis to explain the tolerance to heavy metals shown by populations of *Agrostis* suggests itself, viz., that the plants are able to exclude the heavy metals and thereby escape being poisoned by them. Like so many other obvious, plausible hypotheses, this one is wrong. Plants growing on heavy metal soils in many parts of the world have been analyzed and found to contain concentrations of them which would be toxic to most plants (Turner, 1969). Bradshaw *et al.* (1965) have shown that plants of a copper tolerant population of *Agrostis tenuis* absorbed as much copper from experimental solutions as did plants from a population not tolerant of copper.

Another observation which has a bearing on the mechanism of heavy metal tolerance was made by Jowett (1958). He obtained collections of plants of *Agrostis tenuis* and *A. stolonifera* from sites contaminated by different heavy metals. He tested the growth of the plants in separate solutions, each containing one of these heavy metals, following a technique suggested by Wilkins (1957). It developed that plants from a habitat high in a given heavy metal, and tolerant of that metal, were not necessarily tolerant of other heavy metals. Rather, the plants were more or less specifically tolerant of the particular heavy metal predominating in the habitat of the population to which they belonged.

Together, the observations that plants tolerant of a certain heavy metal absorb it, and that they are not necessarily tolerant of other toxic metals, suggest the following hypothesis first advanced by Jowett (1958) to explain heavy metal tolerance: tolerant plants synthesize chelating compounds which form complexes with the heavy metal ions and thereby render them innocuous. If this basic explanation is correct it means that there must be a number of such complexing agents, each more or less specific for a certain toxic metal but with less chemical affinity for others. No such compounds have been isolated from heavy-metal-tolerant *Agrostis,* but the hypothesis is the most promising one and is in keeping with other information on internal complexing of heavy metal ions by organic ligands (Brown and Tiffin, 1965; Schmid and Gerloff, 1961; Tiffin, 1970).

In the investigations discussed above, populations of single species from different habitats responded differently to the presence of high concentrations of various metals. These findings suggest that the populations represent ecotypes adapted to these unusual soil conditions, and there is much evidence that this is so, both in populations of *Agrostis tenuis* and in other species populations of which inhabit heavy metal sites.

Gregory and Bradshaw (1965) compared *Agrostis tenuis* from old mine tips in Wales and from ordinary pasture lands nearby by growing the plants in solutions containing the heavy metals, one by one, and measuring their growth. Plants belonging to a population from a high-lead habitat were tolerant of lead, but not equally tolerant of other heavy metals. The same was true for other metals: the tolerance of the plants to a given heavy metal was specific. Furthermore, pasture plants were not tolerant of heavy metals. The various tolerances and sensitivities were not lost in cultivation, individuals differed, and seed samples of populations had the same tolerances as vegetatively propagated plants. Clearly, the heavy metal tolerances were under genetic control: the various populations showed "genotypical response of the plant species to the habitat," in the sense of Turesson (1922).

Such genetic differentiation in respect to tolerance to heavy metals exists in other species. For example, Wilkins (1960) examined genotypic variation in respect to lead tolerance in *Festuca ovina,* a perennial fescue. Tolerance to lead was inherited in a dominant fashion, but the situation is not a simple Mendelian one: more than a single gene locus is involved. Bröker (1963) made a detailed study of zinc tolerance in *Silene inflata,* a perennial weed with a range of ecotypes differing in their tolerance to zinc (and copper). Zinc tolerance was dominant; as in *Festuca,* the inheritance of the trait was found to be polygenic.

The heavy-metal-tolerant populations of the Scottish and Welsh mine tailings are of great interest to the evolutionist. For example, McNeilly and Antonovics (1968) studied *Agrostis tenuis* from the copper mine, Drws-y-Coed, in Caernarvonshire. The copper-contaminated habitat from which they obtained representatives of copper-resistant ecotypes probably date from the latter part of the 19th century. Therefore, evolution of and colonization by the copper-tolerant population took place in a very short time, compared with the time scale we commonly associate with evolutionary sequences. Other discussions of the evolutionary aspects of heavy-metal-tolerant populations have been given by Antonovics (1968), Bradshaw (1969), Bradshaw *et al.* (1965), and McNeilly and Bradshaw (1968), among others.

Certain species of plants have unusual affinities for relatively rare elements, most of them heavy metals, which may or may not be essential to them. Such plants are called "accumulator" plants. Mention has already been made in Chapter 4 of *Astragalus,* several species of which are selenium accumulators. Other groups have distributions which correlate well with the content of some element in the soils and rocks; such plants are "indicator plants" and they are used in the prospecting for ore deposits (Brooks, 1968; Cannon, 1960; Chikishev, 1965; Viktorov *et al.,* 1964).

Serpentine Soils.   Under the preceding two headings we have discussed calcium and heavy metals as ecological factors. Both are important aspects of serpentine soils, and so is an additional one: magnesium. In many parts of the world there are distinctive soils which have developed on serpentine rock. In these soils, magnesium predominates over calcium, major nutrients are in short supply and so, in some cases, is the level of molybdenum available for absorption. Lastly, heavy metals, especially chromium and nickel, are often present in high concentrations. The line of demarcation between serpentine and non-serpentine soils is often very sharp, so that one can amble from a serpentine onto a non-serpentine soil in a few steps.

The contrast between the vegetations of serpentine and non-serpentine soils is startling. The plant cover of serpentine soil makes an impoverished impression. Along the ranges of mountains extending along the Pacific coast of America, serpentine vegetation is sparse, shrubby chaparral, with many endemic species which are absent from the adjacent non-serpentine areas. The latter bear a much more varied, more luxuriant plant cover. Descriptions have been given by many authors. A good account of these soils and their floras is contained in the papers on the ecology of serpentine soils given at a 1950 symposium by Kruckeberg (1954), Walker (1954) and Whittaker (1954). Kruckeberg (1969) has a more recent discussion, and further references.

Serpentine soils have several chemical features which may be encountered in non-serpentine soils also, and which therefore are probably not the main ecological factors responsible for the distinctive nature of the vegetation of these soils. However, high concentrations of magnesium, accompanied almost invariably by low levels of calcium, are a definitive characteristic. In view of the importance of calcium for ionic regulation at the plasma membrane (Chapter 6, pp. 117–120) it is not surprising to find that specialized adaptations were necessary if plants were to colonize such soils. In the immediate vicinity of roots, the rhizosphere, the Mg/Ca ratio in solution will be even higher than in the bulk soil. The reason is the elevated concentration of carbon dioxide from their respiration and that of the bacteria of the rhizosphere, because it has been found that carbon dioxide in solution tremendously accelerates the rate of solubilization of magnesium from serpentine (Wildman *et al.*, 1968). It is therefore likely that low calcium concentrations and high Mg/Ca ratios are the principal chemical features which tend to exclude many species from these soils, and to which those plants which do grow there have become adapted. Experimental evidence bears this out.

Martin *et al.* (1953) studied a serpentine area in Lake County, northern California, where over the years neither forage nor small grains had been grown successfully. Of the exchangeable cations on the exchange complex of the soil, only 8–10 per cent were calcium ions—a much lower percentage than is the rule for good agricultural soils. They applied calcium sulfate at rates of 1, 2, 4, and 8 tons per acre and disked it in. Figure 13-6 shows the yields of hay as a function of the degree to which the soil cation exchange complex was saturated with calcium. The calcium concentration in the plant material (vetch, *Vicia* sp., and oats, *Avena* sp.) in-

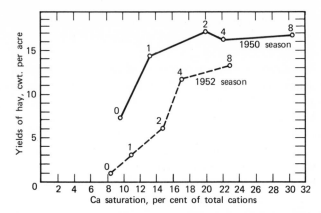

Figure 13-6. Yield of oat-vetch hay (*Avena* and *Vicia*) on a serpentine soil as a function of the degree of calcium saturation of the soil. The figures on the graphs indicate tons of gypsum ($CaSO_4 \cdot 2H_2O$) applied to the soil to achieve the degrees of calcium saturation indicated on the abscissa. After Martin *et al.* (1953), in the Agronomy Journal.

creased as a result of the treatment; in oat straw, it nearly doubled as a result of the highest rate of application of calcium.

Walker (1954) treated samples of serpentine soil in such a way as to obtain a graded series in which the degree of calcium saturation of the cation exchange complex ranged from 5.6 per cent to 34 per cent. (In unaltered field soil, the calcium saturation percentage was 13.5.) He also fertilized the soils with nitrogen, phosphorus, and potassium, and then grew tomato plants, *Lycopersicon esculentum,* and the serpentine endemic crucifer, the jewel flower, *Streptanthus glandulosus* var. *pulchellus,* in pots of these treated soils in the greenhouse.

The results are shown in Figure 13-7. The crop plant, tomato, responded to the calcium levels much as did oat and vetch in the field trial by Martin and co-workers (Figure 13-6). The serpentine endemic, *Streptanthus,* was much more tolerant to low calcium levels in the substrate—even to levels lower than those normally found in serpentine soils. Both tomatoes and *Streptanthus* absorbed more calcium and less magnesium the higher the calcium level of the soil. It is not yet understood what features of the calcium (and magnesium) metabolism render serpentine plants capable of growing there. The ability to absorb calcium from low concen-

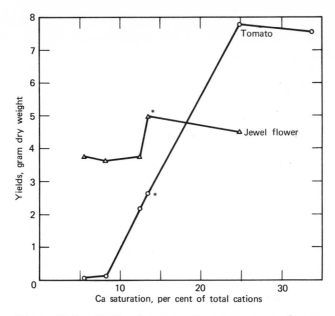

Figure 13-7. Yields of tomato, *Lycopersicon esculentum,* and the jewel flower, *Streptanthus glandulosis* var. *pulchellus,* on serpentine soil adjusted to various degrees of calcium saturation. The asterisks indicate unaltered soil, with calcium representing 13.5 per cent of the exchangeable cations. Plotted from data of Walker (1954).

trations in the substrate is evidently one, and perhaps, the ability to exclude magnesium or to tolerate high internal concentrations of it is another.

There are species which occur on serpentine and also on nearby non-serpentine soils. The question arises whether plants of a given species have sufficient phenotypic plasticity to adjust to both kinds of substrate, or whether the species is differentiated into edaphic ecotypes. Kruckeberg (1954) showed that of 21 species tested, 12 showed definite differentiation into serpentine and non-serpentine races.

Salinity.   Salt restricts the growth of plants over larger areas of the earth than does any other inhibitory substance that they may encounter in the natural environment. Natural vegetation in saline habitats is often sparse, and extremely saline areas are barren. The plants which do grow in saline habitats all possess special adaptations; collectively, they are called halophytes (salt plants), but it should be clearly understood that this term does not refer to particular botanical taxa. There are halophytes which are grasses and other monocotyledons; others are dicotyledons, and in each of these classes halophytes are encountered in a number of families. There are also bacteria and algae adapted to saline conditions, including predominately the marine forms. World wide, salinity is one of the great facts of plant life, and therefore, of all life. Except for the fact of salinity, saline habitats are diverse: there are saline marshes, wet and often cool, and hot, dry saline deserts (Chapman, 1960; McGinnies *et al.,* 1968); there is the salt spray community of the seashore (Boyce, 1954); there are the oceans (Hill, 1963; Raymont, 1963); and the estuaries of rivers (Lauff, 1967).

Saline conditions pose two distinct physiological threats to plants. The low osmotic potentials of saline solutions make it necessary that plants exposed to these media maintain intracellular osmotic potentials lower yet. Otherwise they would be subject to osmotic desiccation because water would move osmotically from the cells into the substrate. In other words, to cope with low external osmotic potentials, plants have to fight fire with fire and maintain internal osmotic potentials which are still lower (Dodd and Coupland, 1966; Harris *et al.,* 1924).

The second difficulty which arises from saline conditions is of a nutritional type which we have already discussed in this chapter. It is the presence of high and potentially toxic concentrations of ions in the medium. Although low osmotic potentials can be due to the presence of any soluble salt or other solutes, the term salinity commonly refers to solutions in which sodium ions predominate. The most prominent anion is usually chloride, as in the oceans which contain about 0.46 M NaCl and lesser concentrations of other salts (Table 2-2).

There is a condition of some salt-affected soils in arid and semiarid regions in which the first factor discussed above, the osmotic one, does not apply. "Alkali" or "sodic" soils are those whose cation exchange complex is occupied to the extent of more than 15 per cent by sodium ions but which do not necessarily contain much soluble salt in their soil solutions. The higher the percentage of sodium among the exchangeable cations, the lower is the percentage of the exchange sites occupied by calcium, potassium, and magnesium ions. In such soils, sodium toxicity, often aggravated by calcium deficiency, is not accompanied by the low osmotic potentials which characterize saline, as distinct from sodic, conditions (Richards, 1954).

Although sodium and chloride are the most prominent potentially toxic ions of saline substrates, other ions which often are found in saline soils may play important, often decisive, roles in the ecology of such areas. Sulfate, bicarbonate, borate, and lithium ions are four which figure prominently in large areas and there are still others. Salinity, then, is characterized by two unique features: low osmotic potential and high concentrations of sodium and other ions which can be toxic. But these factors do not represent the whole range of diversity and adversity which plants in saline habitats may have to be competent to cope with. For example, the concentration of essential nutrients may be low, aeration may be restricted, and photosynthetic gas exchange limited, especially in tidal zones during times when the plants are submerged (Piggott, 1969).

However, looked at from a global perspective, these deficiencies in terrestrial saline habitats are not as crucial as another deficiency: that of water. The largest areas of saline soils are saline because they are in the arid and semiarid zones of the world: they are by-passed by the hydrologic cycle. Being unleached, such soils retain their salts and with time, may become progressively salinized from salt-bearing ground waters and cyclic salt. Evaporation and transpiration by plants remove water but not salt, so that the solution in the soil tends to become more and more saline. The largest areas of salt-affected soils, then, combine two adverse features: salinity and aridity (Dregne, 1963). Consequently, plants native to saline areas must be adapted to whatever adverse conditions these habitats may have in common with other habitats, and in addition, to the unique condition of salinity combining the threat of osmotic withdrawal of water with the toxicity of sodium and other ions present at high concentrations.

The osmotic aspect of this dual dilemma has traditionally been upmost in the considerations of both plant ecologists (Schimper, 1935) and agricultural scientists (Bernstein and Hayward, 1958). The first edition of Schimper's book was published in 1898 and his views exerted an important influence on thinking in this field—too large an influence, as it turned out. His idea was that plants suffer "physiological drought" under saline conditions, even when the soil or other medium is physically wet. This view is embodied in the common practice of agricultural experts on salinity of expressing the degree of salinity of a soil in terms of the electrical conductivity of an aqueous suspension or extract obtained from it (Richards, 1954). This measurement reflects the total concentration of ions in solution but it does not give any inkling of its composition in terms of specific ions.

This osmotic hypothesis, however, has been widely questioned (Eaton, 1927; Greenway, 1962; Lagerwerff, 1969; McNulty, 1969; Slatyer, 1961; Walker, 1960). These and other authors maintain that plant cells, by accumulating and maintaining high intracellular concentrations of solutes, keep internal water potentials sufficiently lower than that of the external medium to effect an osmotic adjustment and thereby prevent physiological drought. Bernstein (1961) has provided firm evidence for osmotic adjustment even in roots. This is crucial since the root is the organ most intimately exposed to the saline medium.

Nevertheless, these findings do not signify that "physiological drought" can be ruled out as a factor when plants are exposed to saline conditions. When the cells of leaves adjust their solute content upwards in response to increased salinity of the medium, i.e., when they effect an osmotic adjustment, the rate at which water moves in may be too slow to effect an adjustment in water potential, and the turgor pressure of the leaf cells may decline. There is in fact evidence that high concentrations of salt cause a decrease in the permeability of roots to water, and hence, a decrease in the rate of its entry into the plant (Kramer, 1969). As a result of this hydraulic resistance, entry of water into leaf cells may be so slow as to give rise to a water deficit even though the cells have generated a sufficiently high internal solute concentration to bring about an osmotic adjustment. This point has been made by several authors (Lagerwerff, 1969; Meiri and Poljakoff–Mayber, 1969; O'Leary, 1969; Riley, 1969). These investigators, however, all worked with bean plants, *Phaseolus vulgaris,* a salt-sensitive species.

Still another osmotic factor has been discussed by Bingham *et al.* (1968) and by Oertli (1968): the capacity of leaf cells to accumulate salt is limited. Salt is absorbed by roots and moves with the transpiration stream into the "outer" space of the leaf (its vessels and tracheids and the cell walls of the mesophyll cells). As the water evaporates, the concentration of salt in the "outer" space may become sufficiently high to cause the turgor of the leaf cells to drop. Determination of the osmotic potential of expressed leaf sap (which includes the extracellular fluid) may well show it to be lower than that of the medium and may lead to the conclusion that osmotic adjustment has taken place, but the point these authors make is that the solution to which the leaf mesophyll cells are exposed is not the external medium. Rather, it is the solution in the "outer" space of the leaf tissue (cf. Smith and Epstein, 1964). The osmotic potential of this extracellular solution may be sufficiently low to cause an intracellular water deficit.

Taken together, these studies show that osmotic adjustment tends to be effected, and that it is a necessary though not always sufficient condition for the growth of plants in saline conditions. We must therefore inquire into the means by which the cells of the plant adjust their osmotic potential downward, below that of the medium. There are only two ways in which the intracellular osmotic potential can be lowered: by the absorption of solutes and by the synthesis of solutes. In both halophytes and nonhalophytes absorption of solutes is paramount, and the solutes absorbed are ions from the external solution. It is in this respect that halophytes differ most markedly from nonhalophytes. Most halophytes absorb sodium from the medium, translocate sodium to the leaves, and tolerate the high concentrations of it which build up in the leaves (Black, 1956; Collander, 1941; Rains and Epstein, 1967; Scholander *et al.,* 1966; van Eijk, 1934). Nonhalophytes accumulate less sodium in the roots, and transfer little to the leaves. Most of the sodium that is absorbed is retained in the roots and lower stem (Bernstein *et al.,* 1956; Jacoby, 1964, 1965; LaHaye and Epstein, 1971; Rains, 1969; Wallace *et al.,* 1965).

Chloride, like sodium, is absorbed and translocated by many halophytes to a much greater extent than in nonhalophytes (Biebl and Kinzel, 1965; Greenway, 1968; Harris *et al.,* 1924; Scholander *et al.,* 1966). As in nonhalophytes, when mineral cations are absorbed in larger amounts than anions, organic acid anions make up the mineral anion deficit (Osmond, 1963; Williams, 1960).

As discussed in Chapter 4, pp 57–58 sodium is an essential nutrient for some halophytes. Potassium is essential to them as it is to all other higher plants. In saline media, the ratio of sodium to potassium is typically high; for example, in sea water it is between 40 and 50 to 1 and it can be much higher yet in saline soils. This means that halophytes need the physiological competence of absorbing adequate amounts of potassium from media containing a large excess of a closely related alkali cation, sodium. This consideration draws attention to the type 1 mechanism of alkali cation absorption (see Chapter 6) which has a high selectivity for potassium vis-à-vis sodium (Elzam and Epstein, 1969a,b; Epstein, 1969; Rains and Epstein, 1967). In leaf tissue of one halophyte investigated in this regard, the mangrove *Avicennia marina,* even the type 2 mechanism of alkali cation absorption took up potassium preferentially, resisting the potentially inhibitory effect of sodium (Rains and Epstein, 1967); in fact, sodium chloride caused some increase in the rate of potassium absorption (Figure 13-8). Calcium plays a crucial role in the responses of both salt-sensitive and salt-tolerant plants to salt (Chaudhuri and Wiebe, 1968; Elzam and Epstein, 1969a; Hyder and Greenway, 1965; Kelley, 1963; LaHaye and Epstein, 1969, 1971).

Unlike salt-sensitive plants, halophytes are able to tolerate the high concentrations of mineral ions which accumulate in their tissues. The mechanisms of this salt toleration, or conversely, the mechanisms of salt damage to salt-sensitive tissues, are unknown. Salts may disrupt the structure of enzymes or other macromolecules (Rauser and Hanson, 1966; Warren and

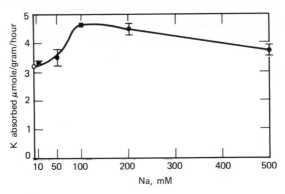

Figure 13-8. Absorption of labeled potassium by excised leaf tissue of the mangrove, *Avicennia marina,* as a function of the concentration of sodium, as NaCl. Concentration of potassium, as KCl, 10 mM, well within the range of operation of mechanism 2 (cf. Chapter 6). Concentration of calcium, as CaSO₄, 10 mM. The potassium was labeled with [86]Rb. After Rains and Epstein (1967).

Cheatum, 1966), may damage cellular organelles (Blumenthal–Gold-schmidt and Poljakoff–Mayber, 1968), may affect photosynthesis and respiration (Boyer, 1965; Nieman, 1962), may inhibit protein synthesis (Kahane and Poljakoff–Mayber, 1968), and may cause ion deficiencies, as discussed above for the pair potassium–sodium. The implication is that halophytes are able to resist or accomodate to these and still other effects that salinity may have on salt-sensitive plants.

The accumulation and toleration of high intracellular concentrations of ions are universals in the mineral metabolism of halophytes. Some but not all of them have additional adaptations. Some halophytes have salt glands in their leaves which excrete salt onto the surface from which eventually it is removed by the action of wind or water. Such is the case in species of tamarisk, *Tamarix,* which are desert shrubs and trees (Decker, 1961; Thomson *et al.,* 1969; Waisel, 1961) and in some mangroves, trees which grow in sea water along tropical shorelines (Atkinson *et al.,* 1967). This subject is discussed in section 5 of Chapter 7. Still another adaptation found in certain halophytes is succulence—a hypertrophy of leaves which causes a dilution of the intracellular solution of salt (Biebl and Kinzel, 1965).

Like other stressful features of the mineral environment, salinity often results in the evolution of races or ecotypes adapted to this condition (Dewey, 1960, 1962; McMillan, 1959; McNaughton, 1966; Tanimoto, 1969; Workman and West, 1969). We therefore have the possibility of breeding salt tolerance into crop species, as well as the option of breeding economic usefulness into salt-tolerant wild plants. Neither of these strategies has yet been tried in any sustained, energetic manner.

Deficiencies.   When we grow peach trees, *Prunus persica,* in the Central Valley of California, or Monterey Pine, *Pinus radiata,* in New Zealand, we are not surprised by the appearance of nutrient deficiencies. These species are not natives of those regions, and it is therefore not to be expected that they should be well adapted to the soils of areas to which they have been transported by man. If the plants are not adapted to the soil, we can adapt the soil to the plants by appropriate fertilization. On the other hand, we tend to consider wild plants well adapted to their native habitats on the basis that if they were not, they would not be there. Their presence does in fact indicate that they are well enough adapted to grow there and to hold their own against competitors, but this does not mean that the soil is an ideal nutritional medium for them and that there are no nutrient deficiencies; on the contrary, nutrient deficiencies are common and wide-spread in native plants in their natural habitats. The plants occupying soils of low nutrient status are those which are least disadvantaged by this condition, in comparison with other plants.

Two factors would be advantageous to plants in any situation in which the level of some nutrient available for absorption is low. The first is superior ability to absorb the element from a medium containing it at a low level, and the second, superior efficiency in the metabolic utilization of the element. This efficiency would result in normal functioning of metabolism at lower concentrations of the element in the tissues than less efficient plants require.

That nutrient deficiencies occur in native plants in their native habitats is well documented from many parts of the world. On the other hand, the specific adaptations enabling plants to cope with conditions of deficiencies—absorptive capacity and efficiency of utilization—are virtually blank spaces on the map. Study of these adaptations represents one of the foremost challenges in the realm of mineral metabolism of plants and plant ecology, with profound implications for environmental science, range management, forestry, and other applied aspects of plant science.

Hellmers *et al.* (1955) studied the mixed vegetation of evergreen shrubs (chaparral) on the slopes of the San Gabriel Mountains of southern Cali-

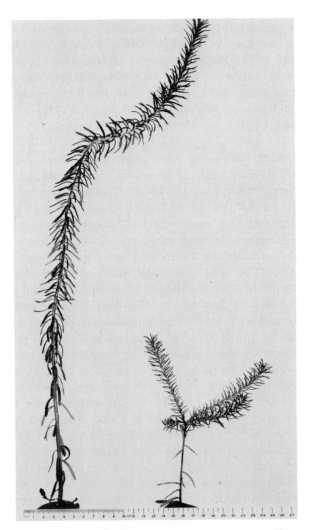

Figure 13-9. *Banksia ericifolia.* (Left) Seedling grown for 6 months with the addition of Hoagland solution. (Right) Seedling grown for 6 months in soil of low fertility on which the species is found in the field. From Beadle (1966).

fornia. The vegetation is sparse, partly as a result of the long, dry summers when no rain falls for about half a year. They addressed themselves to the question of whether nutrient deficiencies also limit the growth of this vegetation and found that this was so. Combined fertilization with nitrogen and phosphorus caused the growth of several species in pot tests to exceed that of the unfertilized controls by as much as a factor of 25. In the field, results were less spectacular but still impressive, the most marked responses approximately trebling the growth of the plants. In another California study, Martin and Berry (1970) applied nitrogenous and phosphatic fertilizers to rangelands, with the result that the growth of range plants was increased up to three-fold. Across the continent, in New York State, Mitchell and Chandler (1939) showed nitrogen deficiency in deciduous forest trees of northeastern United States.

The soils in virtually the whole of the continent of Australia are low in available nutrients, especially phosphate, and native plants no less than introduced agricultural species respond to fertilization (Beadle, 1966). Figure 13-9 shows this for *Banksia ericifolia*. Examples of this kind could be multiplied almost at will, not only for macronutrient elements but for micronutrients as well. Iron deficiency is a limiting factor for the growth of many plants on calcareous soils, as discussed above under the heading, Calcium and pH. In Australia, deficiencies of micronutrients are so widespread and so severe that Anderson and Underwood (1959) have referred to Australia's "trace-element deserts."

Mineral deficiencies also occur in aquatic ecosystems where primary productivity depends on the growth of algae. For example, Goldman (1960) has shown that in Castle Lake in northern California, the growth of the phytoplankton population is limited by a deficiency of molybdenum (Figure 13-10). For a general discussion of micronutrient deficiencies in aquatic ecosystems, see Goldman (1966).

Differences in the capacity of plants to absorb elements present at low concentrations have been studied by several authors. Examples of intraspecific variation in this regard have been given in the preceding chapter. As might be expected, differences among species are even more common and often more pronounced. Table 13-3, from Gerloff *et al.* (1966), gives the concentrations of essential mineral nutrients in samples of thirteen native plants of Wisconsin. General morphological features such as size of the root system may account for some of the differences among species but efficiency of selective absorption is suggested for others. For example, *Pteridium aquilinum* has the lowest calcium concentration among the plants of the Northern dry forest site tested, but the highest nitrogen, sulfur, and chlorine concentrations.

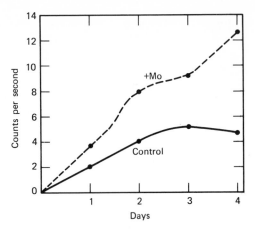

Figure 13-10. Stimulation of the rate of carbon fixation, as measured by incorporation of ^{14}C, by addition of molybdenum at 0.10 ppm to a culture of phytoplankton in water of Castle Lake, northern California. After Goldman (1960). Copyright 1960 by the American Association for the Advancement of Science.

Another approach to the problem of the efficiency of absorption is the experimental one of measuring the absorption of nutrients by different plants from solutions maintained at very low concentrations. This approach has been taken by Loneragan and co-workers in Western Australia. Their findings on calcium absorption by various plants are discussed in section 4 of Chapter 11. Similar experiments on the absorption of potassium (Asher and Ozanne, 1967) and phosphate (Loneragan and Asher, 1967) have also revealed different rates of uptake by different species.

The same experiments have also provided evidence for different efficiencies of utilization. For example, in some species a potassium concentration in the tops of about 100 μmoles per gram fresh weight was the critical concentration above which the yield of the plants did not increase, while for other species approximately double this concentration of potassium was needed for maximal growth (Asher and Ozanne, 1967). The experiments with phosphate revealed instances of phosphate toxicity in some species (Loneragan and Asher, 1967).

Forest trees of different species also differ in the efficiency of utilization of nutrients. Thus, Mitchell and Chandler (1939) speak of some trees as

TABLE 13-3. Comparisons of the Concentrations of the Essential Elements in Samples of Various Plant Species Collected from the Same Site[1]

Date sampled	Plant species	Per cent (oven-dry basis) of						Ppm (oven-dry basis) of						
		N	P	S	Ca	Mg	K	Fe	Mn	Cu	Zn	Mo	B	Cl
Northern dry forest (site 61)														
7-10-59	Aster macrophyllus	1.78	0.24	—	1.00	0.39	4.74	88	328	7.8	51	0.18	54	1475
7-12-59	Epigaea repens	1.02	0.10	0.08	0.44	0.27	0.52	445	1225	7.0	32	0.17	20	161
7-10-59	Gaultheria procumbens	0.87	0.09	0.07	0.96	0.46	0.60	129	697	2.7	25	0.13	27	382
7-10-59	Pteridium aquilinium	2.00	0.25	0.12	0.26	0.22	1.76	39	229	8.5	23	0.22	20	1808
7-10-59	Waldsteinia fragariodes	1.39	0.18	0.06	1.10	0.74	1.59	118	724	4.1	86	0.32	26	550
Southern mesic forest (site 50)														
7-24-60	Adiantum pedatum	1.45	0.23	0.15	0.40	0.37	1.84	97	66	4.2	16	0.31	21	684
7-24-60	Caulophyllum thalictroides	1.83	0.31	0.25	0.82	0.30	2.09	152	96	6.1	23	1.29	28	407
7-24-60	Osmorhiza claytoni	1.17	0.37	0.12	1.77	0.47	3.25	258	38	4.5	15	1.04	31	378
7-24-60	Sanguinaria canadensis	1.49	0.33	0.28	1.77	0.12	5.73	483	103	4.0	22	0.41	45	1866
Dry prairie (site 82a)														
8-24-61	Andropogon gerardi	1.00	0.20	0.07	0.23	0.16	1.02	54	55	3.6	30	0.20	11	776
8-24-61	Euphorbia corollata	1.38	0.32	0.09	0.75	0.30	1.26	69	52	5.5	31	0.65	54	462
8-24-61	Bouteloua hirsuta	0.97	0.19	0.08	0.15	0.18	0.83	56	100	4.6	38	0.22	18	1188
8-24-61	Coreopsis palmata	1.17	0.15	0.12	1.05	0.42	1.19	52	117	7.9	65	0.37	61	957

[1] After Gerloff et al. (1966).

being "nitrogen-deficiency tolerant," referring to their capacity to grow at near maximal rate at low leaf nitrogen levels.

3. INTERPLAY BETWEEN PLANTS AND THEIR MINERAL MEDIA

General Considerations.   In our discussion of the mineral nutrition of plants we have looked upon the mineral medium as given, and upon the plant as responding to the mineral supply available in the medium. The processes of evolution have rendered existing plants fit for the environments in which they are now found, including their mineral environments. Conversely, we may consider "the fitness of the environment," as Henderson (1924) did, and in the present context ask whether the physiological activities of the plant exert a significant influence upon it. Even casual reflection will reveal that this is so.

One of the profound ecological consequences of the immobility of plants, contrasted with the mobility of animals, is this: the plant is a factor in its own environment in a sense that the animal is not. The exchange of materials between the animal and the environment does not alter the chemistry of the environment with which the animal closely interacts. In fact, behavioral patterns have evolved which minimize such alteration, like those whereby animals remove excrement from their immediate surroundings.

The stationary plant cannot escape from the chemical consequences of its physiological activities and those of other plants nearby (Muller, 1969; Tukey, 1969). It must continue to live in the soil from which it withdraws phosphate, nitrate, and other nutrients, and to which it contributes exudates from its roots, leachate from its leaves, and eventually, the leaves themselves, when they fall. There is thus a close chemical reciprocity between the mineral medium and the plant. We have so far studied the influence of the chemistry of the medium on the nutrition of the plant; we shall now consider the effects of the plant upon the chemistry of its medium.

Addition of Organic Matter to the Soil.   Such is the chemical reciprocity between soil and the life it supports that the very development of soil depends as much on the life on it and in it as the development of living things depends on the soil. Jacks (1965) considered that the nearest thing to soil that is formed in the absence of life is a body of physically and chemically weathered particles of rock high on a mountain.

Such a body of finely divided rock is devoid of any mechanism for storing energy. With time, it would become progressively less suited as a habitat for life. Gravitation would compact it, leaving less and less space for water and air, nutrients would be leached out, and all energy input would be dissipated as heat.

But let life appear, probably in the form of hardy algae, lichens, and mosses to begin with, and this sterile scene is transformed into a system capable of absorbing and storing radiant energy. Organic matter, the result of photosynthesis, becomes a substrate for microorganisms and insects. Minerals are absorbed into cells and tissues, and upon the death of the organisms return along with the organic matter into the developing soil. They are now readily available for absorption even by plants which are unable to grow on bare rock. Eventually, grasses and other higher plants follow. The ecological question—why do plants grow where they do?—can therefore be answered for terrestrial plants in part by saying: plants grow where plants have grown before. This is because they grow in soil, and there is no soil except by virtue of prior plant life.

The processes whereby the addition of organic matter makes and modifies soil are numerous. It changes the pH of the soil solution, chelates heavy metal ions in it, supports its microbial life and with it the release of carbon dioxide, accelerates the chemical weathering of minerals, and has effects on the physical condition and water-holding capacity of the soil. The organic materials may be solid plant residues, leachates from leaves, flower nectars, root exudations, and products of decay. Indirectly, animals depending ultimately on the plants also alter the chemistry of the soil through their excretions and the products of their decay after death. These additions of organic matter have the general effect on mineral nutrients of making them more mobile and available for absorption by plants and for removal via run-off and percolation.

Withdrawal and Depletion of Minerals.   Although a good portion of the minerals absorbed by plants eventually returns to the soil at the same site, a fraction does not remain there. The minerals leached from leaves on the tree and from leaf litter are much more subject to the agencies of erosion than they were in their native condition in the soil, and water removes minerals so mobilized by run-off and percolation. The same is true of the minerals in the entire plant after its death.

The removal of nutrient and other minerals by plants often causes profound alterations in the chemistry of the soil. For example, many tropical trees are silicon accumulator plants, containing in their wood and bark several per cent silica (SiO_2) on a dry weight basis. Lovering (1958) calculated that on the basis of 13 tons dry weight of new growth per acre per year (a realistic figure for tropical forests) and 3 per cent silica in this material, the trees pump a ton of silica out of the soil of each acre every $2\frac{1}{2}$ years. While some of it would return to the soil, much of the organic debris containing it would be swept into drainage channels by the frequent torrential tropical downpours. Lovering concluded that this process is a major factor in the genesis of the silica-impoverished, iron oxide-rich lateritic soils of the tropics.

Other trees cause the very opposite alteration in the chemistry of the soil. Figure 13-11 shows a road cut at a site in New Zealand where a large Kauri pine, *Agathis australis*, had grown. Unlike the soil to the left and right, the soil where the tree grew is white; it is almost pure silica. This tree is poor in bases and the leachate from its leaves and litter is very acidic. In the course of time, the dark iron oxides of the soil are solubilized and leached away, leaving the silica. The acidification of the soil kills the

Figure 13-11. Depletion of iron oxides from soil of a site in New Zealand where a Kauri pine, *Agathis australis*, had grown. The iron oxides which give a dark appearance to the soil (left and right) have been leached out where the tree grew (middle), leaving the white silica. From the Soil Bureau, D.S.I.R., New Zealand, through the courtesy of R. H. Jackman.

fine feeder roots, leaving only several large "peg" roots which anchor the tree. Nutrition of the tree is shifted to fibrous roots which grow into the thick layer of litter on the ground. The story of the Kauri pine dramatically demonstrates the chemical and biological consequences of immobility. Kickuth (1970) has discussed the removal of iron oxide from the soil where spruce, *Picea abies,* grows. He attributes this not to acidity, but to the release of (unidentified) chelating substances from the trees.

It is no accident that in the examples given of mineral denudation of soils by plants, the plants in question have been trees, and trees in areas of high or very high rainfall at that. Trees are large and long-lived, and their mineral metabolism, combined with the erosive action of rain, may cause chemical depletion of soils on a very large scale. The roots of smaller plants, especially of annuals, and those of plants in arid and semiarid regions, explore a more shallow layer of soil, and the minerals absorbed are returned to the soil and remain there (see below, Local Cycling of Nutrients).

Interception of Airborne Dust and Aerosols. As wind with a burden of dust and other finely divided matter and of cyclic salt sweeps through forests and woodlands the crowns of the trees act as giant traps. The matter so intercepted, upon wetting by rain or dew, contributes soluble mineral matter to the ecosystem, resulting in a net gain of nutrients by the system, including the soil (Tamm, 1958).

Modification of the Composition of the Soil Atmosphere. Roots respire: they take up oxygen and give off carbon dioxide. Furthermore, they may give off ethylene, and Radin and Loomis (1969) concluded therefore that in soil, they are probably continuously exposed to it. The release of carbon dioxide is an aspect of the addition of organic matter to the soil, discussed above under that heading. Whether the release of ethylene and possibly other volatile organic substances has implications for mineral metabolism and ecology is not known. However, the depletion of oxygen in the vicinity of roots is an important factor in that it slows down aerobic metabolism and hence, the absorption of ions by roots (Letey *et al.,* 1964; Stolzy and Letey, 1964; Wiegand and Lemon, 1958, 1963). This is apt to be particularly important in water-logged soil devoid of air spaces (Greenwood, 1969). Plants whose roots are well aerated by conduction of atmospheric and photosynthetic oxygen from the shoots via open air spaces in the stems are favored in bogs and on inundated soils (Williams and Barber, 1961).

Local Cycling of Nutrients.   The influence of plants on the mineral substrate is not confined to processes resulting in net losses of nutrients from the ecosystem or net gains by it. Nutrients cycle within the system and may be shifted either vertically or laterally as a result of mineral transport by plants. Hibbard (1940) observed that there was a high concentration of zinc in the top layers of soil in several localities in California where trees or grass had grown undisturbed for long periods—much higher than in similar nearby soils where there had been no such long accumulation of plant matter. He attributed this phenomenon to the action of plants whose roots bring up zinc from the lower layers of the soil. With the fall of the leaves and other plant matter, and the eventual death of the plants, the zinc contained in this material had become incorporated and immobilized in the upper layers of the soil. Differences in zinc concentration between top and lower layers of soil were large. For example, under both pine and oak, the zinc concentration in the top 2.5 cm of soil was about 7 times higher than at a depth of about 36 cm.

Other elements also undergo such shifts, and often in a highly selective manner, depending on the selectivity of absorption by the plants. Hopsage, *Grayia spinosa,* and greasewood, *Sarcobatus vermiculatus,* are halophytic shrubs of the deserts and mesas of western North America. The fairly shallow-rooted hopsage accumulates much more potassium than sodium in its leaves, while the deep-rooted greasewood accumulates sodium preferentially and much less potassium. The leaves of both shrubs eventually fall, and the potassium and sodium they contain become incorporated into the top layer of the soil because there is not enough rain to leach them away. Over a period of time the surface soil under hopsage becomes enriched in potassium, and that under greasewood enriched in sodium (Rickard and Keough, 1968). In the area of Washington State which these authors studied, near the Hanford reservation, 20 per cent of the ground area was affected, 7 per cent being enriched in potassium and 14 per cent in sodium.

Under trees, soil properties such as pH and nitrogen content show the influence of the tree canopy, direction of prevaling winds, and other factors, and the distribution of various species of the ground flora is in turn influenced (Fisher and Stone, 1969; Zinke, 1962). Ecologists have described the cycling of minerals in entire ecosystems, especially in forests (Duvigneau and Denaeyer–De Smet, 1964; Ovington, 1965; Rodin and Bazilevich, 1967).

Mineral cycling, without major net losses or gains of nutrients, is characteristic of mature ecosystems. The mineral economy of such systems is in a steady state. The activities of plants solubilize and mobilize nutrients, making them potentially more susceptible to the agencies of erosion. On the other hand, the same processes also make them more readily available for absorption by plants, and evolutionary processes fill the ecological niches thus created. In its totality, the system does not let the mobilized nutrients get away but recaptures them, through absorption, and thereby retains them within the local nutrient cycle.

However, when natural disasters or the short-sightedness of man upset this delicate balance the highly mobile, highly labile nutrients of the system are lost, often on a catastrophic scale. This has been observed time and time again when tropical forest is cleared for agriculture. Suddenly, the lush, dense world of plant life which used to recapture nutrients in short supply, especially phosphate, is cut and burned. The mineral supply which was husbanded by the economy of the living jungle is swept away. In a few years, the land is minerally impoverished and infertile (Nye and Greenland, 1960; Richards, 1952).

Such mineral denudation occurs in temperate regions also, as when forest is clear-cut. This has been shown by the studies of Bormann *et al.* (1968) and Likens *et al.* (1970) at the Hubbard Brook Experimental Forest in New Hampshire. Even a much milder interference with a stable situation, such as tillage of a native soil, can cause large losses, especially of nitrate, because of the speed with which nitrate is released from the organic matter of the soil once oxidation is accelerated by tillage (Stout and Burau, 1967).

Fittingly, with this subject we have come full circle, back to the subject of cycling of elements in nature which is discussed early in this book (Chapter 2). These very large-scale movements of matter ultimately depend on mineral transport through cellular membranes 100 Å across or less. At these membranes discriminating mechanisms select atoms from the random reservoir without and assemble those that go into the make-up of the world of life.

REFERENCES

Anderson, A. J. and E. J. Underwood. 1959. Trace-element deserts. Sci. American 200(1):97–106.

Antonovics, J. 1968. Evolution in closely adjacent plant populations. V. Evolution of self-fertility. Heredity 23:219–238.

Asher, C. J. and P. G. Ozanne. 1967. Growth and potassium content of plants in solution cultures maintained at constant potassium concentrations. Soil Sci. 103:155–161.

Atkinson, M. R., G. P. Findlay, A. B. Hope, M. G. Pitman, H. D. W. Saddler and K. R. West. 1967. Salt regulation in the mangroves *Rhizophora mucronata* Lam. and *Aegialitis annulata* R. Br. Austral. J. Biol. Sci. 20:589–599.

Bamberg, S. A. and J. Major. 1968. Ecology of the vegetation and soils associated with calcareous parent materials in three alpine regions of Montana. Ecol. Monographs 38:127–167.

Beadle, N. C. W. 1966. Soil phosphate and its role in molding segments of the Australian flora and vegetation, with special reference to xeromorphy and sclerophylly. Ecol. 47:992–1007.

Bernstein, L. 1961. Osmotic adjustment of plants to saline media. I. Steady state. Am. J. Bot. 48:909–918.

Bernstein, L., J. W. Brown and H. E. Hayward. 1956. The influence of rootstock on growth and salt accumulation in stone-fruit trees and almonds. Proc. Am. Soc. Hort. Sci. 68:86–95.

Bernstein, L. and H. E. Hayward. 1958. Physiology of salt tolerance. Ann. Rev. Plant Physiol. 9:25–46.

Biebl, R. and H. Kinzel. 1965. Blattbau and Salzhaushalt von *Laguncularia racemosa* (L.) Gaertn. f. und anderer Mangrovebäume auf Puerto Rico. Österr. Bot. Z. 112:56–93.

Billings, W. D. 1957. Physiological ecology. Ann. Rev. Plant Physiol. 8:375–392.

Bingham, F. T., L. B. Fenn and J. J. Oertli, 1968. A sandculture study of chloride toxicity to mature avocado trees. Soil Sci. Soc. Am. Proc. 32:249–252.

Black, R. F. 1956. Effect of NaCl in water culture on the ion uptake and growth of *Atriplex hastata* L. Austral. J. Biol. Sci. 9:67–80.

Blumenthal–Goldschmidt, S. and A. Poljakoff–Mayber. 1968. Effect of substrate salinity on growth and on submicroscopic structure of leaf cells of *Atriplex halimus* L. Austral. J. Bot. 16:469–478.

Bormann, F. H., G. E. Likens, D. W. Fisher and R. S. Pierce. 1968. Nutrient loss accelerated by clear-cutting of a forest ecosystem. Science 159:882–884.

Boyce, S. G. 1954. The salt spray community. Ecol. Monographs 24:29–67.

Boyer, J. S. 1965. Effects of osmotic water stress on metabolic rates of cotton plants with open stomata. Plant Physiol. 40:229–234.

Bradshaw, A. D. 1952. Populations of *Agrostic tenuis* resistant to lead and zinc poisoning. Nature 169:1098.

Bradshaw, A. D. 1965. Evolutionary significance of phenotypic plasticity in plants. Adv. Genetics 13:115–155.

Bradshaw, A. D. 1969. An ecologist's viewpoint. In: Ecological Aspects of the Mineral Nutrition of Plants. I. H. Rorison, ed. Blackwell Scientific Publications, Oxford and Edinburgh. Pp. 415–427.

Bradshaw, A. D., T. S. McNeilly and R. P. G. Gregory. 1965. Industrialization, evolution and the development of heavy metal tolerance in plants. In: Ecology and the Industrial Society. G. T. Goodman, R. W. Edwards and J. M. Lambert, eds. Blackwell Scientific Publications, Oxford. Pp. 327–343.

Bröker, W. 1963. Genetisch-physiologische Untersuchungen über die Zinkverträglichkeit von *Silene inflata* SM. Flora 153:122–156.

Brooks, R. R. 1968. Biogeochemical prospecting in New Zealand. New Zealand Sci. Rev. 26:9–12.

Brown, J. C. and L. O. Tiffin. 1965. Iron stress as related to the iron and citrate occurring in stem exudate. Plant Physiol. 40:395–400.

Cannon, H. L. 1960. Botanical prospecting for ore deposits. Science 132: 591–598.

Chapman, V. J. 1960. Salt Marshes and Salt Deserts of the World. Leonard Hill [Books], Ltd., London. Interscience Publishers, Inc., New York.

Chaudhuri, I. I. and H. H. Wiebe. 1968. Influence of calcium pretreatment on wheat germination on saline media. Plant and Soil 28: 208–216.

Chikishev, A. G., ed. 1965. Plant Indicators of Soils, Rocks, and Subsurface Waters. Consultants Bureau, New York.

Clarkson, D. T. 1965a. The effect of aluminium and some other trivalent metal cations on cell division in the root apices of *Allium cepa*. Ann. Bot. N. S. 29:309–315.

Clarkson, D. T. 1965b. Calcium uptake by calcicole and calcifuge species in the genus *Agrostis* L. J. Ecol. 53:427–435.

Clarkson, D. T. 1966. Aluminium tolerance in species within the genus *Agrostis*. J. Ecol. 54:167–178.

Collander, R. 1941. Selective absorption of cations by higher plants. Plant Physiol. 16:691–720.

Darwin, C. 1859. On the Origin of Species by Means of Natural Selection, or the Preservation of Favoured Races in the Struggle for Life. John Murray, London. (Available in a facsimile edition, Harvard University Press, Cambridge, 1964.)

Davies, E. B. 1956. Factors affecting molybdenum availability in soils. Soil Sci. 81:209–221.

De Candolle, M. A. 1855. Géographie Botanique Raisonnée. Librairie de Victor Masson, Paris. Vols. 1 and 2.

Decker, J. P. 1961. Salt secretion by *Tamarix pentandra* Pall. Forest Sci. 7:214–217.

Dewey, D. R. 1960. Salt tolerance of twenty-five strains of *Agropyron*. Agron. J. 52:631–635.

Dewey, D. R. 1962. Breeding crested wheatgrass for salt tolerance. Crop Sci. 2:403–407.

Dodd, J. D. and R. T. Coupland. 1966. Osmotic pressures of native plants of saline soil in Saskatchewan. Can. J. Plant Sci. 46:479–485.

Dregne, H. E. 1963. Soils of the arid West. In: Aridity and Man—The Challenge of the Arid Lands in the United States. C. Hodge and P. C. Duisberg, eds. American Association for the Advancement of Science, Washington. Publication No. 74. Pp. 215–238.

Duncan, W. G. and A. J. Ohlrogge. 1958. Principles of nutrient uptake from fertilizer bands. II. Root development in the band. Agron. J. 50:605–608.

Duvigneaud, P. and S. Denaeyer–De Smet. 1964. Le cycle des éléments biogènes dans l'écosystème forêt. Lejeunia N. S. No. 28, pp. 1–148.

Eaton, F. M. 1927. The water requirement and cell-sap concentration of Australian saltbush and wheat as related to the salinity of the soil. Am. J. Bot. 14:212–226.

Eckardt, F. E., ed. 1968. Functioning of Terrestrial Ecosystems at the Primary Production Level. Proc. Copenhagen Symposium. United Nations Educational, Scientific and Cultural Organization, Paris.

Elzam, O. E. and E. Epstein. 1969a. Salt relations of two grass species differing in salt tolerance. I. Growth and salt content at different salt concentrations. Agrochimica 13:187–195.

Elzam, O. E. and E. Epstein. 1969b. Salt relations of two grass species differing in salt tolerance. II. Kinetics of the absorption of K, Na, and Cl by their excised roots. Agrochimica 13:196–206.

Epstein, E. 1969. Mineral metabolism of halophytes. In: Ecological Aspects of the Mineral Nutrition of Plants. I. H. Rorison, ed. Blackwell Scientific Publications, Oxford and Edinburgh. Pp. 345–355.

Epstein, E. and R. L. Jefferies. 1964. The genetic basis of selective ion transport in plants. Ann. Rev. Plant Physiol. 15:169–184.

Eyre, S. R. 1968. Vegetation and Soils. A World Picture. 2nd ed. Edward Arnold, London.

Fisher, R. F. and E. L. Stone. 1969. Increased availability of nitrogen and phosphorus in the root zone of conifers. Soil Sci. Soc. Am. Proc. 33:955–961.

Gerloff, G. C. 1963. Comparative mineral nutrition of plants. Ann. Rev. Plant Physiol. 14:107–124.

Gerloff, G. C., D. G. Moore and J. T. Curtis. 1966. Selective absorption of mineral elements by native plants of Wisconsin. Plant and Soil 25:393–405.

Goldman, C. R. 1960. Molybdenum as a factor limiting primary productivity in Castle Lake, California. Science 132:1016–1017.

Goldman, C. R. 1966. Micronutrient limiting factors and their detection in natural phytoplankton populations. In: Primary Productivity in Aquatic Environments. Proc. I.B.P.P.F. Sympos., Pallanza, 1965. C. R. Goldman, ed. Pp. 121–135.

Greenway, H. 1962. Plant response to saline substrates. I. Growth and ion uptake of several varieties of *Hordeum* during and after sodium chloride treatment. Austral. J. Biol. Sci. 15:16–38.

Greenway, H. 1968. Growth stimulation by high chloride concentrations in halophytes. Israel J. Bot. 17:169–177.

Greenwood, D. J. 1969. Effect of oxygen distribution in the soil on plant growth. In: Root Growth. W. J. Whittington, ed. Butterworths, London, Pp. 202–223.

Gregory, R. P. G. and A. D. Bradshaw. 1965. Heavy metal tolerance in populations of *Agrostis tenuis*. Sibth. and other grasses. New Phytol. 64:131–143.

Harris, J. A., R. A. Gortner, W. F. Hoffman, J. V. Lawrence and A. T. Valentine. 1924. The osmotic concentration, specific electrical conductivity, and chlorid content of the tissue fluids of the indicator plants of Tooele Valley, Utah. J. Agric. Res. 27:893–924.

Hellmers, H., J. F. Bonner and J. M. Kelleher. 1955. Soil fertility: a watershed management problem in the San Gabriel Mountains of Southern California. Soil Sci. 80:189–197.

Henderson, L. J. 1924. The Fitness of the Environment. The Macmillan Company, New York.

Hibbard, P. L. 1940. Accumulation of zinc on soil under long-persistent vegetation. Soil Sci. 50:53–55.

Hiesey, W. M. and H. W. Milner. 1965. Physiology of ecological races and species. Ann. Rev. Plant Physiol. 16:203–216.

Hill, M. N., ed. 1963. The Sea. Vol. 2. The Composition of Sea-Water. Comparative and Descriptive Oceanography. Interscience Publishers, John Wiley and Sons, New York and London.

Hutchinson, T. C. 1967. Ecotype differentiation in *Teucrium scorodonia* with respect to susceptibility to lime-induced chlorosis and to shade factors. New Phytol. 66:439–453.

Hutchinson, T. C. 1968. A physiological study of *Teucrium scorodonia* ecotypes which differ in their susceptibility to lime-induced chlorosis and iron-deficiency chlorosis. Plant and Soil 28:81–105.

Hutchinson, T. C. 1970. Lime chlorosis in seedling establishment on calcareous soils. III. The ability of green and chlorotic plants fully to reverse large leaf water-deficits. New Phytol. 69:261–268.

Hyder, S. Z. and H. Greenway. 1965. Effects of Ca^{++} on plant sensitivity to high NaCl concentrations. Plant and Soil 23:258–260.

Jacks, G. V. 1965. The role of organisms in the early stages of soil formation. In: Experimental Pedology. E. G. Halsworth and D. V. Crawford, eds. Butterworths, London. Pp. 219–226.

Jacoby, B. 1964. Function of bean roots and stems in sodium retention. Plant Physiol. 39:445–449.

Jacoby, B. 1965. Sodium retention in excised bean stems. Physiol. Plantarum 18:730–739.

Jefferies, R. L., D. Laycock, G. R. Stewart and A. P. Sims. 1969. The properties of mechanisms involved in the uptake and utilization of calcium and potassium by plants in relation to an understanding of plant distribution. In: Ecological Aspects of the Mineral Nutrition of Plants. I. H. Rorison, ed. Blackwell Scientific Publications, Oxford and Edinburgh. Pp. 281–307.

Jefferies, R. L. and A. J. Willis. 1964. Studies on the calcicole–calcifuge habit. II. The influence of calcium on the growth and establishment of four species in soil and sand cultures. J. Ecol. 52:691–707.

Jenny, H. 1941. Factors of Soil Formation. McGraw-Hill Book Company, Inc., New York and London.

Jowett, D. 1958. Populations of *Agrostis* spp. tolerant of heavy metals. Nature 182:816–817.

Kahane, I. and A. Poljakoff–Mayber. 1968. Effect of substrate salinity on the ability for protein synthesis in pea roots. Plant Physiol. 43:1115–1119.

Kelley, W. P. 1963. Use of saline irrigation water. Soil Sci. 95:385–391.

Kickuth, R. 1970. Ökochemische Leistungen höherer Pflanzen. Naturwiss. 57:55–61.

Kinzel, H. 1963. Zellsaft-Analysen zum pflanzlichen Calcium- und Säurestoffwechsel und zum Problem der Kalk- und Silikatpflanzen. Protoplasma 57:522–555.

Kinzel, H. 1969. Ansätze zu einer vergleichenden Physiologie des Mineralstoffwechsels und ihre ökologischen Konsequenzen. Ber. Deutsch. Bot. Ges. 82:143–158.

Kramer, P. J. 1969. Plant and Soil Water Relationships—A Modern Synthesis. McGraw-Hill Book Company, Inc., New York.

Krause, W. 1958. Boden und Pflanzengesellschaften. In: Encyclopedia of Plant Physiology. W. Ruhland, ed. Springer–Verlag, Berlin. Vol. 4, pp. 807–850.

Kruckeberg, A. R. 1954. The ecology of serpentine soils. III. Plant species in relation to serpentine soils. Ecol. 35:267–274.

Kruckeberg, A. R. 1969. Soil diversity and the distribution of plants, with examples from western North America. Madroño 20:129–154.

Lagerwerff, J. V. 1969. Osmotic growth inhibition and electrometric salt-tolerance evaluation of plants. A review and experimental assessment. Plant and Soil 31:77–96.

LaHaye, P. A. and E. Epstein. 1969. Salt toleration by plants: enhancement with calcium. Science 166:395–396.

LaHaye, P. A. and E. Epstein. 1971. Calcium and salt toleration by bean plants. Physiol. Plantarum. In press.

Lauff, G. H., ed. 1967. Estuaries. American Association for the Advancement of Science, Washington. Publication No. 83.

Letey, J., L. H. Stolzy, O. R. Lunt and V. B. Youngner. 1964. Growth and nutrient uptake of Newport bluegrass as affected by soil oxygen. Plant and Soil 20:143–148.

Likens, G. E., F. H. Bormann, N. M. Johnson, D. W. Fisher and R. S. Pierce. 1970. Effects of forest cutting and herbicide treatment on nutrient budgets in the Hubbard Brook watershed-ecosystem. Ecol. Monographs 40:23–47.

Loneragan, J. F. and C. J. Asher. 1967. Response of plants to phosphate concentration in solution culture: II. Rate of phosphate absorption and its relation to growth. Soil Sci. 103:311–318.

Lovering, T. S. 1958. Accumulator plants and rock weathering. Science 128:416–417.

Lyr, H. and G. Hoffmann. 1967. Growth rates and growth periodicity of tree roots. Internat. Rev. Forestry Res. 2:181–236.

Major, J. and S. A. Bamberg. 1963. Some cordilleran plant species new for the Sierra Nevada of California. Madroño 17:93–109.

Martin, E. V. and F. E. Clements. 1939. Adaptation and origin in the plant world. I. Factors and functions in coastal dunes. Carnegie Institution of Washington Publication No. 521.

Martin, W. E. and L. J. Berry. 1970. Use of nitrogenous fertilizers on California rangeland. Proc. XI Internat. Grasslands Congress. Pp. 817–822.

Martin, W. E., J. Vlamis and N. W. Stice. 1953. Field correction of calcium deficiency on a serpentine soil. Agron. J. 45:204–208.

Mason, H. L. 1946. The edaphic factor in narrow endemism. I. The nature of environmental influences. Madroño 8:209–226.

Mason, H. L. and P. R. Stout. 1954. The role of plant physiology in plant geography. Ann. Rev. Plant Physiol. 5:249–270.

McGinnies, W. G., B. J. Goldman and P. Paylore, eds. 1968. Deserts of the World. An Appraisal of Research into their Physical and Biological Environments. University of Arizona Press, Tucson.

McMillan, C. 1959. Salt tolerance within a *Typha* population. Am. J. Bot. 46:521–526.

McNaughton, S. J. 1966. Ecotype function in the *Typha* community-type. Ecol. Monographs 36:297–325.

McNeilly, T. and J. Antonovics. 1968. Evolution in closely adjacent plant populations. IV. Barriers to gene flow. Heredity 23:205–218.

McNeilly, T. and A. D. Bradshaw. 1968. Evolutionary processes in populations of copper tolerant *Agrostis tenuis* Sibth. Evolution 22:108–118.

McNulty, I. 1969. The effect of salt concentration on the growth and metabolism of a succulent halophyte. In: Physiological Systems in Semiarid Environments. C. C. Hoff and M. L. Riedesel, eds. University of New Mexico Press, Albuquerque. Pp. 255–262.

Meiri, A. and A. Poljakoff–Mayber. 1969. Effect of variations in substrate salinity on the water balance and ionic composition of bean leaves. Israel J. Bot. 18:99–112.

Mitchell, H. L. and R. F. Chandler, Jr. 1939. The nitrogen nutrition and growth of certain deciduous trees of Northeastern United States. The Black Rock Forest Bull. No. 11. Pp. 1–94.

Muller, C. H. 1969. Allelopathy as a factor in ecological process. Vegetatio 18:348–357.

Nieman, R. H. 1962. Some effects of sodium chloride on growth, photosynthesis, and respiration of twelve crop plants. Bot. Gaz. 123:279–285.

Nye, P. H. and D. J. Greenland. 1960. The Soil under Shifting Cultivation. Commonwealth Bureau of Soils, Harpenden. Tech. Communication No. 51.

Oertli, J. J. 1968. Extracellular salt accumulation, a possible mechanism of salt injury in plants. Agrochimica 12:461–469.

O'Leary, J. W. 1969. The effect of salinity on permeability of roots to water. Israel J. Bot. 18:1–9.

Osmond, B. 1963. Oxalates and ionic equilibria in Australian salt-bushes (*Atriplex*). Nature 198:503–504.

Ovington, J. D. 1965. Organic production, turnover and mineral cycling in woodlands. Biol. Rev. 40:295–336.

Parsons, R. F. and R. L. Specht. 1967. Lime chlorosis and other factors affecting the distribution of *Eucalyptus* on coastal sands in southern Australia. Austral. J. Bot. 15:95–105.

Piggott, C. D. 1969. Influence of mineral nutrition on the zonation of flowering plants in coastal salt-marshes. In: Ecological Aspects of the

Mineral Nutrition of Plants. I. H. Rorison, ed. Blackwell Scientific Publications, Oxford and Edinburgh. Pp. 25–35.

Radin, J. W. and R. S. Loomis. 1969. Ethylene and carbon dioxide in the growth and development of cultured radish roots. Plant Physiol. 44:1584–1589.

Rains, D. W. 1969. Cation absorption by slices of stem tissue of bean and cotton. Experientia 25:215–216.

Rains, D. W. and E. Epstein. 1967. Preferential absorption of potassium by leaf tissue of the mangrove, *Avicennia marina*: an aspect of halophytic competence in coping with salt. Austral. J. Biol. Sci. 20:847–857.

Rauser, W. E. and J. B. Hanson. 1966. The metabolic status of ribonucleic acid in soybean roots exposed to saline media. Can. J. Bot. 44:759–776.

Raymont, J. E. G. 1963. Plankton and Productivity in the Oceans. The Macmillan Company, New York.

Richards, L. A., ed. 1954. Diagnosis and Improvement of Saline and Alkali Soils. U. S. Department of Agriculture Handbook No. 60.

Richards, P. W. 1952. The Tropical Rain Forest—An Ecological Study. Cambridge University Press, Cambridge.

Rickard, W. H. and R. F. Keough. 1968. Soil–plant relationships of two steppe desert shrubs. Plant and Soil 29:205–212.

Riley, J. J. 1969. Physiological responses of plants to salinity: plant–water relations. In: Physiological Systems in Semiarid Environments. C. C. Hoff and M. L. Riedesel, eds. University of New Mexico Press, Albuquerque. Pp. 249–254.

Robson, A. D. 1969. Soil factors affecting the distribution of annual *Medicago* species. J. Austral. Inst. Agric. Sci. 35:154–167.

Rodin, L. E. and N. I. Bazilevich. 1967. Production and Mineral Cycling in Terrestrial Vegetation. Oliver and Boyd, Edinburgh and London.

Rorison, I. H. 1960a. Some experimental aspects of the calcicole–calcifuge problem. I. The effects of competition and mineral nutrition upon seedling growth in the field. J. Ecol. 48:585–599.

Rorison, I. H. 1960b. The calcicole–calcifuge problem. II. The effects of mineral nutrition on seedling growth in solution culture. J. Ecol. 48:679–688.

Rorison, I. H., ed. 1969. Ecological Aspects of the Mineral Nutrition of Plants. A Symposium of the British Ecological Society. Blackwell Scientific Publications, Oxford and Edinburgh.

Salisbury, E. 1952. Downs and Dunes—Their Plant Life and its Environment. G. Bell and Sons, Ltd., London.

Salisbury, E. J. 1959. Causal plant ecology. In: Vistas in Botany. W. B. Turrill, ed. Pergamon Press, New York. Vol. 1, pp. 124–144.

Schimper, A. F. W. 1935. Pflanzengeographie auf physiologischer Grundlage. 3rd ed. Verlag von Gustav Fischer, Jena.

Schmid, W. E. and G. C. Gerloff. 1961. A naturally occurring chelate of iron in xylem exudate. Plant Physiol. 36:226–231.

Scholander, P. F., E. D. Bradstreet, H. T. Hammel and E. A. Hemmingsen. 1966. Sap concentrations in halophytes and some other plants. Plant Physiol. 41:529–532.

Slatyer, R. O. 1961. Effects of several osmotic substrates on the water relations of tomato. Austral. J. Biol. Sci. 14:519–540.

Smith, R. C. and E. Epstein. 1964. Ion absorption by shoot tissue: technique and first findings with excised leaf tissue of corn. Plant Physiol. 39:338–341.

Snaydon, R. W. 1962a. Micro-distribution of *Trifolium repens* L. and its relation to soil factors. J. Ecol. 50:133–143.

Snaydon, R. W. 1962b. The growth and competitive ability of contrasting natural populations of *Trifolium repens* L. on calcareous and acid soils. J. Ecol. 50:439–447.

Snaydon, R. W. and A. D. Bradshaw. 1969. Differences between natural populations of *Trifolium repens* L. in response to mineral nutrients. II. Calcium, magnesium and potassium. J. Appl. Ecol. 6:185–202.

Stebbins, G. L., Jr. 1950. Variation and Evolution in Plants. Columbia University Press, New York.

Stolzy, L. H. and J. Letey. 1964. Correlation of plant response to soil oxygen diffusion rates. Hilgardia 35:567–576.

Stout, P. R. and R. G. Burau. 1967. The extent and significance of fertilizer buildup in soils as revealed by vertical distributions of nitrogenous matter between soils and underlying water reservoirs. In: Agriculture and the Quality of our Environment. N. C. Brady, ed. Plimpton Press, Norwood. Pp. 283–310.

Tamm, C. O. 1958. The atmosphere. In: Encyclopedia of Plant Physiology. W. Ruhland, ed. Springer-Verlag, Berlin. Vol. 4, pp. 233–242.

Tanimoto, T. T. 1969. Differential physiological response of sugarcane varieties to osmotic pressures of saline media. Crop Sci. 9:683–688.

Tansley, A. G. 1917. On competition between *Gallium saxatile* L. (*G. hercynicum* Weig.) and *Gallium sylvestre* Poll. (*G. asperum* Schreb.) on different types of soil. J. Ecol. 5:173–179.

Thomson, W. W., W. L. Berry and L. L. Liu. 1969. Localization and secretion of salt by the salt glands of *Tamarix aphylla*. Proc. Nat. Acad. Sci. 63:310–317.

Tiffin, L. O. 1970. Translocation of iron citrate and phosphorus in xylem exudate of soybean. Plant Physiol. 45:280–283.

Tukey, H. B., Jr. 1969. Implications of allelopathy in agricultural plant science. Bot. Rev. 35:1–16.

Turesson, G. 1922. The genotypical response of the plant species to the habitat. Hereditas 3:211–350.

Turner, R. G. 1969. Heavy metal tolerance in plants. In: Ecological Aspects of the Mineral Nutrition of Plants. I. H. Rorison, ed. Blackwell Scientific Publications, Oxford and Edinburgh. Pp. 399–410.

van Eijk, M. 1934. Versuche über den Einfluss des Kochsalzgehalts in der Nährlösung auf die Entwicklung von *Salicornia herbacea* und auf die Zusammensetzung der Salze im Zellinnern dieser Pflanze. Koninkl. Akad. Wetensch. Proc. Sect. Sci. 37:556–561.

Vickery, H. B., G. W. Pucher, A. J. Wakeman and C. S. Leavenworth. 1940. Chemical investigations of the tobacco plant. VIII. The effect upon the composition of the tobacco plant of the form in which nitrogen is supplied. Connecticut Agricultural Experiment Station Bull. 442.

Viktorov, S. V., Y. A. Vostokova and D. D. Vyshivkin. 1964. Short Guide to Geo-Botanical Surveying. The Macmillan Company, New York.

Vlamis, J. 1953. Acid soil infertility as related to soil-solution and solid-phase effects. Soil Sci. 75:383–394.

Waisel, Y. 1961. Ecological studies on *Tamarix aphylla* (L.) Karst. III. The salt economy. Plant and Soil 13:356–364.

Walker, R. B. 1954. The ecology of serpentine soils. II. Factors affecting plant growth on serpentine soils. Ecol. 35:259–266.

Wallace, A., N. Hemaidan and S. M. Sufi. 1965. Sodium translocation in bush beans. Soil Sci. 100:331–334.

Walter, H. 1969. Die Vegetation der Erde in öko-physiologischer Betrachtung. I. Die tropischen und subtropischen Zonen. 2nd ed. VEB Gustav Fischer Verlag, Jena.

Walter, H. 1968. Die Vegetation der Erde in öko-physiologischer Betrachtung. II. Die gemässigten und arktischen Zonen. VEB Gustav Fischer Verlag, Jena.

Warren, J. C. and S. G. Cheatum. 1966. Effect of neutral salts on enzyme activity and structure. Biochem. 5:1702–1707.

Whittaker, R. H. 1954. The ecology of serpentine soils. IV. The vegetational response to serpentine soils. Ecol. 35:275–288.

Whittaker, R. H. and W. A. Niering. 1968. Vegetation of the Santa Catalina mountains, Arizona. IV. Limestone and acid soils. J. Ecol. 56:523–544.

Wiegand, C. L. and E. R. Lemon. 1958. A field study of some plant–soil relations in aeration. Soil Sci. Soc. Am. Proc. 22:216–221.

Wiegand, C. L. and E. R. Lemon. 1963. Corrections in paper "A field study of some plant–soil relations in aeration." Soil Sci. Soc. Am. Proc. 27:714–715.

Wildman, W. E., M. L. Jackson and L. D. Whittig. 1968. Serpentine rock dissolution as a function of carbon dioxide pressure in aqueous solution. Am. Mineralogist 53:1252–1263.

Wilkins, D. A. 1957. A technique for the measurement of lead tolerance in plants. Nature 180:37–38.

Wilkins, D. A. 1960. The measurement and genetical analysis of lead tolerance in *Festuca ovina*. Scottish Plant Breeding Station Report, 1960. Pp. 85–98.

Wilkinson, S. R. 1961. Influence of Nitrogen and Phosphorus Fertilizers on the Morphology and Physiology of Soybean Roots. Ph.D. Thesis, Purdue University.

Williams, C. H. and C. S. Andrew. 1970. Mineral nutrition of pastures. In: Australian Grasslands. R. M. Moore, ed. Australian National University Press, Canberra. Pp. 321–338.

Williams, M. C. 1960. Effect of sodium and potassium salts on growth and oxalate content of *Halogeton*. Plant Physiol. 35:500–505.

Williams, W. T. and D. A. Barber. 1961. The functional significance of aerenchyma in plants. In: Mechanisms in Biological Competition. Symposia of the Society for Experimental Biology. Number 15. F. L. Milthorpe, ed. Academic Press Inc., Publishers, New York. Pp. 132–144.

Workman, J. P. and N. E. West. 1969. Ecotypic variation of *Eurotia lanata* populations in Utah. Bot. Gaz. 130:26–35.

Wright, R. D. and H. A. Mooney. 1965. Substrate-oriented distribution of bristlecone pine in the White Mountains of California. Am. Midl. Naturalist 73:257–284.

Zinke, P. J. 1962. The pattern of influence of individual forest trees on soil properties. Ecol. 43:130–133.

NAME INDEX

Numbers in italics (*340*) refer to listings in the References.

SUBJECT INDEX